I0047302

William H.G. Kingston

In the Wilds of Africa

William H.G. Kingston

In the Wilds of Africa

ISBN/EAN: 9783744752053

Printed in Europe, USA, Canada, Australia, Japan

Cover: Foto ©berggeist007 / pixelio.de

More available books at **www.hansebooks.com**

IN THE WILDS OF AFRICA.

CHAPTER I

ON BOARD THE " OSPREY "—OFF THE COAST OF AFRICA.

DENSE mist hung over the ocean; the sky above our heads was of a gray tint; the water below our feet of the colour of lead. Not a ripple disturbed its mirror-like surface, except when now and then a covey of flying-fish leaped forth to escape from their pursuers, or it was clove by the fin of a marauding shark. We knew that we were not far off the coast of Africa, some few degrees to the south of the Equator; but how near we were we could not tell, for the calm had continued for several days, and a strong current, setting to the eastward, had been rapidly drifting us toward the shore. Notwithstanding that the sun was obscured, his rays found means of heating the atmosphere, so that we felt much as if we were surrounded by a hot damp blanket.

I had already made a trip to the West Indies, and two to this terrible coast; and as I had escaped without an attack of yellow fever, or cholera, when the Liverpool owners of the brig *Osprey*—commanded by Captain Page, an old African trader—offered me a berth as supercargo, I willingly accepted it. We were bound out to the Cape of Good Hope, but had arranged to touch at two or three places on the coast, to trade and land passengers. Among other places we were to call at St. Paul de Loando, to land a Portuguese gentleman, Senhor Silva, and his black servant Ramaon. Our object in trading was to obtain palm-oil, bees'-wax, gold dust, and ivory, in exchange for Manchester and Birmingham goods; and for this purpose we had already visited several places on the coast, picking up such quantities as could be obtained at each of them. We had not, however, escaped without the usual penalty African traders have to pay—two of our men having died of fever, and two others, besides the captain, being sick of it. The first mate, Giles Gritton, and another man, had been washed overboard in a heavy gale we encountered on the other side of the Equator, and we were now, therefore, somewhat short-handed. The first mate was a great loss, for he was an excellent seaman and a first-rate fellow, which is more than could be said of the second mate, Simon Kydd. How he came to be appointed mate seemed unaccountable; unless, as he was related to the owners, interest might have obtained for him what his own merits certainly would not. Taking him at his own value, he had few superiors, if any equals.

I felt much for Captain Page. He took the loss of his first mate greatly to heart, and thus the incapacity of the second contributed considerably to increase his malady. Day after day he grew worse, and I began to fear much that his illness would end fatally. He was as good and kind a man as ever lived, and an excellent sailor.

I had not been knocking about the ocean altogether with my eyes shut, and had managed to pick up a fair amount of nautical knowledge. I did not intrude it unnecessarily; I had a notion that I was regarded with a somewhat jealous eye by those who considered me a mere landsman. I certainly understood more about navigation than Mr. Kydd, but that is not saying much. There were few things which I could not do, from handing, reefing, and steering, to turning in a dead-eye, and setting up the rigging; and few situations in which the fickle winds and waves were likely to place a ship with which I was not prepared to contend. Blow high or blow low, I felt myself at home on the ocean. My father had objected to my becoming a sailor, and had placed me in his counting-house. The sedentary life of a clerk was, however, not to my taste, and I was very glad to abdicate my seat on the high stool on every decent pretext. Still I had done my duty when there, and my conscience was at rest on that score. Misfortunes overtook my father's house; speculations were entered into which proved unsuccessful; and his long-established and highly-esteemed firm got into inextricable difficulties. In vain he and his partners struggled to maintain their credit. The final crash came, and although my mother's marriage settlement saved the family from penury, he had no capital with which to recommence business. I was too young to take his place. One of his partners died broken-hearted, and he had not the energy left to undertake the onerous duties he would have been called upon to perform. He and my mother and sisters retired to a modest cottage in Cheshire; while his boys, of whom I was the third, had to seek their fortunes in the world. He had done his duty by us. He had given us a good education, and ever striven to instil into our minds the principles of true religion and honour. I shall never forget his parting advice when I started on my first expedition. "Ever trust in God, Andrew," he said. "Recollect that you were 'bought with a price,' and 'are not your own.' You have no business to follow your own fancies, or to gratify any

of the propensities fallen nature possesses, even though we do
possess them, notwithstanding what the devil and the world
may say to the contrary.   God has given you a body, but ever
remember that he has given you a mind to regulate that body.
To the animals he has given bodies, and indued them with
instincts which we may say are unerring; whereas man's mind,
in consequence of sin, is prone to err; but then again, in his
mercy, he has enabled man to seek for strength from above to
counteract the effects of sin, and so to regulate his mind that
it may properly guide the body.   I have no faith in high
principles, unless those high principles are kept in order by a
higher influence.   Therefore, Andrew, read your Bible daily
for guidance; go daily to the throne of grace for enlighten-
ment and direction, that you may keep your high principles
bright and ever fit for action.   Do not trust your feelings;
they may mislead you.   Do not trust the world or your companions; they may prove faithless monitors or guides.   Do
not trust, as people say, 'manfully to yourself.'   Self often
proves treacherous."   More to the same effect my father said.
I have given briefly his observations.   I did my best to carry
out his counsel; and through it gained the calmness and
courage with which I encountered difficulties and dangers
which would otherwise have appalled and overwhelmed me.
I was never addicted to talking to my companions of myself,
or my principles and feelings; and I sometimes blame myself
for not endeavouring more perseveringly to inculcate on others
those principles which I knew to be so true and valuable.   I
now mention the subject, because I can say on paper what my
lips have often refused to utter.   But I have said enough
about myself.

We had several other passengers on board, who, notwithstanding the risks which they knew must be encountered on
the African coast, had, for the sake of seeing the country. come

on board with the intention of proceeding on to Cape Town, to which, as I said, we were ultimately bound. I will mention first Captain Stanley Hyslop, a near relation of mine, a nephew of my mother's. He was a military officer, and having sold out of the service, was going to settle in the Cape Colony, where his parents already were. He was accompanied by two younger brothers. David was one of the nicest fellows I ever met. He had been educated as a surgeon, and purposed practising in the country. The youngest, Leonard, or Leo, as we always called him, was an amusing little chap, always thinking funny things and saying them, and yet there was a simplicity about him which was very attractive. He had been sent to school in England, but being considered somewhat delicate—not, certainly, that he looked so—it was recommended that he should return to breathe his native air at the Cape. David was also, I should say, an enthusiastic naturalist, and the hope of increasing his knowledge at the places we might visit, had, besides his regard for me, induced him to take his passage on board the *Osprey*, just as his brother expected to get a few days sporting while the brig remained at anchor. I had seen but little of Stanley, but for David I had always felt a warm regard.

There were, however, two other members of the family, in one of whom, at all events, I must own I felt still more interested, although I knew that it would not do for me in my present situation to exhibit my feelings. My cousin, Kate Hyslop, was a very pretty, engaging girl, who had a short time before left school. She was also full of spirit, while she was right-minded and sweet-tempered. Her younger sister, Isabella, or Bella as she was called, was quite a little girl. She also had been at school; but her parents naturally could not bear to have her left behind, and so Kate had undertaken to complete her education; and from the time we sailed she was most assiduous in her attempts to do so. Sometimes I fancied she gave her almost too much teaching. When her brother, however, made a remark to that effect, she answered that it was important not to lose time, as opportunities might be wanting by-and-by; and when once they arrived in the colony, she knew that there would be so many interruptions and hindrances, and she might have so many other duties to perform, that Bella might not get the due amount of knowledge she wished her to possess. Blow high or blow low, Kate always made Bella learn her lessons. Sometimes holding on by the leg of a table in the cabin during a gale, there the two sisters would be found with their books. Both were capital sailors, as people say—that is, they were never ill at sea; so that they were not inconvenienced as most other people would have been by the tossing and tumbling of the stout brig.

They were attended by an old negro, Peter Timbo by name, who was the most watchful of guardians. He was the captain's servant, and had always accompanied him in his shooting expeditions when he was before staying at the Cape. Timbo, also, from what I heard him say, knew more about his native country than any one on board. He was born at some distance from the sea, not far from the Equator. When he was just growing into manhood, his village had been attacked by another tribe, and he, with several companions in misfortune, had been carried off to the coast. He was there shipped on board a Portuguese slaver, which, venturing to the north of the line, was chased and captured by a British man-of-war. Timbo, having a fancy for a sea life, and being an active, intelligent fellow, had been allowed to enter on board her. After serving for some years, he had been discharged at the Cape; where, after following several pursuits, he had become a servant to my uncle and aunt, Mr. and Mrs. Hyslop. Peter was loquacious and ever merry, and it was pleasant to hear him give way to one of his hearty laughs. He had thick lips, a huge flattish nose, and somewhat high head, covered with thick curling wool, now beginning to show signs of turning gray. Although he understood English perfectly, he still spoke it in a somewhat negro fashion, which often gave piquancy to his expressions; but from the way his master treated him, and from the affectionate care he seemed to take of the younger members of the family, it was evident that he must be a worthy man, notwithstanding his want of personal attractions.

"Ah, Massa Andrew, we nebber know as kind God does what is good for us," he remarked to me one day. "I bery sorryful when slaver people carry me off from my home in Pongo country. I t'ink I go to die, dat dere was no God to look after poor black fellow. I know only of Fetish, and I afraid of Fetish. Den I get among white men, and I see and hear much dat is bad, and still I t'ink dere is no God. Den years pass by, and I hear of de merciful Saviour, who die for me; and I say, 'Dat is just what I want,' and I learn to be Christian. But I will tell you anoder day more about myself; I now go to get ready de cabin dinner."

I told Timbo that I should keep him to his promise, as I was much interested in the short account he had given of himself.

We had four other passengers—Mr. John and Mr. Charles Rowley, and Miss Julia Rowley their sister, who seemed very nice people, but they kept themselves rather aloof from me, as well as from the mate, though they were friendly enough with the passengers, whom they considered their equals. The last person I need name was a young Irishman, Mr. Terence O'Brien, who was of no profession that I could find out, but proposed settling as a colonist at the Cape. I have thus at once run off a brief description of my companions, of the last mentioned of whom, at that time, I knew comparatively little. Having said thus much of them, I will continue the thread of my narrative.

"How is the weather, Andrew?" said Captain Page as I went into his cabin. We had the skylight off, to let in as much air as possible, but yet it felt hot and stifling. He was very pale. His lips were of a bluish tinge, and his eyes were sunken and dim. On a locker close by him sat a young boy with a book before him, from which he was in vain endeavouring to read. I saw that Natty had been crying, for tears bedewed the page. He was the captain's only son. His mother was dead, and rather than leave him on shore to the care of strangers, his father had brought him on this African voyage. "It was a choice of two evils," he said to me one day. "The boy's constitution is good, and we must not let him be exposed to the night air or hot suns up the rivers, and he will probably stand the climate better than most of us." Such indeed had been the case, and Natty had been well, and had until now been full of life and spirits—the favourite of all on board. He and my young cousin Leonard soon struck up a friendship, and were of course always together. For once Natty had left his friend to remain by the side of his father. The captain had been speaking to him, for his voice ceased as I appeared.

I replied to the captain's question, "No signs of a change, Captain Page. We hove the lead, but found no bottom. We must still be some distance off the coast."

"I trust so," was the answer. "Heave the lead every quarter of an hour, and let me know when we are in soundings. Take another cast at once, and then come back."

I told the mate. "Why, I did so not twenty minutes ago," he answered. "What does the old man want us to do it again for?"

"The captain knows this coast well, Mr. Kydd," I answered. "We may be thankful to get an anchor to hold as soon as we get into shallow water."

Seeing the mate did not seem disposed to obey, I took the lead, and calling to two of the hands, prepared to heave it. "No, no," observed Kydd, "that is my work;" and taking the lead from me, hove it carelessly. "No bottom," he answered; "I should think not, indeed, out here."

It appeared to me that as the line ran out the whole length, he could not be mistaken. Returning to the cabin, I made my report to the captain. "Andrew," he said, "sit down; I want to have a few words with you. I am going to that haven whence I shall never come back. I feel that I shall not hold on much longer to life. I have not been a successful man, and leave my boy but ill-provided for. As to my friends, there are none that I can think of who are able to help him; and the few acquaintances I have who could do so, I cannot trust. The thought of what will become of my orphan boy weighs heavily on me. Andrew, you are young and healthy, and may Heaven preserve your life for many years! I have no great claim on you, but Andrew, as you hope Heaven will watch over you, do you keep an eye on my boy. Do for him the best you can. I have seen enough of you to know that you will act wisely and kindly. I do not desire to have him pampered and spoiled by riches, if I could give them, but I cannot bear the thought of his being left friendless and in poverty to fight his way through this often hard and cruel world. You will see to this, Andrew? I am sure you will."

"I will, Captain Page; I promise you," I answered, and I took his cold clammy hand.

Poor Natty was all this time sobbing violently. The truth that he was going to lose his father burst upon him, and that father had ever been kind and indulgent.

"That is well, that is well," murmured Captain Page. "I trust to you to be his human protector, and to One "—and he turned his eyes upward—"who will ever be a Friend of the fatherless."

The captain said a good deal more, and made various arrangements about Natty. Desiring me to get some papers from his desk, he showed me how I could obtain the little property he was likely to leave.

"I wish I could see the brig safely brought to an anchor," he observed after a long silence. "It is a nasty coast at best. With a breeze we could work off it, but while this calm lasts we cannot help ourselves from being carried wherever the current takes us, till we get into water shoal enough for anchoring. I shall be happier when once we can bring up, for if we do not, we may, when we little expect it, be driven on shore; and let me tell you, Andrew, what with the surf and the sharks, few of us are likely to escape with our lives. I know this coast well, and a sandy beach, exposed to the whole sweep of the Atlantic, is even more dangerous than a rocky shore. It must be time again to heave the lead. Go on deck, Andrew, and see how things are."

I found the passengers seated under an awning, which the mate had rigged at their request. He himself was walking up and down the deck, coming the officer in fine style, and endeavouring to make himself agreeable to the young ladies. He evidently anticipated the moment when he should have the command; indeed, he seemed to fancy himself the master already. When I told him that the captain desired me again to heave the lead, he appeared not to hear me, but continued talking to Miss Rowley with the insinuating air he knew so well how to assume. Miss Hyslop took but little notice of him when he addressed her, and turned away, giving her attention to Bella's lessons, or going on with any work she might

(272)

have in hand, for she never was a moment idle. She was admirably fitted for colonial life; indeed, I may say, for any position in which she might be placed. If she had become a duchess, she would not have been an idle one.—I again addressed Mr. Kydd. I told him that the captain wished to have the lead hove.

"The old man is always issuing his orders through you, Mr. Crawford," he answered at length, in a scornful tone. "I know, I should think, what ought to be done, and I will do it And I beg you will not interrupt me when I am talking to ladies." He added the last sentence in a whisper, sufficiently loud, however, for Miss Rowley to hear him.

"As the captain has been too ill to take an observation for some time, I suppose that you know our correct longitude, Mr. Kydd. He, at all events, considers that we are close in with the African coast ; and, as you are aware, it would be a terrible thing to have the brig cast on one of the sandbanks which lie off it," I remarked.

"No fear of that," he answered scornfully. "We shall have a breeze soon, probably, and then we will stand to the westward, and run down to the latitude of Loando. We are not many degrees from that, at all events."

"The captain is a good seaman, and he has his reasons for ordering the lead to be hove," I answered. "If the calm continues, he wishes us to anchor as soon as the water shoals sufficiently."

"Shoals sufficiently !" repeated the mate, in the same scornful tone; "we have no line on board to reach the bottom, I'll warrant." The mate unintentionally spoke loud enough for the gentlemen to hear him.

"Come, Mr. Kydd, I suppose you intend to obey the captain's orders," said Captain Hyslop, coming up to where we were standing. "It seems to me that he has good reason for giving them."

"I believe, sir, that I am chief officer of the *Osprey*, and that I know my duty," said the mate. "It is not customary for passengers to interfere with the navigation of the ship."

"Certainly not, sir," answered Stanley; "but I trust all on board will obey the captain's orders while he is able to give them."

"That will not be for long," muttered the mate in an undertone. "I intend to do what is necessary, and I do not see that there is any use to keep heaving the lead out here almost in mid-ocean."

"But are we in mid-ocean, Mr. Kydd? The captain considers that we are close in with the coast," remarked Stanley.

"Faith, there is going to be a row," I heard Terence O'Brien exclaim to young Mr. Rowley. "See! I would like to be after giving them a poke. It would be rare fun."

"It would not be rare fun if the captain is right," was the answer.

"Am I to report to Captain Page that you decline heaving the lead, Mr. Kydd?" I said at length, seeing that he made no movement to obey the order.

"Do as you like, Mr. Crawford. I am not going to be dictated to by any man on board," replied the mate in an obstinate tone.

"The captain is very ill, as you know, and I fear your conduct will greatly vex him and tend to aggravate his disease," I said, still unwilling to return below. "I hope you will let me heave the lead if you will not do it yourself."

"Are you hired to navigate this ship, or am I?" he said in an angry tone, turning round to me. "I am chief officer, and unless the captain comes on deck to give his commands, I intend to do as I think fit. If you touch the lead, I shall consider it an act of mutiny, and order the crew to put you in irons."

I did not wish to bring things to extremities, and yet I could not bring myself to tell the captain how the mate was behaving. I waited, but waited in vain, to see whether he would change his mind. He still stood with his hands in his pockets, casting defiant looks around. I was in hopes that Stanley and the other gentlemen would interfere; but they remained silent, though somewhat astonished at the mate's behaviour. At last, finding there was no help for it, I went back to the cabin.

"I am sorry to say, Captain Page, that Mr. Kydd seems to consider that there is no necessity for heaving the lead, and refuses to do so at present," I said on entering. "I will do anything you wish, and again carry your orders if you desire me."

"I must go on deck myself then," said the captain, attempting to rise. "Help me on with my clothes, Andrew. I feel very weak, but if he forces me to it, I must go." I assisted the captain to dress, with the help of Natty. "Here, give me your arm, Andrew; it is a stronger one than poor Natty's. I must do it, though it kills me."

I felt the poor captain tremble all over as I helped him along to the companion-ladder. He climbed up with the greatest difficulty; indeed, without my assistance he could not have got along. He at length reached the deck. He could scarcely stand, and was obliged to hold on by the companion-hatch. His face was pale as death. His white hair hung down on each side of his forehead, over which the skin seemed stretched like thin parchment. His lips had lost all colour, and his blue eyes, as he gazed around, had an unnatural brightness.

"Mr. Kydd," he said, "you have compelled me at a severe cost to come up on deck. I order you to heave the lead. And, men," he cried out, "assist the mate to carry out my orders."

Kydd was now obliged to obey. Going to the chains, he hove the lead. I looked over the side to watch him, and saw by the way the line slackened that bottom was found. Just at that moment I heard some one cry out, " See ! see ! What is the matter with the captain ?" I ran aft. He had fallen to the deck. " Oh, father, father ! speak to me !" cried Natty, who was by his side. I lifted up the old man's head. David Hyslop had hastened to him, and was kneeling on the deck holding his hand. " He has swooned," he said. " He should not have left his bed."

" Can you do anything for him ?"

" We will carry him below, but I fear the worst," he whispered.

Just then the sails of the brig gave a loud, thundering flap, and yet there was no wind ; but I felt that a huge wave coming along the ocean had passed under her. The passengers looked at each other with an expression of dismay in their counte- nances, not knowing what was next going to happen ; while David and I, with the assistance of Stanley and Mr. Rowley, began to carry the captain down below. Not without difficulty, as he was somewhat heavy, we placed him on his bed. David again felt his pulse.

" It is all over, I fear," he said in a low voice, so that Natty could not hear. " Bring a glass ! I cannot feel his heart beating." His brother brought a small glass from their sisters' cabin, and David held it over Captain Page's mouth, and again felt his heart. " He is gone," he said. " No human skill can restore him."

Natty, who had been standing outside, now sprang into the cabin. " Oh, tell me !" he said, looking imploringly up at David, " tell me !—is my father likely to get better? Why will he not speak to me ?"

David did not reply, but made a sign to me to lead him out of

the cabin. I saw my cousin close the old man's eyes as I took
Natty by the hand and led him to the main cabin. I thought
I would tell him at once what had happened; but a choking
sensation came into my throat, and I could not utter a word.

"Is father not getting better," he asked, after a time. "Why
did he not speak to me just now?"

"I am afraid he is not getting better," I replied; "but come
on deck." The idea struck me that I would get one of the
young ladies to speak to him, as they would tell him of his
loss with more gentleness than any one else. When we reached
the deck he saw Leo, who ran up to him, and took him aft to
show him a large shark he had been watching swimming about
close astern. I seized the opportunity of speaking to Miss
Rowley, and told her what had happened.

"Oh, no, no; I am sure I cannot speak to the child. I
should not know what to say," she answered. "Just tell him
yourself. I do not suppose boys are likely to be much
affected by such an occurrence."

I could not help giving her a look expressive of the surprise
and pain I felt. Could that elegant young lady be so heartless
and indifferent to the sorrow of others? My cousin Kate was
sitting a little further off, out of hearing of her brother and
Natty.

"The captain is dead," I said, in a low voice; "but his poor
boy does not know it."

"Is the kind old man really gone?" she exclaimed, looking
up into my face, and a tear starting into her eye. "Oh, how
sad for poor Natty! But he must be told; and yet he will feel
it dreadfully."

"Will you tell him then, Kate?" I asked. "It is necessary
to do so at once, and yet it is hard to wound his feelings."

"Yes," she said; "I will try, even though it would greatly
pain me. Yes yes!" she continued. "Come here, Natty,

and sit down by me.—You need not be afraid, Andrew, I will speak gently to him."

I was sure she would. Her sweet countenance showed me how much she felt for the boy. I did not hear what she said, but she took his hand, and looked kindly into his face. He saw the tears in her eyes as she went on talking, and then, at length, he seemed to comprehend the truth, and began to sob violently. I saw her take both his hands, and cast on him a look of sympathy, of more avail just then than any words she could have uttered. Directly after he started up, as if to run to the cabin where his dead father lay; but she held him back by gentle force; and then he sat quiet, and sobbed and sobbed as if his young heart would break; and she again began to speak so soothingly to him, looking so kindly into his face, her tears falling fast, that I knew he was gaining the comfort he needed.

The mate meantime, hearing what had happened, went into the cabin, as if to satisfy himself. When he returned on deck, I saw that he could scarcely conceal his satisfaction as he looked about with an air of authority.

"Men," he said—for the crew had come up, a rumour having reached them of what had occurred—"I am now captain of this brig, and you will have to obey my orders. You understand me. I am not going to have any of the nonsense we had before; what I say I'll have done, and if there's any slackness, look out for squalls."

Captain Page, I should have said, had been accustomed to have prayers every morning and service on Sunday—a practice not common, I am sorry to say, in those days aboard merchantmen. The good old customs of our forefathers had long been given up, when, rough as seamen might have been, there were far more God-fearing men among them than at present; so I have read. I am afraid Kydd alluded to this practice of the captain's, as well as to the kind and gentle way in which he ruled his crew. The men touched their hats in recognition of his authority; but I saw from the looks they cast at him, that they held him in very different estimation to their late master A stricter captain, perhaps, might have kept them in better order. Many of them were somewhat rough hands; but still his kindness had won their hearts, and, rough as they were, they now showed unmistakable signs of sorrow for his death.

When the mate ceased speaking, and turned aft with a conceited air, I saw them talking together, and casting no very complimentary looks towards him. The old boatswain, indeed, Jeremiah Barker, took but little pains to conceal his indignation. No sooner was the mate's back turned than he lifted up his fist with a threatening gesture, which made me fear greatly for the future discipline of the ship. As to expostulating with a fellow like Kydd, I knew it would be utterly useless; and I was afraid that even if Stanley or the other gentlemen spoke to him, he would be as little likely to attend to them as to me.

I must confess that the captain's death and this conduct of Kydd made me forget altogether the almost dying injunctions of the former to anchor as soon as we got into shallow water. The latter also seemed entirely to have forgotten that we were already in soundings.

"Well, sir," he said, coming up to Stanley, "I suppose we must see about getting the old man buried. I am no hand at preaching or praying, and so I will ask you to read the funeral service. We will do all things ship-shape and right."

"Why, sir," exclaimed Stanley, in a tone of indignation, "the poor man's breath is scarcely out of his body! You would not throw him overboard at once surely!"

"We have to manage that sort of thing pretty sharply out in these latitudes," answered Kydd "I shall be wanting his cabin, too; and as it may be two or three days before we reach Loando, we cannot have him buried on shore. We are not far off that place, and I hope we shall be able to get an observation in a short time, and see exactly where we are."

"You are now master of the vessel, and I shall not interfere with your authority," said Stanley; "but I think it would be more decent to wait as long as we can for the sake of the poor little boy there. When his feelings are more calm he would like to see his dead father."

"Oh, certainly, sir, as you please, as you please," said the mate, turning away. "I will give you another hour to indulge your fancy, but I have no maudlin feelings of that sort."

If the look of unutterable disgust which passed over my cousin's countenance could have made Kydd ashamed of himself, he would have hid his face; but he continued pacing the deck and turning his head about as if considering which order he should next issue. I saw Kate at length take Natty down into the captain's cabin, and I thought it best to allow her and the boy to be alone there together at that sad moment. The boatswain then came aft and said that he and the crew wished to see their late captain.

"What is that for?" asked Kydd, and I thought he was going to refuse the request.

"He was our friend, and we would like to have a last look at his kind face," answered the boatswain.

"Well, if the passengers do not mind your going into the cabin, I do not," said the mate, turning aside.

Perhaps he did not quite like the expression of old Barker's countenance. I led the way into the cabin, and the crew came, one by one, following the boatswain

"Well, you was an honest, kind man as ever lived, and that's more than can be said of him who has stepped into your shoes," said old Barker, apostrophizing the captain. "He is less of a sailor than your little finger was; and as to sense, he has not as much as was in your thumb-nail." The remarks of the other men, as they passed by, were still less complimentary to the new master; and had he heard them, he might well have doubted his power of keeping his crew in order. I felt, indeed, very anxious, for though I had thought very little of Kydd, I was not aware how he was despised and detested by the men.

I more than ever wished that a breeze would spring up, that they might have something to do, and that we might get away from this dangerous part of the coast. The calm, however, still continued, and at length the time came for lowering our late captain into his ocean grave. The sailmaker came aft with the boatswain to superintend the operation of enclosing him in a hammock, into which they fastened a pig of iron ballast.

"Them sharks shan't have him, nohow," observed the sailmaker; "for though the bottom may be a long way off, he will reach it pretty quickly, and lie quiet there till the day when we all come up from the land or sea—it will not then matter where we have lain in the meantime—to answer for ourselves. I only wish I was as sure to give as good an account of myself as he is."

"Be quick there!" cried the mate from the deck. "There is a breeze coming up, I have a notion, and we shall have to trim sails. I wish to get this business over first."

Kate had been keeping Natty by her side while this was going forward. Two of the other men now came below and assisted to carry the captain's body on deck, where my cousin Stanley had got his prayer-book, and stood ready. The old boatswain had thrown a flag over the body, now placed on a plank, one end of which projected out of a port. While the funeral service of the Church of England was read, not a sound was heard except the unrepressed sobs which burst from poor

Natty's bosom, and the creaking of the yards and blocks as the brig moved imperceptibly from side to side. Then came the dull, sullen sound of a plunge, as, old Barker lifting up the end of the plank, the body slid off into the water. As I looked over the side I could see the white shrouded figure descending into the depths of ocean. Just as it disappeared, I caught sight of the dark form of a huge shark gliding towards it; but I had hopes that it had sunk far below the creature's reach before he could seize it.

"Stow away that plank," said the mate, the instant the captain's body had been launched overboard. "I wish this breeze would come, though," he added, glancing round.

Still he gave no orders to heave the lead, as the late captain had advised. I knew well enough that to remind him would only make him less likely to do it, so I said nothing, though I kept looking over into the water to see if there was any change of colour which might be produced by our getting nearer the land. Now again came one of those sullen flaps of the sails, showing that though we might seem to be at rest, the vessel was occasionally moved by no gentle force; and I could distinguish, as I looked eastward, a smooth undulation which seemed rolling away in that direction. Still the sky remained obscured as before, and a gauze-like mist hung over the ocean. The atmosphere felt hotter and more oppressive than ever. The passengers remained on deck, for the cabins were almost unbearable. The ladies were trying to read or work, but Kate alone continued to ply her active fingers. Miss Rowley scarcely turned a page, while little Bella kept looking with her large blue eyes at poor Natty, who sat with his head resting on his hands, utterly unable to recover himself. As I looked over the side I observed that the undulations I have spoken of became more and more frequent, on each occasion, as they passed, giving the brig a slow shake, and making the sails flap loudly as before. The crew were talking together, and, led by old Barker, were ranging the cable for anchoring, Mr. Kydd having disappeared below. Suddenly he returned on deck.

"Who ordered you to do that?" he exclaimed in an angry tone. "Did I tell you I was going to bring the ship to an anchor?"

"No, sir," answered the boatswain; "but any one who is acquainted with these parts must know that it is the only thing to be done to save the brig and our lives. For who can tell that we may not be ashore any moment!"

"You are a mutinous rascal," exclaimed Kydd. "I will not allow the brig to be brought to an anchor till I see fit. We are fifty miles off the coast, and more than that, perhaps."

"What, with fifty fathom only under our keel!" exclaimed the boatswain. "What is the meaning, too, of these breakers away in the south-east? Mr. Kydd, we must anchor, and you ought to know it."

I looked out in the direction towards which the boatswain pointed. The sun was already sinking into the ocean, and his rays lighted up a line of foam, or what looked like it, in the south-east.

Kydd, on the boatswain's remark, broke out into a furious passion, and, hurrying into his cabin, appeared again with a brace of pistols in his hand. Placing them in his belt, he walked the deck, muttering incoherently to himself. No one interfered. I felt unwilling to go below, though the steward called me to supper. The sun had long disappeared—the moon rose, and shed a bright silvery light upon the ocean. It was perfectly calm; and as, on looking round, I could see no breakers, nor hear their sound, I at length turned in. I was too anxious, however, to sleep long. On going on deck and again looking out, there I saw, not a quarter of a mile off a black ledge of

rocks rising some feet out of the water.   The brig was drifting
by them at a rate which showed how strong a current was
running.   What was my surprise to see a boat coming off from
the rock.   "What is that?" I asked.

"Why, I have treated one mutinous rascal as I intend to
treat you if you follow his example," answered Kydd, who
heard my question.

I was too much astonished to speak.   After pacing the deck
for a few minutes I went below to consult with Stanley.

"We must put him under arrest," he said at length.   "But
go on deck and learn how the men take the proceeding."

On my return I found the boat alongside.   The crew
climbed on board.   Could they really have executed so
barbarous an order!   Great was my relief to find the boatswain among them.

"You rascals, I ordered you to land him on these rocks!"
exclaimed Kydd, when he caught sight of the old man.

"So we did; and he ordered us to take him off again,"
answered one of the crew.   "We have as good a right to obey
him as you, Mr. Kydd.   If you was to die, like the captain
and first mate, he's the only officer left to take charge of the
brig."

Kydd was a coward.   This answer silenced him, and without uttering a word he went below.

The passengers assembled at breakfast the next morning
with anxious faces.   They knew that something was very
wrong, but could not exactly tell what.   The calm continued.
A thick mist hung over the ocean as on the previous day, the
rocks were no longer in sight, the vessel floated tranquilly on
the treacherous waters.   Kydd had completely recovered himself.   He had the awning spread, and with a smiling countenance invited the passengers to come on deck, and tried to
make himself agreeable to Miss Rowley.   Some time thus

passed. At last I saw the boatswain and several of the men coming up

"Mr. Kydd," said the former, "I have to ask you whether you intend to anchor, and try to keep the ship out of danger or not?"

"Not till the land is in sight, and I see the necessity," answered Kydd quite calmly. He said nothing more for a minute or so. Then suddenly he exclaimed in a furious tone, "But I am not going to be dictated to by a set of mutinous scoundrels." I need not repeat all his words.

Just at that moment I heard that peculiar low, suppressed roar which a seaman knows so well to indicate breakers I begged the mate to listen, telling him what I had heard, but he was deaf to reason, and declared he would only anchor when he saw fit. He seemed to have gone out of his mind, and I felt that I should be justified in assisting the crew in putting him under restraint; but he was in reality as much in his senses as ever, though under the influence of his passion and obstinacy. Just at that moment another roller came in toward the brig from the westward, and the next instant all on deck were almost thrown off their feet. A blow was felt which made her shake fore and aft, and the water, which had hitherto not even rippled against her side, now broke over her in a shower of spray. The passengers started up. Kate clasped her little sister round the neck, and seized the arm of her brother David, who was standing near her. "What is the matter? what has happened?" shrieked out Miss Rowley in an attitude expressive of her terror.

"We are on shore," cried some of the men; "that is what has happened."

Such was too truly the case. The old captain's warnings had been neglected, and his prognostications were thus terribly fulfilled.

CHAPTER II

BOASTFUL as the mate had been, he turned deadly pale as he saw the dangerous position in which the brig was placed. When, however, she lay quiet—the sea not again breaking over her—he recovered himself. The crew meantime, led by Barker, had gone aloft, without his orders, to furl sails, the first thing under the circumstances to be done.

" Get the boats out," he said at length. His voice had lost its usual authoritative tone. " We must warp the vessel off."

" No easy matter to do that," observed the boatswain. " I know what these banks are made of, and it will be a hard job to find holding ground. Which way will you haul her off, sir?"

" The way she came on," answered the mate.

" That was sideways, I have a notion," observed old Barker. " You will not get her off so."

I soon saw, by the manner the brig lay over, that Barker was right; but without sounding round her, it was impossible to judge properly what to do. I suggested that this was the first thing to be done. " Give your advice when it is asked, Mr. Crawford," said Kydd, walking up and down the deck. " Be smart there with the boats!"

While he was speaking, another wave came rolling in **and**

struck the vessel with greater force than the former one, breaking over the fore part of the deck.

"We must get the boats over to the starboard side," said
Barker (the vessel's head was to the north). "They will be
stove in if we attempt to lower them on the outer side."

"What are you afraid of, man?" exclaimed Kydd. "Why,
the sea is as smooth as a mill-pond between these rollers. Am
I to be obeyed, or am I not? Here, lower this boat first. We
will have her round on the other side before the next roller
comes in."

Several of the men hastened to obey him. The boat was
cleared, and two of the crew jumping into her, she was lowered.
Just, however, as she reached the water, before the others
could follow, another far heavier roller came gliding towards
us. "In board for your lives, lads!" I cried out, but the men
either did not hear me or despised the warning. The wave
struck the boat and dashed it with tremendous force against
the counter, sweeping them off towards the shore. They held
out their hands imploring assistance.

"If we get the starboard-quarter boat lowered we shall be
in time to save them, Mr. Kydd," I said; and without waiting
for his reply, Barker and I, with Jack Handspike, assisted by
some of the gentlemen, lowered the boat. Scarcely, however,
had we seized the oars, when we heard a loud shriek, and one
of the poor fellows disappeared beneath the surface. A shark
had taken him. The other, who was at a little distance, saw
his companion's fate, and cried out to us to make haste. We
pulled away as hard as we could lay our backs to the oars, old
Barker steering. But just before we reached the man, his
arms were thrown up, and down he sank. He, too, had
become the prey of one of the rapacious monsters of the
deep. We now returned on board, the boat remaining perfectly quiet on the starboard side. No attempt had been

made in the meantime to sound round the vessel. I offered to do it.

"I have made up my mind to haul her off astern," answered Kydd. "We will carry a kedge out in that direction."

"As you please," I said. "It may be right, but it may be the wrong way."

"It is my way, at all events," was the petulant reply.

It was necessary to get the long-boat into the water to carry out the kedge. Before this could be done, she had to be cleared of all sorts of articles stowed in her. It took us some time. The fate of their companions had thrown a damp over the spirits of the men, and they did not work with their ordinary activity. I could not help looking out seaward now and then, thinking that heavier rollers might be coming in, when our position would be truly dangerous. Where we were all this time we could not ascertain,—whether we were on a sand-bank at a distance from the coast, or on the coast itself. In either case the danger was great. At last we got a kedge out right astern, and the crew manned the capstan. They worked away for some time. It seemed to me that the anchor was coming home. I was sure of it indeed, for not an inch did the vessel move. I meantime had got hold of the hand-lead, and hove it ahead. There was ample water there, at all events, for the brig to float. I then ran with it aft and dropped it over the taffrail. The water was evidently much shallower at that end of the ship. We had been working away all that time, therefore, to haul her still faster on to the bank. I was determined not to stand this any longer.

"Perhaps, Mr. Kydd, you will try the depth of the water astern as well as ahead," I said; "and it strikes me that if we were to attempt to haul her off the other way, we might have a better chance of success."

"Leave that to me," he answered in the same tone as before.

" Round with the pauls, my lads," he sung out. "We will soon have her afloat."

"I am not going to step another foot round the capstan without I know that we are trying to haul off the right way," said Barker, who overheard what I had said.

This remark made the mate furious. The men followed Barker's example. Mr. Kydd swore and stamped about the deck, declaring that there was a mutiny.

"No mutiny, sir," answered old Barker; " but our lives are worth as much to us as yours is to you."

"Take that then!" cried the mate, rushing forward toward the old man and striking him a blow which brought him to the deck. " Who is going to oppose me now?"

I thought the boatswain was killed, for he lay motionless. The crew, indignant at the way one they looked upon as their friend was treated, threw down the pauls, and refused to work any longer. Jack Handspike alone remained firm in entreating them to obey orders. " Mr. Kydd is now master of the ship, and if we do not obey him, whom are we to obey?" he said.

While the dispute was going on, the passengers taking no part in it, the mist which had hitherto hung over the sea slowly lifted, and looking to the eastward I saw a line of coast, fringed with mangrove bushes, and blue mountains rising in the distance. " The land! the land! we are all right!" cried some of the crew. " I for one am not going to stop here and be bullied by an ignorant greenhorn!" cried one. " Nor I," exclaimed another. " Well, mates, let us take the old boatswain, who was our friend at all times, and see what is to be got on shore. Would any of you ladies and gentlemen like to come with us?"

Captain Hyslop now stepped forward. " My men," he said, " I know what you are likely to find on yonder coast, and I entreat you to remain on board till we see if we can get the brig off. The probabilities are that the boat will be upset in the surf as you attempt to land, and if not, when you get on shore there are savage people, who are as likely as not to murder you immediately."

"Oh, that's all humbug!" cried one of the men, "just to make us remain. Mates, are we to go, or are we to stop and get abused by this ignorant fellow?'

The crew, one and all, with the exception of Handspike, were in a state of mutiny. I spoke to them, but they would not listen to me. "Well, you may go with us," they said, "but go we will. We do not want to leave anybody behind." Without attempting even to bring the anchor on board, they lifted the still insensible boatswain into the boat, and in spite of the entreaties of the ladies and Stanley's warnings, shoved off. Kydd not till then seemed to recollect that he had pistols in his belt. Drawing one, he senselessly fired, but the men were too far off to be injured. They answered with loud laughs and gestures of derision, and away they pulled. We had now only one boat left, and she was too small to weigh the anchor. I begged Stanley and David and one of the Mr. Rowleys to come with me in her, however, to sound round the vessel. Kydd by this time was almost beside himself with rage, and did not interfere with us. We found, as I suspected, that the brig had driven broadside on to a long sandbank, an eighth of a mile in width, but how long we could not tell, for the water was deep on the outer or port side of the vessel; ahead it was also sufficiently deep to float her; and should the wind come off shore, I was in great hopes that we might yet forge her off. Astern, however, the water was far more shallow; and, indeed, the senseless efforts which Kydd had made had contributed to drag her still further on. It all depended, however, upon the wind coming from the eastward. A westerly wind must inevitably prove our destruction, a-with the sea which broke against her in that perfect calm, it was clear that the breeze would have the effect of driving her further on, and sending the sea completely over her. Our position was a truly fearful one. Stanley, however, who was no seaman, did not seem to dread it so much, but Handspike and Timbo fully agreed with me that we should be prepared for the worst. Deserted by the crew, even should the wind come off the shore, we could with difficulty make sail, and then it would be a hard matter to navigate the vessel. We only, hoped, however, that they would return on finding the unattractive appearance of the coast. The mist clearing away to the west, the rays of the sun glanced almost horizontally across the waters, over which they cast a ruddy glow, showing us the boat just as she reached the shore. I went aloft with a spy-glass to watch her, and could make out a number of dark figures hurrying down to the beach. She stopped for some time when at no great distance, and the people in her seemed to be holding a conversation with those on shore. She then pulled on, and directly afterwards I saw her surrounded by the dark figures, who seemed to be running her up the beach. Presently, to my horror, I perceived some of the crew running, and the blacks apparently pursuing them. Now one was struck down, now another. It was too evident that the infatuated men were being murdered by the savages. Soon all pursuit ceased; and here and there I could see figures stretched their length and motionless on the sand. Then I made out a crowd of blacks dancing and leaping, so it seemed to me, round the boat. A new alarm seized me. I was afraid that they might attempt to come off, and treat us as they had done the crew. Anxious to watch them, I did not descend till the shades of night, which rapidly came on, hid them from my sight. I then returned on deck, and taking Stanley and David aside, told them what had occurred.

"We must defend ourselves to the last," he answered, "if they do come. It will be better to die fighting than let them get on board. What do you advise?"

"We have nearly a dozen muskets," I said, "and with our two guns we may make a stout defence. I do not think they would wish to encounter our firearms, even though they possibly have some themselves."

"I am afraid that fellow Kydd will be of no use to us," observed Stanley. "He seems beside himself. We will hear what Timbo says, however. He knows more of these people than any of us."

Timbo was standing at no great distance, and Stanley called him up. I told him what I had seen.

"Not surprised," he observed. "De white men make dem slave, and so when dey catch de white men dey kill dem. Dat's it; but dey no come off at night. No fear of dat. Dey t'ink we one slaver; and if we fired a gun, dey no come off at all."

This information was cheering, as we thought we could rely on Timbo's knowledge.

"Would you consent to go on shore and gain their friendship?" I asked. "If they know that we are not their enemies, they may possibly be disposed to help us; for as to getting off the brig, I fear greatly it is not to be done."

He hesitated. "Yes," he said at length; "I go to-morrow morning. I talk deir lingo; and if dey come from up de country, as I t'ink, I make friends wid dem."

I agreed to accompany him, with David and the younger Mr. Rowley. Darkness at length came on; and as the mist settled once more over the ocean we were unable to see many fathoms on either side of the vessel. We made these arrangements without consulting Kydd, for his conduct had been such that we felt it would be useless: indeed, when I looked round

I could not distinguish him on deck. All this time the brig lay tolerably quiet, for though the sea every now and then struck her, and I feared sent her even more on to the bank, yet it did not break over sufficiently to wash anybody off the deck; the after part, indeed, remained perfectly dry. Here the ladies had collected, with the two boys, while the five gentlemen passengers, Jack Handspike, Timbo, and I, busied ourselves in getting up the muskets and ammunition for them and the guns. "We are going to fire," I heard Stanley say, and soon afterwards Timbo appeared with a hot poker from the galley fire, and our guns were discharged in succession. "Dat keep de niggers away," he observed, returning to the galley. I was surprised that Kydd made no inquiry when the guns were fired. As I was going aft I saw a figure come up the companion-hatch. I could make out that he had a number of packages under his arm. I was sure it was the mate, and my suspicions were aroused, though I could scarcely tell what he was going to do. I pointed him out to Stanley, who was standing near the mainmast. "We will follow him, at all events," he answered. As we got aft we saw him leaning over the quarter, and evidently engaged in hauling up the boat.

"Mr. Kydd, what are you about?" exclaimed Stanley, seizing him by the arm. "Are you going to leave the brig?"

"I am captain, and who dares question me?" was the answer.

"You shall not deprive us of our only boat, at all events," said Stanley. "If you leave the vessel, it must be on a raft, or swim for it."

Kydd made no answer, but continued leaning over the side. We saw that he was dropping something into the boat. It seemed that he was about at that instant to throw himself over, when Stanley seized him and dragged him back. As he did so Kydd let go the painter, and before I could spring forward and seize it, the boat had drifted away from the vessel. I would have jumped overboard and swam to her—I was on the point of doing so—when David, who had followed us, stopped me.

"Stay, Andrew!" he exclaimed. "We are surrounded by sharks. I saw three just before dark. You would be their prey in an instant."

Meantime Kydd was struggling with Stanley, who however quickly overpowered him.

"I was not going to take the boat," said Kydd, "whatever you may fancy. I am captain of this vessel, and I have a right to do what I like. It was through your fault that the boat got away, and you are answerable for that. Let me go, I say!"

Stanley released Kydd, who slunk away without uttering another word.

"This is not a time for disputes," said my cousin. "We must be ready for resistance should the blacks come off to us; though I hope that Timbo is right in supposing that they will not venture from the shore till daylight."

So short a time did the occurrence I have described take, that the ladies were scarcely aware of what was happening till it was over.

"What is the matter, Stanley?" asked Kate.

"Nothing to alarm you, my dear sister. I trust all will yet be well. There is every sign of the calm continuing; and perhaps in the morning, when the wind comes off the land, we may get the brig afloat. What do you say, Andrew?"

"I hope we may," I answered, "as she has not struck very hard."

"Had not you and Bella better go below, Kate, with Miss Rowley, and Leo and Natty will attend on you! We men must remain on deck to do what is necessary should any fresh emergency arise."

Kate begged to remain also, but David, and the Rowleys joining him, persuaded the young ladies at length to retire to the cabin. Timbo followed them to light the cabin lamp, and I saw them, as I looked through the sky-light, seated at the table, Kate having a large book before her, which I recognized as the old captain's Bible. She was reading from it to her companions, the two boys and Bella listening with earnest looks, though Miss Rowley seemed to be too much alarmed to pay any attention. The young Irishman and the two Rowleys now exerted themselves as much as the rest of us in making preparations to defend the vessel.

" If there were boarding-nettings, we should find them useful," said Stanley. " Mr. Kydd. have you any on board?"

" No, sir," was the answer. " We do not carry such things; and, for my part, I think all this preparation is useless. The blacks are not likely to come off to attack us, and if they do, we could very soon drive them back again."

" If we are properly prepared we may," said my cousin; and we all continued the work we had in hand.

Besides the firearms we had a few ship's cutlasses; and at Timbo's suggestion we fastened all the knives and axes we could find to some long spars, to use them as boarding-pikes. We ran lines also along the sides between the rigging to answer in a measure the purpose of boarding-nettings; and before the morning broke, we were as well prepared as we could expect to be to resist an attack. We were looking out for the rising sun, when I felt a light wind fan my cheek. I said nothing, but again I felt it blow stronger.

" We shall have the wind off the shore soon," I cried out, " and we must be ready to trim sails to make the most of it."

" Who is issuing orders on board this vessel?" I heard

Kydd exclaim. "Mr. Crawford, I am the man to say what is to be done."

"If you will tell us what to do, we will take good care to do it," Stanley said to me, in a low voice. "There is little use in listening to that fellow."

The breeze came stronger and stronger; and by the time the first streaks of early dawn appeared over the land, there was a strongish breeze blowing, hot, and smelling of the arid sand and damp mangrove marshes."

"Faith, there is but little of the spices of Araby," I heard Terence O'Brien observe to one of his friends.

"Those who know how to handle ropes, come and help me to trim the sails," exclaimed Kydd. "Handspike, you are the only man under my orders. You go to the helm."

We all set to work to trim the sails. Senhor Silva and his servant, who had hitherto not done much, now joined with a will. The canvas blew out, and the yards creaked and strained, but not an inch was the vessel moved. Kydd then ordered us to run fore and aft; but the light weight of a few people on board the stout brig produced no perceptible effect.

"Had we the boat, and could we carry an anchor out, we might get the brig off," I observed to Stanley. "But, I fear, now it is hopeless, unless, indeed, we were to build a raft. With that we may do something, though there will be no slight risk in the undertaking."

"If you think it can be done, we will do it," said Stanley.

"Certainly," I said, "it is our only chance."

"Then it shall be done," he exclaimed. "Mr. Kydd, we wish to build a raft to carry out the anchor."

Kydd was about to reply, but the captain's look silenced him. All hands now set to work to collect all the spare spars and planks to be found. We got up also a number of small casks from below, in which palm-oil was to be stowed; and this assisted us greatly.

"Massa," said Timbo, coming up to Stanley, "me t'ink it better to have two raft. Suppose no get de brig off, den we want dem to get away. Suppose de niggers come off, den what we do? We not stay here for eber."

"A wise suggestion, Timbo," said his master. "Crawford, will you undertake to build another raft? Mr. Kydd seems busy with the one forward."

Senhor Silva and his servant had, they told us, once assisted in building a raft to escape from a wreck, and were well able to lend a hand. While the rest of the party were collecting materials, I went aloft, anxious to see what the negroes on shore were about. The mist which usually hangs over the land at early dawn had by this time disappeared. With my glass I could distinguish the boat on the beach, and a number of people moving about. As, however, they did not seem preparing to launch her, I hoped that we might have time, at all events, to get our rafts ready; and quickly again descended with the satisfactory intelligence. Believing that there was but little prospect of getting the vessel off, we did not scruple to use the hatches and bulkheads, and, indeed, to rip off the inner planking. It would require, we saw, two rafts of considerable size to carry so many people with any degree of safety even in smooth water. Still, what other prospect had we of saving our lives? I had not for a moment allowed my mind to dwell more than I could help on our too possible fate; indeed, it would almost have unmanned me to contemplate the hardships to which the young ladies must inevitably be exposed even at the best. However, we were doing all that men could do under the circumstances, and that kept up our spirits. Kydd had become somewhat humbled by this time, and worked away like the rest of us, without taking any leading part; indeed, several of the rest of the party were far more expert in constructing the rafts than he was.

The water, as I said, remained smooth inside of us. We now set to work to launch our rafts. Kydd took charge of the one forward; I of the after one, at the construction of which I had assisted. Having cut away the bulwarks, we worked them over the side with the capstan bars, and then lowered them as gently as we could with ropes. Mine, I found, was somewhat the largest, and floated higher than the other out of the water. We had now to fit masts and sails to them. Fortunately there was a number of spare oars on board, so that our time was not occupied in making fresh ones. I however thought it well to have one long one to serve as a mast. The important business of provisioning our rafts had next to be attended to. We first got up four water-casks, which we secured in the centre of the raft. Round them we formed a strong railing, with a raised platform, on which a few of the party could sit well out of the water, which I feared, as soon as there was any sea, would wash over the main part.

I saw Kydd hurrying on with his preparations. "Now, Miss Rowley," he said, "I hope you will entrust yourself to my charge. I ought to know better how to manage a raft than those landsmen," and he cast a glance at me; "and I promise to take good care of you and your brothers."

I did not hear what the young lady said, but directly after wards I saw her being lowered down on to Kydd's raft. Her brothers and the young Irishman followed.

"Come, Handspike; we want you," sung out Kydd, standing up on the raft.

"No, no," answered Handspike. "The landsmen, as you say, will want my help, and I must go aboard the other."

While this was going on, I saw that Timbo had gone aloft.

Presently he came gliding down by a backstay on deck. "Quick! quick! Massa Andrew," he exclaimed. "No time to lose! De niggers coming off in de boat! If we stop and fight, dey take away de rafts. If we sail off, dey come aboard vessel, and stop and steal and get drunk, and we get away."

Kydd overheard him. "Shove off!" he cried out to his companions. They obeyed him; and immediately the raft was clear of the vessel, he began to hoist his sail.

"Stop! stop!" I cried out. "Take more of our party on board! Senhor Silva and his servant will go with you!"

He paid no attention to my shouts, but continued hoisting his sail, though I saw the gentlemen on board were expostulating with him.

"We must all go, then, on the one raft," I said. "I trust it will hold us, although it was treacherous of the mate to go away, leaving the party thus unequally divided."

"I am sorry our friends are under no better charge," said Stanley. "But, Andrew, we are ready to place ourselves under your and Handspike's guidance. Timbo, too, will be of no slight service; so that we need not complain of what has occurred. We have no time to lose, though."

Jack and Timbo now going on to the raft, assisted the rest of the party to descend. I was the last to leave the unfortunate brig. As I looked round I did not see Natty. "Where can he be?" I exclaimed. I sprang up the side. My young charge had fallen on the deck, and lay concealed from those on the raft by the bulwarks in the fore part of the vessel. "Hold on for a moment," I cried out; "I will bring him down to you." I lifted the poor boy up in my arms. A falling block or spar, I conjectured, had struck his head and stunned him. Had I not discovered his absence, how dreadful would have been his fate left alone on board the brig. To my great joy he soon recovered. Jack Handspike received him in his arms as I lowered him down, and I following, without delay we shoved off, and passed under the brig's stern. The blacks could not see what was occurring, and would therefore, I hoped, not hurry themselves in coming off, so that we might have a considerable start of them should they pursue us. The raft was, as may be supposed, deeper in the water than I could have wished; at the same time, in that smooth sea, it was well capable of supporting us all. My hope was that we should be picked up by some cruiser or passing merchant vessel, and that we might not have long to remain on it. Still, the risk was a fearful one, but it seemed better than venturing to the shore after we had discovered the savage disposition of the natives. If they had murdered the seamen, there was no reason to suppose that we should escape the same treatment.

The mate's raft, being lighter, had already got a considerable distance ahead. Our sail, however, was larger than his; and as we had hands enough to lower it quickly, we could venture to carry it longer in the increasing breeze. We got out the oars also, which contributed to urge the raft through the water. We thus, in a short time, had nearly overtaken the mate and his companions. Few of us spoke much. We were all too anxious for talking. Senhor Silva advised that we should alter our course, as soon as we had got out of sight of the brig, to the southward, hoping that we might be picked up by some vessel bound to Loando, the nearest European settlement on the coast. One thing was certain, that should the wind shift to the eastward we should have no choice, but should be compelled to run back for the land.

We had placed Kate and Bella on the most secure part of the raft, with the two boys, while we spread a piece of awning, which projected a little way over their heads, thus

affording them some shelter from the hot rays of the sun.
The water remained smooth, and was bright and clear; and
could we have forgotten that it might at any moment be tossed
into huge waves, there was little to give us a sense of danger.
Jack Handspike was at the helm, and tended the sheets while
the rest of us pulled; I kept an eye on the halliards, ready
to let go should the breeze increase too much for our sail. We
had brought a telescope, through which, every now and then,
I took a glance astern to ascertain whether the negroes had
reached the brig. We were gradually getting to a distance
from her, so that our white sails would have looked almost
like specks on the ocean, unless seen through a spy-glass, and
those that remained on board we hoped the savages would not
know how to use. Presently I saw the bright flash of a gun,
and, a few seconds after, the sound came booming across the
water; then, once more looking through the glass, I caught
sight of several dark objects moving above the bulwarks. There
was no doubt that the blacks must have reached the vessel;
but whether or not they had discovered us remained uncertain. All we could do was to use our best exertions in
getting away from them, by rowing as hard as we could and
keeping our sails spread to the breeze. By this time we had
come abreast of the other raft. I hailed her and told what
I had seen.

"Never fear," cried out Kydd. "We will drive them back
if they do come."

He exhibited several muskets which he had placed on his
raft. We also had taken a couple, and a small quantity of
ammunition.

We had got some little way ahead of the other raft, when
I proposed hauling down the sail, not to run away from her.
I was about to do so, when the wind, which had hitherto been
getting somewhat lighter, fell altogether, and we were left on

a perfect sea of glass, the other raft being about a quarter of a mile away from us. The heat was very great; and as we had been rowing all day, we felt scarcely capable of further exertion. We had also, we hoped, got beyond the reach of the negroes, as it was not likely they would follow us so far out to sea. Timbo asserted that they were black fellows from the interior, as he did not think the coast natives would have murdered the crew. As we had brought an ample supply of provisions, we took our meals regularly. Timbo had provided a small charcoal stove, with which we could boil water, and make our tea and coffee—a great luxury under the circumstances. We had, however, to economize our fuel, of which there was but a small quantity. Considering all things, our spirits rose wonderfully; and I believe every one of us hoped before long to fall in with a vessel and be taken on board.

" Our friends on the other raft seem to be making themselves merry," observed Stanley. "Listen. They are singing!"

So indeed they were. The sound of their voices, though so far off, reached us across the smooth water. We had brought some cloaks, with which we wrapped the young ladies up; and they lay down on the platform I have described, under the awning, to sleep, the remainder of us dividing ourselves into watches. The watch below, as we called it, placed themselves on the other side of the platform, to seek such rest as could be found. I know, when it was my turn to lie down, I slept as soundly as I had ever done in my life. The two boys lay down close together; but during the night I heard poor Natty sobbing. He had awoke, it seemed, and recollected his loss. It was sad to hear him in the still silence of the night out there on the ocean. Poor fellow! he at length sobbed himself to sleep again.

I woke up, feeling a gentle moving of the raft, and, rising to my feet, found that the night wind had again come off the shore, though it seemed rather more to the northward than before. We again hoisted the sail, as we were not far enough out to be in the track of any traders.

The night at length came to an end; and when the dawn once more broke, we found the same mist as on the two previous mornings hanging over the ocean. The young ladies and the boy were still sleeping. We looked round, but could nowhere discover our companions. That was, however, what might be expected, as the mist greatly circumscribed our view. I was standing by Timbo's side.

" I fear dis calm weader not last much longer," he observed to me. " I hope we soon get aboard ship; for if it come on to blow, den we in bad way."

" We must pray to Heaven to protect us," I said.

" Yes, Massa Andrew. If Heaben no protect us, den it be bery, bery bad indeed."

" We must not, however, alarm the young ladies," I observed ; " so do not express your fears, but let us pray that a vessel may be sent to relieve us. Now, I think we had better prepare breakfast. It will cheer our spirits."

Soon after this Kate and little Bella appeared from under their awning.

" My father would have had prayers, I think," said Natty to me, in a low voice.

" He would, I am sure; and so will we," I answered ; and before going to our meal, we offered up a prayer to Heaven for our protection, and Kate read a chapter from her Bible, which she had not forgotten to bring.

The hours after this sped slowly on. Once more the mist lifted. We looked round for the raft. It was nowhere to be seen.

" I trust no accident has happened to it," said Stanley. " It would be a sad fate for the Rowleys and that pretty girl."

I could not suppose this, and yet I could not account for its disappearance should Kydd have continued steering the course we had agreed on. On sweeping the horizon with my glass, I made out a small sail in the distance to the southward. It was, however, so far off, that, in consequence of the slight mist which still remained, I could not be certain whether it was the top-gallant sail of some ship rising above the water or the bow-sail of the raft. I gave the glass to Jack Handspike.

"To my mind it is the raft," he said. "The lighter sails of a large vessel would not look so clear as that does."

If Jack was right, there could be no doubt that the mate had purposely altered his course for the sake of getting away from us. I could not help thinking that he was fully capable of such treachery. Soon after this, again sweeping the horizon with the glass, my eye fell on the topsails of a vessel far away to the north-west. I pointed it out to Jack, and both he and Timbo were of opinion that she was standing toward us on a wind, and that if we continued running as we were doing, she would before long be up with us.

CHAPTER III.

UR spirits, which had naturally been at a low ebb, were greatly cheered by the sight of the strange sail. She had evidently a strong breeze with her, stronger than we should like when it reached us, as it probably would do before long. Already, indeed, it had freshened, and the sea had got up considerably. This made us more than ever anxious to be seen and taken on board. Gradually her topsails rose above the horizon. We watched her anxiously. Although we were not seen, Timbo and Leo could not resist an impulse to stand up and wave towards the stranger. She was standing steadily to the southward, gradually edging in towards the land. Our hopes increased of cutting her off. We made her out to be a large topsail schooner—a rakish-looking craft. Nearer and nearer she drew. Still she came on so fast that we began to fear that we should not get sufficiently to the westward to be seen, for though we could make her out clearly, and could now see her hull, we were so low in the water, that unless those on board were keeping a bright look-out, they might easily pass us.

"What do you think, Timbo? Shall we get up with her?" asked Stanley.

4

"Not quite sure, massa. If dey look dis way, den dey see us; but if dey not look dis way, den dey pass to westward one mile or perhaps two mile."

At length Jack Handspike gave a loud shout. The schooner was coming up to the wind. Her foretopsail was thrown aback, and she lay hove to. "We are seen! We are seen!" we exclaimed, one after the other. Presently a boat was lowered; she came gliding over the water towards us. As she approached we saw that she had a crew of dark, swarthy men, evidently not English. They hailed us in a foreign language. Senhor Silva replied, and a short conversation ensued.

"They are my countrymen," he said, for he spoke English well. "The schooner is, I understand, a Portuguese man-of-war, and you will be kindly treated on board."

"We are indeed fortunate," said Stanley.

"Oh! say rather that God has been very merciful to us," said Kate, looking out towards the beautiful vessel which rose and fell on the fast increasing seas at no great distance from us.

"The officer desires to know whether you would like to be towed on board or would prefer getting into the boat," said Senhor Silva.

I was naturally anxious to preserve the raft, and begged that we might be towed; but Stanley requested that his sisters and the boys, at all events, should be taken into the boat. Senhor Silva joined them. We now proceeded rapidly towards the vessel. I saw Timbo and Jack eyeing her narrowly.

"She seems to be a fine man-of-war schooner," I observed, "and a craft of which the slavers must have no little dread. We thought the *Osprey* a clipper, but yonder schooner, I suspect, could easily have walked round her."

"Not know 'xactly," observed Timbo. "She may be man-of-war schooner, but she very like some slavers I have seen."

"Senhor Silva surely must know," I observed, "and he told us positively that she was a man-of-war."

When we got near the schooner the boat cast off, Senhor Silva saying that he would go on board, and send her back for us.

"I wish I had gone with them," observed Stanley on hearing this. "I do not like their appearing on board a strange vessel without David or me to protect them."

"Oh, but Leo will do that," said David. "He is quite escort enough for them till we can get alongside."

As we approached still nearer the schooner we hauled down our sail. In a short time the boat returned and towed us alongside. The crew of the stranger were looking out eagerly at us over the bulwarks, and ropes were now thrown to assist us in getting on deck. An officer stood at the gangway and politely welcomed Stanley, Senhor Silva who stood by interpreting for him. Kate was seated on a chair, with Bella by her side.

"Oh, they are very kind and polite," she said to her brother as soon as he went up to her. "This is indeed a fine man-of-war."

She was certainly a remarkably fine vessel, and I saw that she mounted six broadside guns and a long gun forward; but as I had not been on board many English men-of-war, and never any foreigners, I was not well able to judge of her. She had a numerous crew, of every colour and shade, from the fair European down to the dark tint of the darkest African. Our stores and the various articles we brought on the raft were now hoisted on board, and the structure which had cost us so much pains to build was cast adrift. The officers, I observed, all wore jackets and straw caps, which I fancied was not usual for officers of men-of-war; but probably on account of the heat of the climate the usual custom was departed from.

Senhor Silva and the captain of the schooner were walking up and down the deck conversing eagerly. At length Senhor Silva stopped as he was passing me, and said, "I have found an old friend in the captain of the *Andorinha* (the *Sea Swallow*), and we are happy to meet each other again. He begs that you and our other friends will consider yourselves as welcome and honoured guests on board. I have told him that we have lost sight of the other raft, and he promises to keep a look-out for her. He has already given directions to have cabins prepared for you, and begs that you will make yourself as thoroughly at home as possible."

This was indeed satisfactory news. Timbo, Jack, and Ramaon were sent forward, where they were well received by the crew; for although Jack could not make himself understood, nor understand what was said, Ramaon was always ready to interpret for him. The wind, which had been for some time increasing, now blew half a gale, and we had great reason to be thankful that we had got on board so fine a craft. The captain insisted on giving up his cabin to Kate and Bella, and Stanley and David had another prepared close to them, while a third was devoted to the accommodation of Senhor Silva and I, the two boys being placed in another rather more forward. Not only were we comfortably accommodated, but a handsome dinner was, soon after we got on board, placed on the table. The captain announced himself as Senhor Marques da Costa. He was very polite, and a good-looking man, though somewhat dark even for a Portuguese. This, I concluded, arose from having been a long time on the coast. He understood but little English, so we had to carry on our conversation chiefly through our friend Senhor Silva. He, however, never seemed tired of interpreting for us. When the captain heard that we wished to proceed to the Cape, he expressed his regret that his duties required him to remain on the coast. He could not, he said, indeed promise to land us, for some little time, at Loando, but he begged to assure us that we were heartily welcome on board. Several of the officers sang very well, and after dinner guitars were produced, and they sang numbers of their national songs: somewhat die-away sort of melodies I thought them, but Kate said they were very pretty, and expressed a wish to learn the guitar. Directly one of the officers undertook to instruct her, and presented her with a handsome instrument, which he said he hoped she would keep in remembrance of her visit to the *Andorinha*. The time thus passed very pleasantly on board. Still having some doubts from what Timbo had said about the vessel, I asked Jack, whom I met the next morning, what he thought of her.

"Well, sir," he answered, "the people seem a free-and-easy set, rather fond of gambling—but that's the way with these foreigners; and most of them wear long ugly knives stuck in their belts, which is not the fashion with English seamen; but these Portuguese are odd fellows, and that is how I accounts for it."

With Timbo I had no opportunity for some time of speaking. Next morning I saw that the Portuguese flag was flying from the schooner's peak, while a pennant waved from her mast-head. Certainly the officers did their best to amuse their fair guests and us. Next day, after dinner, some of the men were called aft to dance their national dances, but I can't say much for them. I saw that one or two of the men were always aloft on the look-out, and while the crew were engaged as I have before described, one of the look-outs gave a shout from aloft, and presently two of the officers went up the rigging with glasses at their backs. I saw them looking eagerly to the southward. Presently they returned on deck and reported their observations to the captain. The breeze, which

had before been fresh, had by degrees been falling, and now
failing us altogether, the schooner lay becalmed with her sails
flapping against the masts.   From this I concluded that a sail
had been sighted—a slaver possibly.   The officers continued
talking together, while one of them, who had gone aloft, remained there, his eye constantly fixed in the direction in
which I supposed he had seen the stranger.   I was about
to go aloft with my spy-glass, when Senhor Silva came on
deck.

"The captain says that passengers going up the rigging will
interfere with the duty of the ship," he observed; "you must
remain on deck."

I thought this was very odd, but of course obeyed.   The
schooner lay without moving on the calm ocean.   Some time
passed.   The officers continued pacing the deck, looking even
more anxious, I fancied, than before.   At length, as I swept
the horizon with my telescope, I observed a white sail rising
above it.   I looked again, and made out the royals and part of
the top-gallant-sails of a square-rigged vessel.   I shut up my
glass quickly, however, as I saw the captain looking somewhat
angrily towards me.

"You had better go below," said Senhor Silva, coming up
to me.   "Ask no questions, and do not say what you have
seen.   It will be better for you to do as I advise, and before
long I will explain matters to you."

As I had no inclination to go below, I begged to be excused
doing so; indeed, I was anxious to learn the character of the
stranger, and to observe what was going forward.

"Well, do as you like," said Senhor Silva; "but I tell you
your presence on deck may possibly annoy our friends."

The stranger approached rapidly, bringing up the breeze
with her.   Presently the captain issued some orders to his
crew, and a number of them went aloft with buckets of water,

with which they drenched the upper sails. In a short time some cat's-paws began to play over the ocean, our royals swelled out to the breeze, and the helm being put up, we stood away to the northward. Still the vessel in the south-west, having far more wind, quickly overhauled us. Our lower sails were now wetted, and every inch of canvas the schooner could carry was packed on her. I soon discovered that, instead of pursuing, we were pursued by the stranger. This, if the schooner we were aboard was a man-of-war, seemed unaccountable. Portugal was at peace, so I fancied, with all the world; besides which, the stranger did not appear very much larger than the schooner—a craft which, if she was of the character Senhor Silva had asserted, was not likely to run away. In a short time I made out the stranger to be a brig with taunt masts and square yards—remarkably like a man-of-war. As she drew nearer I saw, to my astonishment, the glorious old flag of England waving from her peak. I looked and looked again. I could not be mistaken. The schooner, now beginning to feel the wind, made rapid way through the water; which, stirred up into wavelets, hissed and bubbled under her bows as her stem clove a passage through it. Faster and faster we went, as the breeze, which had now overtaken us, increased, and, filling our sails, made the yards and masts crack and crack again. The countenances of the officers, as they saw the speed at which we were going, brightened considerably, and I saw them smiling as they gazed astern at our pursuer. Presently a puff of smoke issued from the bows of the brig, and the sound of a gun was heard across the ocean. Another and another followed. The Portuguese only laughed, and made mocking gestures towards the brig. I was glad that Kate and her brothers were below, for they naturally would have been anxious at seeing what was going forward. The *Andorinha* was undoubtedly a fast craft, and there seemed little

probability, if the breeze continued, of the brig overhauling us
That she was a British man-of-war, I had no longer any doubt.
What then could be the schooner? It was now late in the
day, and I saw that there was every probability of her escaping. Still, unless she was a slaver, I could not account for the
anxiety of her crew to avoid communication with the British
man-of-war. The Portuguese crew made every effort to keep
ahead, by throwing water on the sails as soon as they dried.
Sails were also rigged close down to the water on either side,
and several of the crew went below with shot, which they
slung in hammocks in the hold, under the idea, I believe, that
their weight, as the vessel pitched into the seas, would urge
her forward. Two of the officers were at the helm steering
her, every now and then exchanging remarks as to the best
course to be pursued. The brig, I saw, was also doing her
utmost to come up with us, and had also rigged out studding-
sails on either side, with lighter sails above the royals, often
called sky-scrapers, as well as sails hanging from the lower
studding-sail booms. The Portuguese colours were flying at
the peak of the schooner, but I observed that the pennants
had been hauled down. Again the brig fired, but without any
other effect than making the captain utter a low scornful
laugh, and drawing from the crew gestures of contempt.
When I first saw the brig I had hopes that we should be able
to get aboard her; for, polite as Senhor Silva and the Portuguese captain were, I could not help wishing, for my fair
cousins' sakes at all events, that we were in better company.

Night was drawing on. It threatened to be dark, for there
was no moon, and I saw the mist rising which so often hangs
over the water in those latitudes, near the coast. Still, astern
I could distinguish the brig standing on in our wake with all
the sail she had hitherto carried, in spite of the still increasing
breeze. The Portuguese captain and his officers stood care-

fully watching their spars strained to the utmost by the almost cracking canvas, every now and then glancing astern at their pursuer. I kept my eye fixed on her. Now it seemed to me that she was again coming up with us. My hopes revived that she would bring the schooner to, and settle the doubts as to her character. As I was looking at her, I saw what looked like a vast cloud floating away from her mast-head. Some of the Portuguese saw it too, and cheered loudly. Her top-gallant-sails, if not her topsails, had been blown away, probably with their respective masts; but the thickening gloom prevented us seeing the exact nature of the damage she had received. The Portuguese no longer feared being overtaken, but still they continued standing on as before. A few minutes afterwards we altogether lost sight of the brig. The mist, as I expected, came on, and at length the steward announcing supper, being very hungry, I went below to partake of it. The Portuguese captain and Senhor Silva were in very good spirits, and courteous as usual. I had said nothing about the brig, and was about to mention her appearance when Senhor Silva stopped me.

"There is no use talking about that matter, Mr. Crawford," he observed. "The young ladies will not be interested by it, and—you understand me—I will explain matters by-and-by."

Of course after this I said nothing, and we all parted, when we retired to our berths, very good friends. The next day no sail was in sight. My cousins were on deck, and the officers treated them with the same attention as at first. With Timbo I had not exchanged words, but I got an opportunity at last of speaking to Jack Handspike without being observed. I asked him if he had seen the man-of-war brig, and what he thought of the matter.

"Yes, I did see her, and a rum thing I thought it for another man-of-war of a friendly nation to run away from her. To tell you the truth, Mr. Crawford, I have a notion that this here craft—"

What he was going to say I could not tell, for at that moment one of the Portuguese officers passing, took my arm, and led me to where Senhor Silva was standing.

" Our friends do not like to see you talking to your people," he observed in an undertone. " Remember they do not know who we are, and they have some suspicion as to our character."

I thought the excuse a poor one, but yet was unwilling to give any offence, and therefore refrained from again addressing either Jack or the black.

For two days the schooner continued out of sight of land; but the third morning when I came on deck I found that she had been headed in towards it, and as soon as the sea-breeze commenced she ran in under all sail towards the mouth of a river which opened out ahead of us. On either side were dense woods of mangroves, appearing to grow directly out of the water, while on our starboard hand was a glittering sand-bank, and stretching across the river appeared a line of white breakers, which I fancied must completely bar our ingress. David came on deck at that moment. I pointed them out to him. " Surely we cannot be going in there?" he said. Just then Senhor Silva came up to us and said the captain begged that we and all idlers would go below, as we were about to cross the bar, and that as occasionally the seas broke on board in so doing, it might be dangerous to remain on deck. We could but obey. What could take us into the river? I wondered. Presently I felt the vessel rise to a sea, then she pitched into it, then rose again, and in a few minutes she was gliding on in smooth water. I thought we must be inside the river, but again I felt her rise and once more pitch two or three times, then again she glided on as before. From this I knew that we must have passed over two bars, such as are frequently found at the mouths of the rivers on the west coast of Africa. "What can the vessel be about?" said David. I could not enlighten him; and at length, wishing to satisfy our curiosity, we made our way on deck. We were running up the river, with thick woods on either side. It had the appearance of a long lake, for we had already lost sight of the sea, though I knew by the current in which direction it was. In a short time we caught sight of a number of low cottages and sheds standing in a cleared space at a little distance from the banks. The crew sprang aloft and furled sails, and in a few minutes the schooner was brought to an anchor. Several canoes now came alongside, and in one of them was a fat black fellow with a cocked hat and red jacket, and a piece of stuff which looked very like an old flannel petticoat fastened round his waist. The captain bowed very politely to him, as did his officers, and he returned the salute in the same fashion. I asked Senhor Silva who he was?

"Oh, he is King Mungo," he said; "a very important person in these regions. The schooner has come here on a diplomatic mission, and though he is an ugly-looking savage, we must treat him with every respect."

After the first greetings were over the captain ushered King Mungo and three of his sable attendants, dressed in old nankeen jackets and tarry trousers, into the cabin. Kate's astonishment was naturally very great when she saw them. His majesty bowed to her with profound respect; and I saw him afterwards, whenever he had the opportunity, casting glances of admiration at her. Senhor Silva accounted to Captain Hyslop, as he had done to me, for our entering the river.

"If we are to wait any time, I should like to go ashore and see the nature of the country," said Stanley. "We shall probably be able to get a little sport."

Senhor Silva hesitated, and then addressed the Portuguese

captain. "King Mungo declines to guarantee your safety, and without that it would be madness to go into the interior," he answered.

"But we can keep along the banks of the river, and we may find some sport there," said Stanley.

Again Senhor Silva brought forward many reasons for this being inadvisable. "To say the truth," he added, "as I before explained to our young friend here, my countrymen do not altogether trust us, and it would not be wise to offend them."

This answer did not satisfy Stanley, but he made no remark. Wine and spirits were now placed on the table. His majesty, I observed, after taking a glass or two of the former, applied himself with warm interest to the latter beverage, which soon produced a visible effect. His eyes rolled, and he began to talk away in a thick, husky voice. Senhor Silva again whispered a few words to Stanley, who thereon recommended Kate and Bella to retire to their cabin. It now appeared to me that the captain and King Mungo were warmly engaged in bargaining, judging by their gestures and way of speaking. The captain pressed more spirits on his guest. He would, it seemed, have continued drinking till he was unable to move, had not one of his attendants whispered in his ear, and at length snatched the glass out of his hand. The bargaining now once more went on, and seemed to be concluded to the satisfaction of both parties. At length his majesty rose, and supported by his attendants, made his way on deck, whence he was lowered in no very dignified state into his canoe. He was followed on shore by the captain and two of his officers, and a boat's crew well armed. I observed that the schooner's guns were run over to the side nearest the village, which they thus completely commanded. As he was shoving off Stanley begged that he and I might be allowed to accompany him. David evidently wished to go, but told me that he would remain for the protection of his sisters.

"I do not quite like the look of things," he said; "and take care that you and my brother do not go far from the shore."

I said I would be cautious, and persuade Stanley to follow his advice. Scarcely had we landed when there appeared, coming through the woods, a long line of men, women, and children, walking one behind the other. As they drew nearer I saw that they were bound together with rough ropes fastened tightly to their necks by collars. At intervals at their sides walked several savage-looking blacks, with muskets on their shoulders and thick whips in their hands. There were a dozen or more huts built of bamboo, the walls and roofs covered with the leaves of the palm-tree. Some were of good size, from twenty to fifty feet in length, and of considerable breadth. At the further end of the village was another, three or four times the size of the largest. Stanley and I made our way towards it, but the disagreeable odour which proceeded from it as we approached almost drove us back. We persevered, however, and on looking through the door our indignation was excited to find that it was full of human beings—a dense mass, packed almost as closely as they could exist. They were sitting down in rows, and on a nearer examination we discovered, to our horror, that they were secured to long bars which ran across the building. Below were rough benches on which they might sit, but they could only move a foot or two to the right hand or the left. There were men, women, and children. Many of the poor little creatures were crying bitterly, while their mothers were moaning and weeping, as they tried to comfort them. Some of the men were trying to sing, as if to show indifference to their sufferings, but the greater number sat supporting their heads on their knees, with

looks expressive of despair.   Outside were several savage-
looking negroes armed with muskets, who every now and then
took glances through openings in the side into the interior, to
observe, apparently, if any of the prisoners were trying to
escape.

"Why, these poor beings must be slaves; and, Andrew, the
schooner must be a slaver," exclaimed my cousin.

"There is no doubt about the matter," I answered. "I
have for some time suspected it; nay, I was almost certain of
the fact when she ran away from the English man-of-war.
What do you advise, Stanley?"

"That we leave her immediately," he answered.

"But where are we to go?" I asked.

"Anywhere, rather than remain on board so abominable a
craft," he replied.

"That may be very difficult, if not impossible," I remarked.
"We cannot leave her in this place, and I am afraid that the
captain would not venture near any English settlement to
land us.

"We must try him, however," he said. "We must bribe
him. I would pay any amount I can command to be quit
of her."

We agreed to keep Kate in ignorance as long as possible. Just
then two white men appeared on horseback, swarthy, ill-look-
ing fellows, one tall and thin, and the other short and paunchy,
both dressed alike in wide-brimmed straw hats and nankeen
jackets and trousers. We found that they were the principal
slave-dealers on the coast, having, as we afterwards discovered,
several barracoons at numerous other stations, and parties constantly engaged in capturing and purchasing slaves. The
party of slaves who had just arrived were made to halt, and sit
down on the ground under the shade of the barracoons. After
this several men opened the front of the building, and led out

the slaves, linking them together as the others had been. In this state they were marched down to the water's edge, where two dozen or more large canoes had collected. As soon as these were filled they pulled away towards the schooner. I counted the blacks as they passed, and at least two hundred human beings, including several small children, were carried on board the vessel. The captain of the slaver touched me on the shoulder and pointed to the boat, signifying that we were to return on board. We of course obeyed—indeed, what else could we do!—though we intended to beg Senhor Silva to request him to land us at the nearest European settlement, either Portuguese or French, if he would not take us to an English one, which, of course, we could scarcely expect him to do.

As soon as we reached the vessel the anchor was hove up, and, towed by several boats and canoes, she proceeded down towards the bar. We found our friends in great agitation on board on discovering the character of the vessel. Kate was almost in tears.

"Poor creatures! Where are they going to carry them to in that dark hold? Why, there is scarcely room I should think for one hundred, instead of the number who have been placed below."

"They are but a small portion, I fear, of his intended cargo," I answered. "From what I have heard, many more than those who have already been brought on board will be stowed away. A large vessel like this will carry between five and six hundred human beings. I trust, however, that the captain is more humane, and will be content with those he has already obtained."

"I wish we could manage to let them go again," said Leo. "What right have people to carry off their fellow-creatures, even though they are blacks. I am sure they did not come willingly, for I saw many of them crying, and refusing, till they were beaten, to go down into the hold."

"If you could think of a plan, I would help you," said Natty. "I wish we could manage to restore them to their friends and to their native villages."

I was pleased to see this feeling in the boys, although it was hopeless; for, unless captured by a cruiser, the poor blacks were not likely ever again to visit their native land, or to set foot on shore until they had reached the coast of Brazil. I had seen something of the slave-trade on my former visits to Africa, and was well acquainted with the whole system.

When crossing the bar, we were all as before ordered below. The wind was blowing off the land, and with a strong breeze we dashed through the breakers. I felt by the way the vessel pitched that they were of some height, and I confess I was glad when at length I found that she was well outside, and once more gliding through the waters of the Atlantic. Stanley now addressed Senhor Silva, and begged him to urge the captain to land us at the nearest European settlement.

"I will do what I can," was the answer; "but I am sorry to tell you that, as we have all now been let into too many of his secrets, he purposes carrying us across to the Brazils."

This information made Stanley very indignant.

"My friend," said Senhor Silva, "there is no use exhibiting any anger; but if you will leave matters in my hands I will do the best I can for you."

I can scarcely describe the horror and annoyance we all felt on finding out the character of the vessel we were on board. During all hours of the day and night, but especially at night, the cries and groans of the unfortunate slaves reached our ears. Once my curiosity induced me to look into the hold, but the horrible odour which proceeded from it, and the sight of those upturned faces, expressive of suffering and despair, prevented me ever again desiring to witness the sight.

Once more we were close in with the land. Senhor Silva came to us in the cabin. "I am glad to say that I have made arrangements with the captain to land you," he said. "There is another barracoon near this, from whence more slaves are to be brought off, and if you wish at once to go on shore you can be conveyed there. A heavy surf is however setting on the beach, and I am afraid that there is some risk. It is a wild place, too, and you will probably have many hardships to endure before you can reach any European settlement."

"Oh, we would go through anything, so as to get out of this vessel!" exclaimed Kate. The same sentiment was echoed by the rest of us.

"I fully sympathize with you," said Senhor Silva, "and will inform the captain of your determination. I will lose no time, lest he should change his mind. He knows that I hate this traffic in which he is engaged as much as you do."

We at once prepared to quit the slaver, and on going on deck found the boat alongside. The captain and his officers were collected at the gangway to bid us farewell, but we could with difficulty restrain our feelings of abhorrence in spite of the politeness with which they treated us. Notwithstanding the unprepossessing appearance of the shore, we thankfully hurried into the boat. Timbo and Jack followed us. Ramaon stood on the deck. His master called to him. He replied in Portuguese.

"The scoundrel!" said Senhor Silva. "He has been tempted to turn slaver, and tells me he has entered aboard the vessel as a seaman. I am well rid of him then."

I was glad to hear these expressions from our friend, because I was afraid, from his intimacy with the slave captain, that he himself was engaged in the traffic. The slaver remained hove

to while we pulled towards the shore. As I saw the heavy
surf breaking ahead of us, I felt great anxiety for what might
occur. The boat, however, was a large one, and the cockswain
was an old seaman, who seemed calm and collected as he stood
up and surveyed the breakers through which we had to pass.
The crew kept their eyes fixed on him as they pulled on.
Now we rose to the summit of a sea; now they stopped rowing; now again they urged the boat forward, bending to their
oars with might and main. On we dashed. The waters
foamed on either side. A huge sea came rolling up astern.
Once more we stopped and allowed it to break ahead of us.
Again the helmsman urged the crew to pull away. We dashed
on, and the next instant rushed up on the sandy shore. Some
twenty or more blacks were there to receive us, and dashing
into the water, they seized the boat and dragged her up,
and before another sea broke we were high up on the beach.
The crew assisted us to run forward, Stanley helping Kate,
while David took little Bella in his arms, and sprang over the
bows on to the sand. The rest of us followed, Jack catching
hold of Natty and Timbo of Leo, and carrying them up out of
reach of the water. I saw Senhor Silva putting some money
into the hands of the cockswain. " Now," he said, " we are on
shore, we must consult what is next to be done." Our clothing,
and the small amount of articles we had saved from the wreck,
together with numerous packages brought by Senhor Silva,
were next handed out and piled together high up on the beach.

A little way off we saw a few huts and a large barracoon,
similar to those on the banks of the river from which the
slaves had been embarked. On the shore were hauled up a
number of canoes. Scarcely had we landed when a troop of
slaves were seen issuing from the barracoon, and led by their
captors down to the beach. Several were put on board the
boat, which at once shoved off and pulled for the schooner.

The canoes were now launched, and in each a dozen or more negroes were embarked. The boat passed through the surf in safety; then one canoe followed, then another. The third had scarcely left the shore when a huge sea came rolling in. We trembled for the unhappy beings on board. Those who were paddling her must have seen their danger; but their only hope of escape was to paddle on. It was vain, however. The sea struck her, and in an instant over she went, and all those on board were thrown into the raging surf. The crew, accustomed to the water, struck out for their lives, swimming to the nearest canoe ahead; but the unfortunate slaves, unable to swim, were quickly engulfed. Some cried out for help; but others sank without a struggle, perhaps glad thus to terminate their miseries. Out of all those on board the canoe, which must have contained some twenty human beings, only three or four escaped. One reached the shore; the others were taken on board by the canoes ahead. Notwithstanding this the remainder shoved off, and passing through the surf, put their cargoes on board. They then returned, and the schooner, letting draw her head sheets, stood out to sea.

CHAPTER IV.

E sat on the shore under the shade of some tall trees on the outskirts of the forest, which came down in an apparently impenetrable mass nearly to the coast. Our eyes were turned towards the slave-schooner, which now, under all sail, was standing on, with her freight of living merchandise, at a distance from the shore. We were thankful to be out of her; yet our position was a trying one. We could not tell what dangers and difficulties were before us. In front was the dark rolling sea, which broke in masses of foam at our feet; behind us was the thick forest, through which on one hand a creek had forced its way into the ocean, though its mouth was impassable for boats on account of a sand-bank which ran across it; while on the other side was a clear space, in which stood the barracoons and huts of the native slave-dealers. The blacks had taken little notice of us, leaving us to our own devices, probably, till we might be compelled to appeal to them for assistance. Close to us were piled up the articles we had saved from the wreck, as well as others which Senhor Silva had purchased from the captain for his own and our use.

We had been silent for some minutes. " What is to be

done, Stanley?" said David at length. "Are we to proceed to the north, or south; and how are we to travel? We cannot carry all those things, that is certain."

"It must depend on whereabouts we are, the direction in which we proceed," answered Stanley. "The slave captain took good care to keep us in the dark as to that point; but perhaps Senhor Silva can inform us."

"Indeed, my friend, I am sorry to say I cannot," said the Portuguese. "It is only now that I breathe freely, and can assure you that although I appeared on friendly terms with the captain of yonder vessel, I hate the work in which he is engaged as much as you do; and though by a heavy bribe I induced him to land us, he would not tell me where he purposed putting us on shore, lest we might reach some settlement, and give notice of his being on the coast before he can leave it altogether for South America. Though he has already four hundred slaves on board, he will probably, if he can find them, take two or three hundred more before he considers he has his full cargo."

"Dreadful!" exclaimed Kate. "I would rather go through any dangers on shore than have remained longer on board that terrible vessel."

"So would I," said Bella. "I fancy I still hear the cries of those poor little black children." Timbo and Jack shook their fists at the vessel.

"Oh yes; Natty and I often talked of how we could set them free!" exclaimed Leo; "and only wished that the English man-of-war would come and catch them. If I become a sailor, I would rather be engaged in hunting slavers and liberating the poor blacks than in fighting Frenchmen, or any other enemies."

"One thing I would advise is, that we leave this coast and proceed to the highlands in the interior," observed Senhor

Silva. "You saw that range of blue mountains as we approached the shore, though they are now hidden by the trees?
They form the Serra do Crystal. They are but thinly inhabited, and though travelling along them will be rougher
work than on the plains, yet we shall enjoy fresh breezes and
a more healthy climate than down below."

"To the mountains, then, in the first place let us proceed,"
said Stanley, springing to his feet. "After that we can
decide which way to take; but, for my own part, I should
prefer moving towards the south. We shall be going homewards, and may be better able to send a message to our
friends at the Cape. It is a long distance, but we shall, no
doubt, hear from them if we have patience, and, in the meantime, maintain ourselves in the most healthy region we can
find. There is, at all events, no lack of game, and we shall
probably be able to obtain fruit and vegetables sufficient for
our wants."

"An excellent plan!" exclaimed David. "We shall thus
be able to add largely to our knowledge of natural history;
and if Kate and Bella do not object to live a savage life for
so many months, I think we can make our stay not only satisfactory, but in many respects delightful.'

"I am glad to do whatever you wish, my brothers," said
Kate; "and I think I shall enjoy the life you propose very
much. I wish I had a few more books to teach Bella from;
but we must make the most of those we have: and I will
undertake to cook for you and tend the house, for I suppose
you do not intend living out in the woods all the time?"

"Oh no, no," said David. "Wherever we settle to remain
we must at once build a house, where you and Bella can live
in comfort, and where we can stow our stores and collections
of natural history."

Of course I agreed to my cousin's plan; and, indeed, I

thought it, under all circumstances, the most advisable. Even should we reach one of the Portuguese settlements, we might not be able to find a vessel to carry us to the Cape; besides which, they are mostly unhealthy, and it would be far better travelling along the mountains than having to spend any length of time at one of them. I was afraid, however, that Senhor Silva would not so readily agree to this plan, as he might be anxious to reach Loando. I was relieved when I heard him say,—

"Well, my friends, I approve of your proposal; but we must not wait here an hour longer than is necessary. At night we shall find unhealthy vapours rise from yonder river, and the sooner, therefore, we get away from its banks the better."

"But we have no horses or waggons to carry our goods," I observed, looking at the pile of property before me. "Even if each of us were to take a heavy package, we could not carry it."

"I will see to that," said Senhor Silva. "I think I can secure the services of some of those negroes, although they may not be willing to venture far into the country. Mr. Crawford, will you come with me, and we will see what can be done?"

I started up with my gun in my hand, for I did not like the appearance of the black savages. I remembered the way the poor crew of the *Osprey* had been treated, and thought it possible, if we were taken unawares, that we might meet the same fate.

"The case is very different here," said Senhor Silva in reply to my remark as we walked along. "Those poor men fell victims to the treachery not so much of the blacks, as of some of the white slavers who had but a short time before carried off a number of their kindred and friends. I heard the story on board the schooner. They had enticed them down to the coast on pretence of trading, and then surrounding them, had captured some forty or fifty of their number, and carried them off on board their ship. Those who had escaped, very naturally vowed vengeance against the first white men they might meet, of course not distinguishing between English and Portuguese. Thus the unfortunate crew of the brig became their victims. They would, had we landed before they had had time to ask questions, very probably have put us all to death. We have had, indeed, a providential escape."

We found that the slave-dealers and most of their followers had already taken their departure—probably to avoid rendering us any assistance. They had only come down to the coast to embark their captives, and had gone back again, my companion supposed, to obtain a fresh supply. We found, however, about a dozen men, who came out when Senhor Silva called them in their own language. When he assured them that we were friends, and that we would treat them honestly, they agreed, without hesitation, to act as our bearers as far as the Crystal Mountains. Beyond them they declined going, saying that they had enemies on the other side who would certainly, if they found them, kill them, or carry them off as slaves, or, they added, " very likely eat us, for they are terrible cannibals." As soon as the arrangement was made, they all came leaping and hooting and rushing against each other, like a set of school-boys unexpectedly let loose for a half holiday, or a party of sailors on shore after a long cruise.

While the blacks were arranging our property into fit packages for carrying, the two boys and I accompanied David to the mouth of the river, which, as I said, was lined with mangrove bushes, a ledge of rocks which ran out some way enabling us to get a view up the stream. We had thus an opportunity of examining those curious trees. Innumerable roots rose out of the water, lifting the trunk far above it, and from its upper part shot off numerous branches with bright green foliage, which grew in radiated tufts at their ends. Many of them were bespangled with large gaily-coloured flowers, giving them a far more attractive appearance than could be supposed, considering the dark, slimy mud out of which they grew. From the branches and trunk, again, hung down numberless pendulous roots, which had struck into the ground, of all thicknesses—some mere thin ropes, others the size of a man's leg—thus appearing as if the tree was supported by artificial poles stuck into the ground. David told me that the seeds germinate on the branches, when, having gained a considerable length, they fall down into the soft mud, burying themselves by means of their sharp points, and soon taking root, spring upwards again towards the parent tree. Thus the mangrove forms an almost impenetrable barrier along the banks of the rivers. On the other side of the stream, indeed, we saw that they had advanced a considerable distance into the ocean, their mighty roots being able to stem even the waves of the Atlantic. Near where we stood the ground was rather more open, and we saw the black mud covered with numberless marine animals, sea-urchins, *holothuria*, or sea-slugs, crabs, and several other creatures, many of brilliant hues, which contrasted curiously with the dark mud over which they were crawling. The roots of the trees were also covered with mussels, oysters, and other crustacea. But the most curious creature was a small fish which I had before seen, called by sailors Jumping Johnny. David called him a close-eyed gudgeon (*periophthalmus*). He was of the oddest shape, and went jumping about sometimes like a frog, and sometimes gliding in an awkward manner over the mud. We were watching one of them when Leo cried out, " Why,

the fish is climbing the tree—see, see!" And so in reality
he was, working his way up by means of his pectoral fins,
David supposed in search of some of the minute crustacea
which clung to the roots. Jumping Johnny, having eaten as
much as he could swallow, or slipping off by accident, fell
back into the mud, when we saw issuing sideways from under
the roots a huge crab. David said he was of the *Grapsus*
family. Suddenly he gave a spring, and seized the unfortunate
Johnny in his vice-like gripe, and instantly began to make
his dinner off the incautious fish, who, as Leo said, would
have been wiser if he had kept in the water, and not attempted to imitate the habits of a terrestrial animal.

As we looked up the stream we saw numerous birds feeding
along the banks. Among them were tall flamingoes, rose-
coloured spoonbills, snow-white egretts, and countless other
water-fowl.

"I am glad we have been able to witness this scene here,"
said David, "where we can benefit by the sea breeze; for such
deadly miasmas rise from these mangrove swamps, that the
further we keep off from them the better."

While we were watching we saw a canoe, paddled by half a
dozen blacks, dart out from the mouth of a creek which had
been concealed by the thick trees. We drew back, not knowing whether the people in the canoes might prove friends or
foes. Another followed at a little distance, and proceeded up
the stream. They were impelled by paddles with broad blades;
and the sound of voices reached our ears as if they were singing.

"I do not think they can be enemies, or they would not be
so merry," said Natty.

"I hope not," I observed.

"If we could stop them we might hire their canoes to convey us up the river."

"It might be dangerous to do so," said David, "on two accounts: they might prove treacherous, and the miasmas rising from the stream might also possibly give some of us fever. I think we had better let them go on their way, and proceed as Senhor Silva proposed."

Returning, we found the party ready to start. We told Senhor Silva about the canoes.

"I think you did wisely to let them go," he remarked. "Unless we were under the protection of their chief man, or king, as he is called, we could not tell how they might behave. We must use great caution in our intercourse with these people. When we have shown them that we are friends, and desire to do them all the good in our power, we, I hope, shall find them faithful; but they have become so debased by their intercourse with the white people, and especially, I am sorry to say, with my countrymen, who often deal treacherously with them, that they cannot be depended on. They in return, as might naturally be supposed, cheat and deceive the whites in every way."

Our path first led through the forest near the banks of the river, of which we occasionally got glimpses. It was here of considerable width, bordered by mangrove bushes. In one or two places there were wide flats covered with reeds. Suddenly, as we passed a point of the river, I saw drawn up what had much the appearance, at the first glance, of a regiment of soldiers, with red coats and white trousers.

"Why, where can those men have come from?" I cried out.

David, who was near me, burst into a laugh, in which his sisters and the boys joined. "Why, Andrew, those are birds," he answered. "A regiment, true enough, but of flamingoes; and see! they are in line, and will quickly march away as we approach."

A second glance showed me that he was right; and a very curious appearance they had.

"See! there is the sentinel."

As he spoke, one of the birds nearest to us issued a sound like that of a trumpet, which was taken up by the remainder; and the whole troop, expanding their flaming wings, rose with loud clamours into the air, flying up the stream. We went on, and cutting off a bend in the river, again met it; and here our bearers declared that they must stop and rest. We accordingly encamped, though Senhor Silva warned us that we must remain but a short time, as we wished to reach some higher ground before dark. A fire was lighted for cooking; and while our meal was preparing, David and I, with the two boys, went down nearer the banks to see what was to be seen. We observed on the marshy ground a little way off a high mound, and creeping along, that we might not disturb the numerous birds which covered the banks or sat on the trees around, we caught sight of another mound, with a flamingo seated on the top of it, her long legs, instead of being tucked up as those of most birds would have been, literally astraddle on it.

"That is one of their nests," whispered David. "The bird is a hen sitting on her eggs. Depend upon it, the troop is not far off. See, see! there are many others along the banks. What a funny appearance they have."

Presently a flash of red appeared in the blue sky, and looking up, we saw what might be described as a great fiery triangle in the air sweeping down towards us. On it came, greatly diminishing its rate, and we then saw that it was composed of flamingoes. They hovered for a moment, then flew round and round, following one another, and gradually approached the marsh, on which they alighted. Immediately they arranged themselves as we had before seen them, in long lines, when several marched off on either side to act as sentinels, while the rest commenced fishing. We could see them arching their necks and digging their long bills into the ground, while they stirred up the mud with their webbed feet, in order to procure, as David told us, the water-insects on which they subsist. They, however, were not the only visitors to the river. The tide was low, and on every mud-bank or exposed spot countless numbers of birds were collected—numerous kinds of gulls, herons, and long-legged cranes—besides which, on the trees were perched thousands of white birds, looking at a distance like shining white flowers. They were the *egretta flavirostris*. Vast flocks of huge pelicans were swimming along the stream, dipping their enormous bills into the water, and each time bringing up a fish. They have enormous pouches, capable of containing many pounds of their finny prey.

" Could we kill one or two we should get a good supply of fish for supper," said David; " for the pelican stows them away in his pouch, where they remain not only undigested, but perfectly fresh, and not till it is full does he commence his meal. However, as we have no canoe, even were we to kill one we could not get him."

While we were looking on, a huge bird, descending from the sky, it seemed, pounced down into the water, quickly rising again with a large fish in his mouth.

" Ah, that fellow is the fishing-eagle of Africa—the *Haliœtus vocifer*," said David. " His piercing eye observed his prey when he was yet far up in the air. See how like a meteor he descended on it! Now he flies away to yonder rock; and there, see! he has begun to tear his fish to pieces. How quickly he has finished it—and listen to that curious shriek he is uttering, and how oddly he moves his head and neck. It is answered from those other rocks. The birds are calling to each other, and from this the fishing-eagle has gained his name of *vocifer*." Leo was for shouting and making them fly off. " No, no; let them feed," said David. " We have frightened the flamingoes once; and how would you like to be disturbed in your dinner? We must get Kate to come and look at them."

While we were watching the birds, an enormous head emerged from the water at a short distance from us. Leo and Natty, who were a little in front, started back, Leo exclaiming, "What can it be? What a terrific monster!" A huge body rising after the head, the creature swam slowly up the stream.

"Why, that must be a hippopotamus," observed David, watching the creature in his usual calm way.

"It looks to me the size of an elephant," exclaimed Leo. " Run, run, run! If he were to attack us he would swallow us up in a mouthful."

"I do not think it has even noticed us," said David. " It will be time enough to run when the creature lands. See! there is another."

As he spoke, a second and then a third hippopotamus appeared, following the first. The creatures, indeed, had truly terrific countenances; their backs in the water looking, as Leo had declared, nearly as large as those of elephants.

"But see, there are some other creatures nearer!" cried Natty. " Oh, what are they? What fearful jaws!"

He pointed to the bank close below us, and there we saw, just scrambling out of the water, three huge crocodiles. There was no mistaking them. We knew at once by their long snouts and terrific jaws, their scaly backs and lizard-like tails, their short legs and savage eyes. They seemed in no way afraid of the hippopotami, which they kept watching as they swam by.

"I little expected to get a sight of these monsters," said David. " But see! they take no notice of us, and we need not be afraid of them."

I had my gun, and instinctively levelled at the head of the nearest hippopotamus.

"Do not fire," said David. "Even if you were to kill the beast we could not get him, and it would be cruel to slaughter him without any object in view. He intends us no harm; we ought to allow him to enjoy the existence the Creator has given him."

The hippopotami swam by and dived, and presently we saw them rise to the surface with a quantity of weeds in their mouths, which they chewed leisurely as they swam on. The crocodiles meantime crawled up on the bank and lay basking in the sun, enjoying its warmth, and looking at that time, at all events, as if they had no evil intentions. It was a curious scene, and gave us an idea of the vast amount of animal life to be met with in that region.

"I think it would frighten Kate, brave as she is, to see those huge monsters," said Leo.

"Oh, no," answered David. "Bella might be somewhat alarmed; but I am sure Kate would be as much interested as we are in witnessing this curious sight. We will get her to come, but warn her beforehand what she is to expect."

We accordingly hastened back to the camp, but found we had been so long absent that it was now time to proceed; and the bearers taking up their loads, we continued our march. Senhor Silva assured Kate and Bella they need not be disappointed at missing a sight of the flamingoes, as they would have many opportunities of seeing troops of those magnificent birds, which are found in vast numbers throughout that region.

The woods as we proceeded appeared full of life. Birds flitted among the boughs, and monkeys of all sorts sprang here and there, chattering and hooting as we passed. Soon after this we emerged from the wood and entered a beautiful prairie —a natural clearing covered with grass or low shrubs and flowers. As yet we had fallen in with no inhabitants. " Oh, but see!" exclaimed Leo. " There are some huts ahead. Shall we go and pay the people a visit?" The boys ran on. I thought Senhor Silva would have called them back, but he allowed them to proceed. At all events, he knew that if the huts were inhabited, the people were likely to prove friendly. The boys stopped before the seeming huts, and began to examine them. We saw them walking round and round, and they then finally climbed to the top of one of them. After apparently satisfying their curiosity, they came back towards us.

" They are not huts," exclaimed Natty, " but curious mounds, three or four times as high as we are."

" What do you say to those mounds or clay-built domes being the houses of ants, and built entirely by themselves?" said Senhor Silva.

As we approached we saw a dozen or more such mounds, scattered about at short distances from each other. Having got to a secure distance from the last, two of our bearers put down their loads, and advancing towards it with the poles they carried, began to attack it with heavy blows, knocking off one of the small turrets on the side. Instantly a white ant was seen to appear through the opening thus made, apparently surveying the damage done. Immediately afterwards, hundreds of other ants came to the spot, each carrying a small lump of clay, with which they began to repair the damage; and even for the short time we remained, they had made some progress. We could discover, however, no outlet or opening in the mound; nor, except at the hole made by our bearers, were any ants seen. We, however, could not remain to watch the progress of the work. Just as we were going, one of our bearers, much to my regret, commenced a still more furious attack on the citadel, exposing the whole centre to view, when it appeared crowded with thousands and tens of thousands—so it seemed—of ants, who issued forth with pincers stretched out, evidently intending to attack us. David caught up one of the ants to examine it; but we were all too glad hurriedly to make our escape. We found the creature, on examining it, to be a quarter of an inch in length, with a flat hard head, terminating in a pair of sharp horizontal pincers, something like the claws of a crab.

Several, who, in spite of our flight, caught hold of us, bit very hard, and did not fail to draw blood. Senhor Silva, as we marched on, gave us a very interesting account of these white ants, with the habits of which he was well acquainted, as he told us he had had one of the mounds cut completely in two, so as to examine the interior. The under part alone of the mound is inhabited by the ants; the upper portion serving as a roof to keep the lower warm and moist for hatching the eggs. His description put me somewhat in mind of the Pyramids of Egypt. The larger portion is solid. In the centre, just above the ground, is the chief cell, the residence of the queen and her husband. Round this royal chamber is found a whole labyrinth of small rooms, inhabited by the soldiers and workmen. The space between them and the outer wall of the building is used partly for store-rooms and partly for the purpose of nurseries. A subterranean passage leads from a distance to the very centre of the building. It is cylindrical, and lined with cement. On reaching under the bottom of the fortress, it branches out in numerous small passages, ascending the outer shell in a spiral manner, winding round the whole of the building to the summit, and intersecting numerous galleries one above the other, full of cells. The outer end of the great gallery, by which the mound is approached, also branches off into numerous small ones, so as to allow a passage into it from various directions. As the ants cannot climb a perpendicular wall without difficulty, all their ascents are gradual. It is through this great passage that they convey the clay, wood, water, and provisions to their colony.

To give you a correct idea of the way these curious mounds are built and stocked with inhabitants, I should tell you that the perfect termites are seen at certain seasons in vast quantities covering the earth, each having four narrow wings folded on each other. They are instantly set upon by their enemies —reptiles of all sorts, and numerous birds—who eat such quantities, that out of many thousands but few pairs escape destruction. There are besides them in their fortress vast numbers of labourers, who only issue forth with caution to obtain provisions and materials for their abodes. When these discover a couple of the perfect termites who have escaped destruction, they elect them as their sovereigns, and escorting them to a hollow in the earth which they at once form, they establish a new community. Here they commence building, forming a central chamber in which the royal pair are ensconced; while they go on with their work, building the galleries and passages which have been described, till the mound has reached the dimensions of those we have seen. The king in a short time dies, but his consort goes on increasing in bulk till she attains the enormous length of three inches, and a width in proportion. She now commences laying her eggs, at the rate, it is said, of nearly sixty in a minute. This often continues night and day for two years, in which time fifty million eggs have been laid. These are conveyed by the indefatigable labourers to the nurseries, which are thus all filled. When hatched, they are provided with food by the labourers. There is another class, the soldiers. These are distinguished by the size of their heads, and their long and sharp jaws, with which they bravely attack any intruders. When any unwary creature appears to attack their abode, first one comes out to see what is the matter. He summons others, and directly afterwards vast numbers issue forth, doing battle with the greatest courage. When any of them are knocked over, instantly recovering themselves, they return to the assault with a bravery and courage surpassed by no other creature in creation. The labourers meantime are exerting themselves to the utmost to repair the damages which have been effected in their fortress. Those who have watched their proceedings state that in a single night they will repair a gallery, which has been injured, of three or four yards in length. We were thankful that in our attack on the termites' fortress we had escaped with only a few bites ; but probably had we remained longer in their neighbourhood we should have received far more severe injuries.

Travelling on for several days, we emerged into some open ground, where we prepared to encamp. We selected a spot somewhat above the plain, and our bearers at once set to work to cut down poles. These they planted in circles, and interwove them with branches of palm-trees, forming walls which afforded sufficient shelter from the night wind; then bringing the tops close together, they thatched them over with leaves of the same tree. We of course all assisted, and in a short time a number of small circular huts were formed sufficient to accommodate the whole of the party. A quantity of wood was collected, to keep up blazing fires to preserve us from the attacks of wild beasts. We were at a sufficient distance, however, from the skirts of the forest, not to be taken totally unawares. Still, it was considered necessary to place guards round the camp, two of our party and two of the blacks remaining on the watch all night. Before darkness closed in, we saw numbers of monkeys in the trees, watching us with curious looks, leaping from bough to bough, and chattering and grinning, wondering apparently who the strangers could be who had thus ventured into their domain. The two girls had a hut to themselves. We had formed a second wall of sticks round it, so that should any wild beast approach unseen, it could not force an entrance, which Senhor Silva told us had sometimes occurred. The moon rose in an unclouded sky, and cast a mild light over the scene. In the distance were the lofty mountains, on either side the dark woods, and far away to the west was the ocean we had left behind. It was a beautiful scene, such as I had not expected to witness in that region, and we were all more than ever thankful that we had escaped from the slaver. Still, I could not banish from my mind the spectacle I had witnessed on board, and my thoughts went back to the unhappy beings crowded on the slave-deck of that fearful craft. I was reminded that we were in Africa by the cries which proceeded ever and anon from the surrounding forest. Now there was a loud roar, with a suppressed muttering, which it would be hard to describe, and which I afterwards learned to distinguish as the voice of the monarch of the woods; not that he often ventures here, for his rule is disputed by the tremendous gorilla, the creature who had only a short time before been discovered in this region. We were, however, we concluded, on the most southern verge of his territory, and we therefore scarcely expected to encounter one. We kept our fires blazing through the night, and thus avoided any attacks from lions or panthers, or any other wild beasts.

The morning broke brightly, though we could see the mist hanging over the far distant coast. Birds flew about among the trees and across the prairie in all directions, uttering their varied notes; and the monkeys came forth, skipping from bough to bough, muttering and shrieking at us as on the previous evening, as if they had not as yet satisfied their curiosity. While Kate, assisted by Timbo and Jack, prepared breakfast, I accompanied Stanley and David, with the two boys, to shoot some birds for our next meal. I had heard so much of ser-

pents and wild beasts, that I expected every instant to see a
snake wriggling its way through the grass, and about to fasten
its fangs in our legs, or to twine its fearful coils round our
bodies.   I could not help also looking anxiously at every
bush, expecting to have a lion or a panther spring out on us.
David acknowledged that he had a similar feeling.   Stanley,
however, laughed at our apprehensions, assuring us that snakes
were not nearly so common as were supposed, or how could
the almost naked blacks make their way through the country,
though he acknowledged that lions and panthers were in some
places justly dreaded; " But then," he observed, " we can the
more easily defend ourselves against them.   A well-aimed
bullet will settle the fiercest lion we have to encounter."

We had good sport, and shot several varieties of birds.
Among them was a partridge, of a gray colour; and David
said that they were its loud calls we had heard in the forest
the evening before, summoning its mate.   He had observed
them sleeping side by side on a branch of a tree where they
have their home, and the bird which was first there did not
cease calling till its mate arrived.   We also shot several parrots, of a species known as the African damask parrot.   They
are pretty birds, and their habits are very interesting.   Had we
not positively required them for food, I should have been unwilling to kill them.   We had seen numbers flying towards a stream
which ran into the river we had passed on the previous evening.
They there assembled, making a great deal of noise, and huddled
and rolled over each other, frolicking together, and dipping their
feet into the water, so as to sprinkle it over the whole of
their bodies.   Having enjoyed an ample bath and amused themselves for a time, they flew off to the forest whence they came.
There we saw them sitting on the branches, cleaning their
feathers.   The operation over, they flew off in pairs, each pair
seeking its own nest or roosting-place, separate from the

others.   David said that this species is noted for conjugal
affection, for they never separate till one or the other dies,
and the survivor then pines to death for its mate.   The boys
were very anxious to catch one alive for Bella, but we could
not succeed in so doing.   Coming near a dead tree, we saw
several hollows, evidently formed by art.   Leo climbed up to
one of them, and putting in his hand, drew out a beautiful
little bird, with a throat and breast of a glossy blue-black,
having a scarlet head and a line of canary-yellow running from
above the eyes along the neck.   The back also, which was
black, was covered with yellow spots.   Here David brought
his knowledge to bear; and said, from its habits, he should call
it the carpenter bird.   When the birds pair, they fix on a tree,
the wood of which has been sufficiently softened by age to
enable them to work upon it with their bills.   They then take
out a circular opening, about two inches in diameter and about
two deep.   Next they dig perpendicularly down for about four
inches, the last hollow made serving as their nest.   They line
it softly, and the female, laying her eggs, is able to hatch them
without much risk of an attack from birds of prey.

"I suppose monkeys do not eat birds," observed Leo; "or
I suspect our little friend would very soon be pulled out of its
nest."

"Just as you have done, Leo," observed Stanley; "and probably the poor little bird took you for a chimpanzee, or perhaps even for a gorilla."

"But neither chimpanzees nor gorillas eat animal food,"
observed David.   "They live upon roots, fruits, and leaves;
and do not amuse themselves by bird-nesting."

I need not mention the other birds we shot, but, pretty
well loaded, we returned with our prizes to the camp.   Breakfast over, we packed up and proceeded on our journey, leaving
our huts for the occupation of the next comers.

CHAPTER V.

E travelled on for two days, and still the mountains were not reached; but they grew higher and clearer as we advanced, and we had hopes of getting to them at last. My young cousins bore the journey wonderfully well. When we came to difficult places, her brothers and I helped Kate along, making a seat for her with our joined hands. We could thus make but slow progress, and she entreated us to allow her to walk, declaring that she was not at all fatigued; while Timbo or Jack carried Bella on their back, and with long sticks in their hands trudged on merrily. We caught sight of several wild animals. On two or three occasions buffaloes crossed our path, but at too great a distance for a shot. We killed, however, a wild boar, which afforded a fine meal for our party. Natty and I were a little in advance, when we came to a large village of white ants, such as I have before described. We were examining them, when I saw a troop of gazelles come bounding across the prairie towards us. The wind blew from them to us, and as we were behind the hill, they did not observe us. Our larder at the time was ill-supplied, and so I was anxious to kill one. I rested my gun on one of the turrets of the hill. I was not much of a shot.

but I was improving. The herd came by within thirty yards of us. Just then the leader caught sight of the rest of the party, who were coming up. I saw that I must now fire, or lose my chance. I took aim at the nearest—a doe, with her young one by her side. The mother escaped, but the little creature fell to the ground. In spite of my hunger I felt almost sorry for what I had done, when, running forward, the dying animal turned up its large languishing eyes towards me as it stretched out its limbs quite dead. I am afraid it was but a clumsy shot at best, as I ought to have killed the larger animal. Natty and I, placing it on my pole between our shoulders, bore it in triumph to our friends, who received us with shouts of satisfaction. Stanley also shot a beautiful little squirrel and a number of birds—indeed, a good sportsman in health, with a supply of ammunition, need never, in that part of Africa, be without abundance of animal food; but some of the natives, who have no firearms and are very improvident, often suffer from famine even in that land of abundance.

The buffalo of Tropical Africa—*Bos brachicheros*—is about the size of an English ox. His hair is thin and red, and he has sharp and long hoofs, his ears being fringed with soft silky hair. His chief ornaments are his horns, which gracefully bend backwards. In shape he is somewhat between a cow and an antelope. A herd feeding at a distance had very much the appearance of English cattle grazing in a meadow. They differ greatly from the Cape buffalo, to be met with further south.

Evening was approaching, when the head man of our bearers spoke to Senhor Silva, who instantly called a halt. The black's quick ears had detected sounds in the distance. " He thinks there are elephants out there," said Senhor Silva, pointing ahead. We were then in a thinly-wooded country, and a charge from those monsters would have been dangerous. We saw, however, a clump of trees on one side, behind which Senhor Silva advised that we should take post till we had ascertained the state of the case. The blacks were eager for us to attack them, hoping to enjoy a feast off the huge bodies of any we might kill. As it might expose the young ladies to danger should we do so, even Stanley resolved to let them pass by unmolested. I have not yet mentioned the leader or head man of the bearers. His name, he told Senhor Silva, was Chickango ; but Jack and Timbo called him the Chicken. He was an enormous fellow, and ugly even for an African; but there was a good-humoured, contented expression in his countenance, which won our confidence. His costume was a striped shirt, and a pair of almost legless trousers; while on the top of his high head he wore a little battered straw hat, such as seamen manufacture for themselves on board ship—indeed, his whole costume had evidently been that of a seaman, exchanged, probably, for some articles which he had to dispose of. Chickango, signing to us to remain behind the clump of trees, advanced towards the spot where he expected to find the elephants. Suddenly he threw up his arms, and began shouting at the top of his voice. His cries were answered by similar shouts from a distance; and presently, beckoning to us to come on, he hurried towards the spot whence they proceeded. Passing through a belt of wood, we came in sight of an encampment of blacks seated round their fires. There were upwards of one hundred human beings—men, women, and children. A few of the men were dressed in cast-off European garments, with rings round their arms and legs, their woolly heads being mostly uncovered. Chickango advancing, explained, we concluded, who we were; and we received a hearty welcome from the party. The chief, an old man, sat in their centre, attended by his wives. He was distinguished from his companions by an old battered cocked hat, ornamented with beads. He wore, besides, a checked shirt and a regular Scotch kilt, which had somehow or other found its way into his territory. Senhor Silva then explained to us, through Chickango, that he and his party had come from a considerable distance up the country, where they had gone to collect *caoutchouc*, or india-rubber, the packages of which lay piled up near the centre of the camp. They had collected it some distance up the country, where the vines which produce it grow in considerable quantities. In South America it is obtained from a tree; but in Africa from a creeper of great length, with very few leaves growing on it, and those only at its extremity. They are broad, dark green, and lance shaped. The larger vines are often five inches in diameter at the base, with a rough brown bark. The mode of obtaining it is to make an incision in the bark, but not in the wood, and through it the milky sap exudes. A small peg is then fixed in each hole to prevent its closing, and a cup or calabash secured underneath. When this is full, a number of them are carried to the camp, where the substance is spread in thin coatings upon moulds of clay, and dried layer after layer over a fire. When perfectly dry, the clay mould is broken and the clay extracted from the interior. The *caoutchouc*, though originally white, becomes black from the smoke to which it is exposed while drying. It is in this state brought down to the coast and sold to the traders.

"Oh yes," said David. "This is the material with which Mr. Mackintosh makes his waterproof coats. He found that it could be re-dissolved in petroleum; and by covering two pieces of woollen or cotton stuff with the liquid, and uniting them by a strong pressure, he formed a material through which no water can penetrate. Some time afterwards, Messrs. Handcock and Broding discovered that, by the addition of a small quantity of sulphur to the *caoutchouc*, it acquired the property of retaining the same consistency in

every temperature without losing its elasticity. A further
discovery was made by Mr. Goodyear, who, by adding about
twenty per cent. of sulphur, converted it into so hard a substance that all sorts of articles can be manufactured from it
for which tortoise-shell had hitherto been chiefly used—
indeed, it is difficult to say what cannot be made out of it."

Besides india-rubber, the blacks had several huge lumps of
ebony and a small number of elephants' tusks, which they had
either purchased from other natives further in the interior, or
were carrying down to the coast to sell for the original owners
on commission. The ebony was brought from the hilly country, where alone the ebony-tree grows. It is one of the finest
and most graceful of African trees. The trunk, five or six
feet in diameter at the base, rises to the height of fifty or
sixty feet, when fine heavy boughs branch forth, with large
dark green and long and pointed leaves hanging in clusters.
Next to the bark is a white sap wood, and within that the
black wood. This does not appear till the tree has reached a
growth of two or three feet in diameter, so that young trees
are not cut down. The trunk and even the branches of the
mature tree become hollow. It generally grows in clumps
of three or four together, scattered about the forest.

Nearly all the negro tribes on this part of the coast have the
spirit of trade strongly implanted in them; and I cannot help
thinking that it is so for the purpose of ultimately bringing
about their civilization, which the nefarious slave-trade has so
long retarded. That trade is one of the sins which lies at England's door, and she should endeavour to make amends for the
crime, by using every means in her power for the spread of
Christianity and civilization among the long benighted Africans.
We observed that the men, women, and children were very
busy in the camp—the women cooking and making arrangements for the night, while the children were collecting fire-

wood from the neighbouring thickets. Poor little creatures, I was afraid that some of them might be carried off by panthers or other beasts of prey who might be prowling about in the neighbourhood; but their parents seemed to have no such fear. We were anxious to obtain some more bearers to carry Kate and Bella, as also to assist us in conveying our goods up the mountain. Chickango undertook to make the arrangements, and after a good deal of talking with the chief and then with the people, he pointed out four young men who expressed themselves ready to accompany us. These arrangements being made, we encamped on a somewhat higher spot a short distance from our friends, and soon had huts built such as I have before described. Though we heard the cries of wild beasts in the forest, none ventured near us, as we kept up blazing fires all night.

Next morning, even before our party were stirring, the old chief and his followers were on foot, preparing to continue their march towards the sea-coast. The men, however, sat still, with their bows in their hands, talking to each other while the women were employed in packing up their goods in baskets, which they suspended at their backs, with their children in many instances on the top of them. All the elder children also had burdens, but the men walked along with a haughty air, carrying nothing but their arms in their hands. Saluting us with loud cries, they proceeded towards the west.

We meantime had been employed in packing up, but instead of making Kate and Bella carry burdens, we prepared a litter to carry them. Passing through a dense forest, we saw before us the mountain range we hoped soon to gain. Near the banks of the stream we passed a grove of curious trees with short stems, on either side of which projected huge long leaves with feather-like branches on the top. Amid them was an immense number of clusters of nuts, each larger than a pigeon's egg. Chickango ordered one of the men to climb up and bring down a cluster when he saw us looking at them. On pressing the nuts even with our fingers, a quantity of oil exuded; and Senhor Silva told us that the tree was the *Cocos butyracea*, the oil extracted from which is exported in large quantities from the neighbouring rivers, chiefly to Liverpool. We calculated that the tree had fully eight hundred nuts on it; and as each contains a considerable quantity of oil, it may be supposed how large an amount a single tree produces. I had seen something of the trade on my former visit to the coast, when I was at the Bonny river. We took chiefly English manufactures to exchange for the oil, and a few bales of glass beads from Germany. On entering the river we covered in the deck with a mat roofing, to protect us from the sun and the tropical showers; but before we could begin trading we had to pay a heavy duty to the old king of the territory, of muskets, powder, tobacco, calicoes, woollen caps, and, what he valued still more, several dozens of rum The dealers then made their appearance, and received advances of goods to purchase oil in the interior, for the Bonny itself does not produce the oil. Our next business was to erect a cask-house on shore, in which to prepare the puncheons for the reception of the oil. This was brought down in small quantities by the traders; and it took us nearly four months to obtain about eight hundred puncheons, which our vessel carried. The palm-oil or *pulla*, when brought to us, was of a rich orange colour, and of the consistency of honey. To my surprise, the morning after the first quantity arrived I found a basin full of it on the breakfast-table, and learned that it was the custom to eat it instead of butter; and very delicious it was. By the time it reaches England, it has, however, obtained a disagreeable taste, totally different from what it possesses when fresh. The palm-oil is about the most valuable production of this part of Africa; and the natives are beginning to discover that its collection is far more profitable to them than the slave-trade.

To return to my narrative: we encamped at a short distance from the thick wood, by the side of the stream I spoke of, hoping early next morning to begin our ascent of the mountains. We might have proceeded further, but the spot was so tempting, that, although we had a couple of hours of daylight, we agreed to stop where we were. The blacks soon had the huts erected and fires lighted—an operation they would not have undertaken had their wives been present to do it for them. As we were all very hungry, we immediately commenced our evening meal, some birds we had shot not taking long to cook; while we had a good supply of biscuits, which we had brought, with tea as our beverage.

"This is just such a pic-nic as we had in our last holidays," said Bella, looking round with a smiling countenance. "You remember, Leo, it was by the side of a stream; and you went and caught some fish, and we had them cooked before the fire."

"Oh yes; and I will try and get some fish now," said Leo. "Natty, you will come and fish with me as soon as supper is over."

To this, of course, Natty agreed; and Jack produced a ball of twine, while I fortunately had some fish-hooks in my pocket, which I brought from the wreck. While we were laughing and talking, suddenly a loud roar reached our ears, which made Kate start and little Bella turn pale, while a loud hollow sound, as if a drum had been beaten, followed the roar. Leo declared it was more like distant thunder. Our blacks started to their feet, many of them with looks of terror, uttering the word—Ngula. Stanley seized his gun. "That must be a gorilla!" he exclaimed, examining the lock.

"I hope so," cried David. "It would be worth coming here to see the monster."

"No doubt about its being a gorilla," said Senhor Silva, "but you must be cautious how you approach him. Chickango says he will go with you. He is a good hunter; and, I judge by his looks, a brave fellow."

The ugly black nodded his head, and pointing to the forest, advanced towards it. David and I also took our guns.

"Now be steady," said Stanley. "I will fire first, and if I fail to kill him, David, do you fire; and, Senhor Silva, tell our black friend that he must make the third shot; and Andrew, you must act as a reserve in case of accidents,—but I hope not to miss him."

Stanley and David kept together, while the black and I advanced a little on one side. Turning my head for an instant, I saw Leo and Natty following us. I signed to them to go back, but they seemed resolved to take a share in the expected fight. Each was armed with a long pike, which I knew would have been of about as much use as a tooth-pick should they be attacked by the creature. We made our way between large boulders to the edge of the forest, which seemed almost too thick to be penetrated. I had never felt so excited. My sensations were something like those, I fancy, of a soldier going into his first battle; but from what I had heard of the gorilla, I knew him to be almost as formidable an antagonist as the best armed man. For some time as we advanced into the forest there was a perfect silence, yet we were certain that the monster could not be far off. The trees grew closer and closer together; and as the edge of the forest was turned towards the east, we soon found ourselves shrouded in a thick gloom. Still, so eager were we to meet the beast, that, instead of halting, as might have been wiser, we continued to push onwards. Suddenly a terrific roar was heard proceeding from a spot not

many paces ahead. Had it not been uttered, we might have
gone close up to the creature without perceiving him. Just
then we saw the branches waving to and fro, and a huge monster moving on all fours appeared amidst them. Suddenly he
rose up on his hind legs, holding on to a bough with one hand,
and then striking his breast, from which a loud hollow sound
came forth. He uttered another terrific roar, and grinned
fiercely at us. "Oh, what a terrible giant!" I heard Leo
exclaim behind me. I dared not turn my head or speak to
urge the boys to run back. My attention was rivetted on the
huge gorilla, for I now saw before me that monster of the
African woods. Again he uttered a fearful roar, and beating
his breast and gnashing his teeth, he began to move towards
us. He was not many paces from Stanley, who was a little
in advance. "Steady, friends!" cried our leader. I held my
breath with anxiety; for should my cousin's gun miss fire, it
seemed impossible for him to escape being seized by the
tremendous creature. Then I saw his rifle raised to his
shoulder. There was a flash, which lighted up the monster's
face and the surrounding branches, and then with a terrific
roar I saw it spring forward. Just as I dreaded that Stanley
was about to be seized by its sharp claws it stopped, and, with
a groan almost human, fell forwards on its face, crashing
amidst the bushes, and rolled over on the ground. Even then
I expected to see it rise again and attack us, but the bullet
had gone through its huge chest; and though it made several
convulsive struggles, by the time we reached it it lay perfectly
quiet. Chickango struck it with his spear, but it did not
move, and then he plunged it into its breast.

"Have you really killed him?" cried Natty and Leo, running up to us. "We would have fought him, that we would!"
exclaimed Leo, jumping on the gorilla's body.

Chickango at the same time seized one of its huge paws, and

pulled and shook it violently, and then set up a triumphant shout as a compliment to Stanley on his victory.

"I wish we could carry him to the camp," said Leo. "It would show Kate and Bella that they need no longer be afraid of the monster."

"I expect a sight of it would not much tend to allay their fears," said David, "for it would rather show them what sort of fierce beasts we may expect to find in our neighbourhood."

"What! do you mean to say there are any more of them?" exclaimed Natty. "When Senhor Silva was talking about him the other day, he called him the king of the forest; and so I fancied, of course, that he had no rivals."

"Where he exists we shall probably find others," said David; "though their habitation does not reach further south than we now are: indeed, I did not expect to meet them in this latitude. They chiefly inhabit the country about the Gaboon and other rivers to the north of us."

We found, on measuring the gorilla, that it was within a few inches of six feet in height, while the muscular develop-ment of its arms and breast showed that it could have seized the whole of us in its claws, and torn us to pieces without difficulty; but the art of man and the death-dealing rifle were more than a match for it. Still, as it lay extended on the ground, I could not help feeling as if we had killed some human being—a wild man of the woods, who might, under proper treatment, have been tamed and civilized. David laughed when I made some remark to that effect.

"I suspect, if we were to catch a baby gorilla, and feed it on milk, and bring it up in a nursery, it would prove almost as savage and fierce as this creature," he answered. "He can feed himself and fight in defence of his liberty, but he could never make a coat to cover his back, or light a fire to warm himself, though he might have seen it done a hundred times.

There is no real relationship between a man and an ape, however much similarity there may be between the outer form and the skeleton. In man there is the mind, which, even in the most debased and savage, is capable of improvement, and the soul, which nothing can destroy. In the ape there is instinct, and a certain power of imitation which looks like mind, but which, even in the tamest, goes no further. The most enlightened mode of instruction and the utmost patience will never teach an ape to read or talk; while we know that human beings who have been born deaf and blind and dumb have, by a wonderful process, been instructed in many of the glorious truths which can give joy and satisfaction to the soul of man."

As it was already late, and it would delay us greatly should we attempt to carry it to the camp, we agreed to leave the gorilla where it lay and return for it the next morning. We saw Chickango cautiously looking behind him as we turned our backs on the forest; and he gave us to understand by his gestures that he was afraid a lion or leopard, or some other wild beast, might be following us.

My cousins came out to meet us on our return. The roars of the gorilla had aroused unusual fears in their hearts, and our absence had been so prolonged that they had become anxious for our safety. We kept a strict watch all night; for although we did not again hear the gorilla—indeed, had there been one in the neighbourhood, he would by that time have gone to rest—the sounds of other wild animals frequently reached our ears.

We were up early next morning—the instant there was light—for Kate had made us promise to show her the gorilla. "I may never have an opportunity of seeing another," she said. "I should like to be able to say when we get to the Cape that I have actually beheld one in his native wilds."

As neither Jack nor Timbo exhibited such curiosity, we left

them in charge of the camp with the black men, to pack up, while we proceeded towards the forest. We advanced cautiously, Stanley and I going ahead, with David and Senhor Silva on either side of the young ladies, and the boys bringing up the rear, Chickango acting as scout, a little in advance on one side of us. Every now and then we halted, whenever we observed the branches disturbed. Now a huge ape of the ordinary species might be seen grinning down upon us, and then scampering off among the boughs; or a troop of monkeys would come chattering above our heads, not so easily put to flight. Birds of gay plumage flitted before us from bough to bough; and a huge snake, which had been coiled round a branch, giving a hiss at us, went off among the underwood into the depths of the forest.

"And now, girls, be prepared for a sight of Leo's giant of the woods," said Stanley, turning round when we approached the spot where he had killed the gorilla. "But, hillo! the ground looks alive."

The trunk of a tree lay near. By climbing on it we got a view of the spot where the gorilla had fallen; but, as we looked towards it, scarcely a particle of the monster could be seen. The skin was there and the huge bones and monstrous skull, but nearly all the flesh had been eaten away by myriads of ants, which swarmed about it. So engaged were they in their work of destruction, that they did not attack us.

"Why, they must be drivers," said David, "the *bashikouay*, as the natives call them. They have gained their English name by driving every other species of the animal creation out of their way."

They were not much larger than the common English ant, of a dark brown colour. David, jumping down, caught one, and showed us that he had a sharp head, terminating in a pair of horizontal nippers—very like those of the warrior ants.

In taking one up another had caught hold of his little finger, and gave it a nip which drew blood. Senhor Silva told us that they usually traverse the country by day and night, in trains nearly half a mile long, though only a few inches wide, and, as it passes under the grass, presents the appearance of a huge snake. They also, like the warrior ants, have soldiers who march by the side of the regular column, and the instant any danger appears hurry forward, when the column is either halted or turned backward. Should the difficulty be removed, it again advances. One of their most curious proceedings is the formation by the soldiers of a perfect arch, into which thousands of them weave their bodies, expanding across the whole width of a path where danger is apprehended. Under the arch the females and the labourers who bear the larvæ then pass in comparative safety. It is formed in the following manner. One ant stands upright, and then another climbs up and interlocks its feet with the fore-feet of the first, and then another climbs up, somewhat in the fashion of acrobats. Another couple form the base of the arch on the opposite side, and then others, stretching themselves longways, form what may be called transverse beams, to keep the two sides connected. When thus formed, the creatures hold together so tenaciously, that the whole could be lifted off the ground without breaking. If attacked, they spread themselves on the ground over a space of thirty or more feet, across which neither man nor beast can pass with impunity. It is difficult to force a horse through them ; and a dog will never venture, unless the space is sufficiently narrow to enable him to cross by a bound. He knows well that, should he fall, they would set upon him ; and, before many hours were over, in spite of his strength, entirely consume him. They have been known to attack horses and cattle shut up in a confined space, and to reduce them to skeletons in less than a couple of days. They sometimes enter a

dwelling-house through a small hole, and literally take possession, proceeding across the floor, over the walls and ceilings.

"When I resided in the Brazils." said Senhor Silva, as we
stood surveying the ants at work, " I was one morning seated
at breakfast with my wife and little boy, when I heard outside
the house a great commotion, and in rushed a black servant
carrying the cage of our favourite parrot in one hand and
grasping a number of pet fowls in the other ; while our negro
girl, hurrying in from another direction, and catching up the
lapdog, cried out, " See ! they come—they come ! Fly, senhor.
Fly, my dear mistress—fly, or you will all be eaten up."
Looking down to the ground, towards which she was directing
her alarmed gaze, I saw that it was covered by countless
numbers of white ants, which came swarming in through a
small hole in the wall. I can only liken the appearance of
the insects to a stream of water suddenly bursting into the
house, so rapidly did they make their way through the opening. It was too late to think of stopping it, for the room
was in a few seconds full of them. My wife, taking the advice
of the girl, seized our boy by the hand and fled into the garden.
I followed quickly, for already I felt the ants biting at my
feet. Not for some hours were we able to return, when we
found that our invaders had devoured every particle of food
in the house. They did us, however, an essential service, by
destroying all the mice and cockroaches, as well as other insects
which they encountered, so that on that account we were much
obliged to them ; but there are many instances on record of
their destroying human beings unable to move on account of
sickness, and with no one to assist them. Formerly, it is
said that criminals secured by shackles were laid in their
way ; happily, however, this terrible custom no longer
exists, even among the most savage tribes. They, in most
cases, as in ours, effectually rid a house of mice, and take

but few minutes to devour one, leaving only its bones and hair."

We were glad to leave the wonderful insects to their repast on the dead gorilla, and, returning to our camp, found our bearers ready for starting.

We toiled on all day, ascending the sides of the mountain range. Now we had to plunge into a valley thickly covered with trees, and then to ascend the opposite side, now to proceed along the edge of lofty precipices. Sometimes the ascent was so steep that we were obliged to use our hands as well as our long poles to make our way up it. I was thankful that bearers had been provided for the young ladies; for although they had spirit enough to attempt whatever we did, yet they must inevitably have been much fatigued had they been compelled to walk. Leo and Natty, however, trudged on bravely in our midst; and often indeed, when ascending steep places, took the lead. Chickango, who knew the way, having often before traversed it, was of great use. He also kept a watchful eye on either side of the path, especially when we were crossing valleys, lest a leopard or lion might spring out on us, or any huge serpent might lie across our path. At length we reached a lofty plateau, or table-land, which Chickango informed Senhor Silva extended a long way to the south. Over this, therefore, we resolved to travel, till we could find a suitable spot in which to fix our abode. We purposed remaining there till we could send a messenger towards the Cape Colony, hoping that he might fall in with either traders or explorers or missionaries, several of whom were settled in Damara or Namaqua land. The further we travelled south, the cooler and more healthy we should find the climate. We had no wish, either, to remain longer than necessary in the gorilla region.

CHAPTER VI.

EVERAL days had passed away. Our progress had, of necessity, been slow; but it was a satisfaction to feel that we were going towards the south, and getting nearer to where we might hope to meet with assistance. We had all kept our health, and even my young cousins seemed in no way to have suffered; indeed, they looked stronger and better than they were when they landed. Our bearers, however, had for some time shown a disinclination to proceed. They told Senhor Silva that they had come further than they had bargained for, and evidently began to doubt our intentions. They knew very well that their countrymen were carried off in great numbers by the whites; and stories had been told them about the cruelties practised by those white men, and that they even collect people merely to slaughter and eat them. Although they did not perhaps suspect us of such intentions, yet altogether, in spite of the bribes we had to offer them, they thought it wiser to return to their own people. Senhor Silva promised them that as soon as we could find a spot on which to settle, if they did not wish to remain with us, they should be paid and allowed to depart.

Chickango and Timbo had by this time become great

friends. They were able to converse freely together; and Timbo told me that he was doing his utmost to instruct his countryman.

"Timbo tell de Chicken all about England and Cape Town, and de oder countries of de world, and de big ships, and de rich white men; and, more dan dat, I tell him dat he got soul, and dat white man and black man hab de same God; and if he stay wid us, we treat him like one broder. You see, I no t'ink he go away now."

Not without the greatest difficulty, however, could Chickango persuade his countrymen to proceed further with us. The hills over which we were travelling were covered thickly with wood, so that often we could see but a short distance either on one side or the other. Now and then we came to openings, whence we looked down on the wide-spreading country on either side, partly hilly or undulating, and then stretching away in an even plain, intersected by rivers, till lost to sight. Stanley and Senhor Silva, with their guns, were ranging the country on either side.

"Listen!" cried David, who was walking by my side. "What noise is that?"

I listened.

"It sounds like the roar of breakers on a rocky shore," I observed.

"No," he said; "it must be a waterfall."

Hurrying on, we saw before us a wide lake-like expanse on one side, and on the other a cloud of spray floating in the air As we drew nearer, a broad stream appeared, rushing over a ledge of rocks and falling into a deep chasm below, after which it ran towards the south and east.

"This would be a grand place to settle on," said David. "Where there is water in this region we are sure to find abundance of game; and it will assist us in defending our selves against any attacks of the natives, should they prove hostile."

I agreed with David, and we anxiously looked out for the appearance of Stanley and Senhor Silva, to learn whether they were of the same opinion. When Kate and Bella overtook us, they were delighted with the scene, and agreed that it was just the place where they should like to settle.

In a short time Stanley arrived. He was as well pleased as we were with the appearance of the country around. Senhor Silva had no objection to fixing our abode there, though he would have preferred moving on, in the hope before long of reaching Portuguese territory. Chickango, however, assured us that the country to the south was more difficult to pass over than that we had traversed, and that without men to carry our provisions and goods we could not perform the journey. The matter was settled by our bearers refusing to proceed further. Senhor Silva asked Chickango whether he intended to return with his people or to remain with us. He hesitated; then he seized Senhor Silva's hands, and gave rapid utterance to an harangue.

" He say we good people, he stop. He my broder now. Hurrah !" exclaimed Timbo.

Although Chickango had resolved to remain with us, he could only induce his countrymen to delay their departure for a few days, in order to assist us in putting up our huts. They at once set to work to construct our usual shelter for the night, which would serve until we could erect a more permanent abode. We fixed upon a spot considerably raised above the head waters of the stream, which would defend us on one side from wild beasts, while the ground sloping downwards on the other would enable us to fortify it against either human beings, or lions or leopards. Those creatures will, without difficulty, leap over the highest fences ; and if erected

on level ground, no ordinary means are capable of keeping
them out.   I should observe that there are no tigers in Africa;
their absence, however, as Leo remarked, being more agreeable
than their company.   Stanley and Senhor Silva had been very
successful in their hunt, and had brought back a good supply
of birds and young deer, besides three or four smaller animals.

By Chickango's advice, we built our huts in the fashion of
his people—that is to say, facing each other, so as to form a
street, with their backs to the outside of our little fortress.   As
the river side was altogether enclosed, one strong door at the
other end was sufficient for all the houses.   For the sake of
air, however, we built our huts separate from each other, and we
thus had windows on all sides.   The poles were of bamboo,
and the walls strong pieces of bark, secured by ropes composed of creepers.   The framework roofing was also formed of
bamboos, with thick palm-leaves at the top, kept down by
ropes.   At the inner end was a shed for cooking; and our
street was sufficiently wide to enable us to light a fire at night
in the centre, to prevent the unwelcome intrusion of wild beasts.
Our habitation, though not very imposing, was sufficiently
strong to keep out the wet and rain, and, at the same time,
was tolerably cool.

The two young ladies had one house to themselves; Stanley, David, and Senhor Silva another; the boys and I a third;
Timbo and Chickango had one to themselves; and Jack was
left alone in his glory, he taking a small one at the entrance,
and having charge of the gate.

"You may depend upon me," he said.   "I will always
sleep with one eye open; and if any strange black fellows
come near us, or any savage beasts, I will be up and have a
crack at them before they know where they are."

The bearers, having performed their contract very much to
our satisfaction, received from Senhor Silva a piece of calico,

a knife, and some tobacco, as their payment, with a few beads for their wives, either present or prospective, with which they seemed highly pleased. When they were about to take their departure, Chickango addressed them. What he said we did not understand, but the result was that they agreed to stop two or three days longer and assist us in hunting, whereby they themselves were to benefit by a share of the spoil. They remained at night in the huts they had previously occupied, while we took possession of our new abode. Besides our sleeping houses, there was a large one intended for what Leo called our banqueting-hall. In the centre we constructed a long table, at which we could all sit, with two chairs at the end for the ladies, Stanley, as our chief, having his seat between them, somewhat in the fashion of ancient days, Jack and the two blacks taking their places at the further one. Our bed-places were formed of bamboos raised from the ground. Senhor Silva politely devoted some of his calico to making curtains for those of the young ladies. He had also brought some mosquito-curtains, which he presented to them; for we found that even in that higher region we were not free from those pests of a hot climate.

As I gazed round our new location, I could not help wish·ing that it was the permanent abode of civilized men. Far as the eye could reach, forest and prairie stretched away into the interior, capable of supporting a dense population; and from the health we had hitherto enjoyed, I saw no reason why even whites should not inhabit it; or, at all events, a civilized black community might there, I hoped, be some day established.

As soon as our black friends had agreed to remain, they set off, headed by Chickango, for the purpose of exploring the banks of the stream, to ascertain in what direction we should commence our hunt the following day. They had not been

long absent, when Chickango came hurrying back in a state
of excitement, and called to Senhor Silva.

" They have discovered an hippopotamus higher up the
stream, and beg that we will go out at once and assist in
killing it."

" What ! can they wish to eat one of those ugly brutes ?"
said Leo. " If they are like those we saw the other day, it
will be a hard matter to kill them."

" Nothing comes amiss to them," said Senhor Silva ; " and
we must not disappoint them."

Senhor Silva, with Stanley and Chickango, accordingly
started off, the two boys and I accompanying them to see the
sport. Chickango led us some way up the stream, where, on a
rock among the trees which lined the island in the centre, we
saw a huge monster. He turned his eyes towards us ; but
from the indifference with which he regarded our approach, it
was evident that he was unaccustomed to the sight of man
Chickango shouted out.

" What does he say ?" asked Stanley.

" It is an hippopotamus. You must fire, and hit him under
the ear, and you are sure to kill him," said Senhor Silva.
" The blacks want the creature for food, and you must not
disappoint them."

The water by the side of the banks above the fall was shallow, flowing amongst numerous rocks. Stanley carried a long
pistol in his belt.

" Here, take my gun," he said. " I can hit the creature
with this ; and if I fall, it will not be of so much consequence."

Springing forward, he levelled his pistol, and the huge
beast rolled over into the water and was carried down the
stream. The report, however, brought out several others from
among the trees on the river's bank. They came swimming
down towards the fall. I was surprised they did not make

towards us, and could not help feeling anxious for Stanley's safety. He stood his ground, however. Two or three had passed before he had again loaded. He then took aim at a third. He missed! The whole herd now made for the falls. The body of the first rolled over and over, but the others plunged downwards in a way which showed that they were well accustomed to the feat; and we saw them swimming down the centre to the lower part of the stream. As the last was passing, Stanley took steady aim, and by the way the creature moved, it was evident that it was severely wounded.

The blacks now shouted out again, and led the way down to the lower part of the waterfall. We all followed. How they proposed getting the bodies of the hippopotami out of the river I could not tell, and fully expected that they would soon be lost to sight. There was, however, an eddy, which probably the blacks had observed, and into this both the huge animals were drawn. Still they were at a considerable distance from the land. The blacks, as soon as they reached the banks, began cutting away at a grove of reeds, a species of palmyra. As soon as they were cut, a layer was thrown on the surface of the water. Another layer was placed crossways on this; and so on, till the raft was of sufficient thickness to bear the party. No binding was required, as the reeds were thus sufficiently united for the purpose. With some long poles and some ratan vines cut from the forest, three hunters embarked. Throwing their ropes round the head of the first animal they got up to, they soon towed it ashore, where their companions secured it, while they shoved off for the other. The second was scarcely dead, though unable to defend itself. They secured it to the raft, when it gave a convulsive struggle, and then opened its enormous jaws, which were certainly big enough to swallow one of the men at a mouthful. It was its last effort, however, for it merely grasped the edge of the raft, and the blacks, shoving on, soon brought it to land.

I had now, for the first time, an opportunity of examining an hippopotamus thoroughly. It is a most singular looking animal, which may be described as intermediate between an overgrown hog and a high-fed bull, without horns and with cropped ears. It has an enormous head. Each of its jaws is armed with two formidable tusks, and those in the lower, which are the largest, are nearly two feet in length. The nostrils, ears, and huge eyes are placed on nearly the same plane, thus allowing the animal to make use of its three senses and of respiration, at the same time exposing but a very small part of its body. It is but little inferior in size to the elephant, though its legs are very much shorter; indeed, the belly in the full-grown one almost touches the ground. The hoofs are divided into four parts, unconnected by membranes. By this means it is able to spread out its clumsy-looking toes, and to walk at a quick pace even through mud or in very deep water. The skin is from one to two inches thick, and completely bullet-proof, except behind the ear and near the eye, where it is thinner; and it has a few hairs only on the muzzle, the edge of the ears, and tail. When out of the water it is of a purple-brown hue. In the young animal it is somewhat of a clay yellow, and under the belly of almost a roseate hue; but seen in a clear pool it is a sort of dark blue, or light Indian-ink hue. As we looked at its head we agreed that few animals have more hideous or terrific countenances.

"Why, he would swallow Natty and me up at a mouthful," said Leo, as he tried to lift up the jaws of one of the huge animals.

"Take care! he will bite!" cried out Natty; and Leo, letting his stick drop, sprang back with an expression of horror in his countenance which made us all laugh.

We left the blacks cutting up their prizes, for which, through Chickango, they expressed themselves duly grateful to Stanley.

We found that the young ladies, aided by Timbo, had prepared a sumptuous repast of wild-fowl and venison, to which we now added some hippopotamus steaks. The meat was somewhat coarse-grained, but tasted not unlike beef. Our black friends consumed it eagerly. During supper we discussed our plans for the future. Chickango assured Senhor Silva that he hoped to obtain a messenger to proceed to the south, although he himself would not venture to go alone. He took his meals with us; indeed he was, in many respects, a civilized black. He knew perfectly well how to behave at table; and used his knife and one of the wooden forks Jack and Timbo had manufactured with perfect ease.

At length our black friends, loaded with as much hippopotamus-meat as they could carry, in addition to the various articles they had received as payment, took their departure. We should have been better pleased had they continued with us, as we might then have proceeded further south without the assistance of strangers.

I have hitherto said very little about Natty Page. He had greatly recovered his spirits after the loss of his father, and now showed that there was a great deal in him. He and Leo and little Bella were the life of our party. They, happily, were not troubled with thoughts of the difficulties and dangers before us, and enjoyed the present to the utmost.

"Do you not think, Andrew," said Natty to me, "that if we were to build a canoe we could explore the river and make our way to the south far more easily than by land? Meantime, it would assist us in our hunting expeditions; and we should be able to go fishing or shooting birds, although I should not much like to meet with any of those fierce monsters the captain killed the other day."

"An excellent idea, Natty," I answered. "I will propose it to Captain Hyslop, and I am sure he will agree with you."

Stanley was well pleased with the suggestion, and it was at once agreed that we should carry it into execution.

"I, however, never built a canoe, and should scarcely know how to set about it, although I understand the management of one thoroughly," said Stanley. "I must trust, therefore, to others."

"No fear, captain," observed Timbo. "Jack, Chickango, and I soon do de work. First t'ing find big tree; and Senhor Silva got axes, so we soon cut it down."

Before the day was over we found a large tree, not more than three hundred yards from the bank of the river, which was likely to answer our purpose. The trunk was perfectly straight, the wood soft, and about twelve feet in circumference. The axes our Portuguese friend had among his stores were, however, rather small for the purpose; but yet, if carefully used, we hoped, with perseverance, to have the tree felled in the course of a day. Jack Handspike undertook to act as chief architect, although Chickango and Timbo, I suspect, knew more about the actual work than he did.

"Now, boys," he sung out, "the first thing we have got to do is to place the craft in the right position for launching, so just see that the tree falls towards the river."

Senhor Silva interpreted Jack's remark to Chickango. He nodded, and forthwith cut from the surrounding trees a number of vines, as creepers of all sorts are called. These, with my aid and that of the two boys, he formed into a strong rope. He then mounted the tree by throwing a band round it and his waist, till he reached the branches, carrying the end of the rope with him. This he secured to the top. Descending, he made signs to us to carry it to a distance towards the river, where he secured the opposite end to another tree.

Jack and Timbo, who were expert axe-men, then began cutting away near the ground. First they made a deep notch on the river side, scoring the tree all round. David and I stood by ready to take their places, while Stanley and Senhor Silva went in search of game.

"But what are we to do?" exclaimed Leo. "We do not want to be idle!"

"No, young masters, nor need you," said Jack. "We shall want spars and oars, so do you go and look out for some small trees fit to make them out of, and cut them down."

"That will be capital," cried Natty. "We will soon have a mast and yard ready for you, and as many paddles as we can pull."

The young ladies, meantime, remained in the house, that Kate might teach Bella, and, when the lessons were over, get dinner ready for us. We worked away with a will, the sound of the axes never ceasing, for as soon as Jack and Timbo were tired, David and I stepped into their places.

"See, we shall soon have the trunk through!" cried Jack. "Run and help Chickango, and haul away as hard as you can. We will have the tree down in a jiffy in that clear space."

We gave a loud cheer as we saw the tall tree bending towards us, and hauling with all our might as we ran from it, down it came with a crash. Then, as if it had been some huge creature with long feelers ready to seize hold of us, we rushed at the branches with our axes, and began hacking away at them. We had now to cut off a piece of the trunk of sufficient length for the canoe. Jack wanted to make it thirty feet long; but Timbo advised that it should not be more than twenty feet, that it might be the more easily managed in the stream. As we had no saw, this had to be done with our axes, and, of course, occupied more than half as much time as getting down the trunk. The boughs, also, had to be cut up and cleared away, that we might have an open road to the river. By the time this was done night had come on, while hunger made us all ready to return to the house.

The boys were very proud of the tree they had cut down for a mast. They had barked it completely, and shaped it partially, and now came towards us bearing it on their shoulders in triumph.

"Do you not think we might saw the thick end off?" cried Leo, after he had gone a little way with us. "It is wonderfully heavy, I can assure you, and I do not think so long a mast can be required."

"Better cut it in half at once and make two masts," said Natty. "It is somewhat heavy to carry up to the top of the hill."

"Come, young masters, I see what it is you want," said Jack. "You have cut down the spar, and done it well, and you think that stronger men ought to carry it. Timbo and I will relieve you of it, and you may run on ahead and say we are coming."

However, the boys, after all, were not very willing to give up the spar of which they were so proud, and carried it on a little way further in spite of their friends' offers. At length Jack quietly put his shoulder under one end, and Timbo took the other, and fairly lifted it off their backs. It was high time, for their knees were beginning to shake, and their faces looked very red with their exertions. The mast was indeed a great deal too long for the canoe, and required more than a third cut off.

We found that the young ladies had, as usual, made ample preparation for our supper, and Kate had found time to give Bella her usual lessons. Her instruction was imparted certainly under difficulties. Her only books were a Bible, a small History of England, a Johnson's Dictionary, and a work on natural history. The latter was especially useful to all of us, as it gave a very fair account of many of the animals we were likely to meet with. Senhor Silva had laid in a good stock of paper, pens, and ink. Kate herself was so well acquainted with geography, that she was able to draw maps, and teach her sister without difficulty. History, too, she seemed to have at her fingers' ends, so that Bella not only learned about England, but most other countries in the world.

Next day we all went back to our work. We began first to shape the outside of the canoe—a task we performed with our axes, and at this four could work at once. By Jack's advice we planed off the upper side of the tree, so that the plan of the canoe could be drawn off on it by exact measurements. We first drew a straight line down the centre, and from this measured off the two sides with the greatest care. In the same way the stem and stern were measured with a plum-line. We then turned the log over, and having levelled that side, marked off the keel, thus having it truly in the centre. Natty and Leo had remained to assist in turning over the log.

" Why, that is exactly how I should cut out a model-boat!" exclaimed Leo. " If we had a saw we could shape the bows and stern much more easily, just as I always used to do."

" But you see, young gentleman, we must make use of what tools we have," observed Jack. " By sticking at it, I dare say we shall not be as long cutting out this here canoe as you would have been making a little model."

" Let me see," said Leo. " No; I remember it took me a good month before I got it ready for painting, and even then, I own, from some unaccountable cause, it was somewhat lopsided."

" Maybe you did not use the plum-line, Master Leo."

observed Jack.   "You see there is nothing like that for getting things perpendicular, though I cannot say exactly the reason why."

"There I have you, Jack, then," said Leo.   "It is on account of the centre of gravitation, and a weight let down on the earth always falls perpendicularly to the plane of the earth."

"That may be philosophy, as you call it, Master Leo, but I cannot say as how I am much wiser than I was; only you will see we will get our canoe to sit fairly on the water—neither heeling over to one side nor t'other."

Having got all our measurements correct, we once more put the canoe on an even keel, and then commenced chopping away round the intended gunwale, so as to have the upper works done first.   By Jack's advice she was sharp at both ends, like a whale-boat, that we might the better back out of danger if necessary.

"Come, you are getting on so fast with the canoe, that we shall not have the spars ready if we do not set to work," said Natty.   "Come along, Leo;" and the boys ran off with their axes on their shoulders in high glee.

They had not been gone long when we heard their voices crying out, "Come, come!—quick, quick!"   Stanley, David, and I hurried on with our guns, which we kept ready for use, and soon reached the boys.   They were too excited at first to speak.   "A wild man!" cried Leo.   "A fierce-looking fellow! I thought he was going to run after us, but he did not, and I do not know if he is still there."

"But was he a wild man?" said Natty.   "'He was walking along on all fours, and then he went up a tree. If he had been a man I do not think he would have done that."

"Probably he was a big ape," said David; "another gorilla."

" No, no; not a gorilla," answered Natty; " but I think he was an ape.   He was not so big as the fearful one the captain killed and the ants ate; but he is a big fellow, notwithstanding."

This account of course excited our curiosity, and we all hurried on, hoping to find the creature which the boys had seen.   They led us some way into the forest.

" We shall frighten him if we make a noise," whispered Natty.

" But I say he is a wild man, and I do not think he will be frightened," said Leo.   " Only take care; if he has companions they may rush out and surprise us."

" Whether man or beast, we will be cautious," said Stanley, advancing in woodland fashion, concealing himself as much as possible behind the trunks and undergrowth.

The boys kept close to his side.   Presently they stopped, and pointed to a tree standing by itself in a little open glade. The lowest branch was about twenty feet from the ground, and on looking up we saw spread above it a curious roof of leaves like an umbrella, while seated on a branch with one arm round the tree was a huge ape.   His feet were resting on the stump of a lower branch, while his head was so completely covered by the roof of his nest that it almost looked like a Chinaman's huge hat.   Presently we heard him give a peculiar sound, something like " hew "—" hew," which was answered from a little distance, and looking round, we discovered another roof with an ape seated under it.   We guessed that it was the female, by her having a funny-looking young ape clinging to her, which she held, as a nurse does a baby, in one arm.   We had advanced so cautiously that neither of the animals saw us.   They were smaller than the gorilla; the hair seemed blacker and longer, and more glossy.

" Do not kill the creatures," said David.   " They will do us no harm, and we do not want them for food."

This remark was made just in time to save the life of the old ape, at whom Stanley was aiming.

"You are right," he answered. "I should like to know more about them, however."

"Perhaps Chickango or Timbo can tell us," answered David.

As it was not far off, the boys agreed to go and get them, while we watched the spot. Before long the two blacks came creeping up. Chickango watched them for a little time. Then he spoke to Timbo, who whispered to us:

"He say dat is *Nshiego Mbouve*. He got bald head, wide mouth, round chin, and—see! beard like one old man! He not nearly so strong as gorilla. Dey stay dere; no fear, not run away now."

With this information we returned to the canoe. Timbo advised the boys to keep at a distance from the animals; for should they discover that they were watched, they might come down and attack them. Being somewhat tired with our work, and having made considerable progress, we retired earlier than usual to the Castle; for such was the name we had given our abode. Chickango and Timbo, however, remained behind, keeping their guns with them, and saying that they would give a few more touches to the canoe.

We had scarcely reached the house when we heard a distant shot. Leo and Natty, who had just given an account of the animal they had seen to the young ladies, and were still somewhat excited, ran out to ascertain who had fired. We heard them shouting out—

"There they are! and they're bringing a little chap along with them!"

"It is a young ape," cried Natty.

"No; I tell you it is a small savage—a boy," exclaimed Leo. "See! why, he is walking along!"

This announcement, as may be supposed, made us all rush to the door. Sure enough, the two blacks were seen dragging along a young ape with a handkerchief tied over its head; and even then it was turning first on one side and then on the other, endeavouring to bite its captors.

"I am afraid they must have killed the old one," said David, "or they would not have caught that young creature. That must be the little ape we saw with its mother. No, we did not tell them to let the animals alone; and they do not understand the humane feelings which, at all events, ought to influence us. They probably were surprised we did not kill the creatures at once."

The blacks now came up with their prize.

"We killed de big mother," said Timbo. "Chickango say he go back and fetch her when we make fast de little one, which we bring as playmate for Missy Kate and Bella."

"I doubt if the young ladies will be pleased with their intended companion," observed David.

"Oh, but he will do as a chum for us!" cried Leo. "He is a brave little chap; I like the spirit he shows, doing his best to bite you."

The young nshiego was at once secured in Chickango's hut, for he undertook to take charge of the creature and tame it. David, hearing that the mother was shot, was eager to go and examine her. We accordingly all set off with some poles on which to convey the body. We found on measuring it that it was about four feet high. The skin was black, and many parts of the body were covered with thin blackish hair. It was a far less powerful animal than the gorilla, though its arms were rather longer in proportion to its size. One of its characteristics was its bald head. Its mouth was wider, and the nose less prominent than that of the gorilla. We found nothing but leaves in its inside, which were apparently the food on which it lives. Our young doctor was anxious to secure its skin; and the blacks wished to have its flesh for eating, but to this even Jack demurred.

"No, no!" exclaimed Jack. "I would as lief almost eat one of your people."

This made Timbo very indignant.

"Dis beast no man," he exclaimed; "no mind, no soul. Why not eat him? Chickango say he bery good food."

It was finally agreed that Chickango should cook it outside the Castle, if he wished it, and that he and Timbo should be welcome to feast off it. Senhor Silva and David's curiosity prompted them to taste some of the animal, which they declared to be very delicate, and not unlike venison. They, however, were very unwilling that Kate and Bella should hear of it.

"You know we eat small monkeys without scruple, and I cannot therefore see why we should not eat the flesh of a big one; in reality, I suspect it is the best of the two," observed the young doctor.

Our amusement for some time every evening was endeavouring to civilize our young prisoner, the little nshiego. Leo at once called him Chico, because Chickango had caught him, and *chico* in Spanish means "little."

The mother's skin had been drying on some trees outside the Castle. No sooner was it brought in than the creature recognized it, and, running towards it, placed its hands on the head, and finding that it did not move, broke out into a plaintive cry which sounded like "Ooye! ooye! ooye!" and then it looked up in our faces as if seeking for commiseration. At length it ran up to the doctor, and appeared to appeal to him to restore its mother. Jack, who stood by, watched it with an eye of pity. The little creature seemed to understand his feelings; and at length the sailor took it in his arms and

caressed it, while Timbo carried off the skin and hid it in his
hut. Chico after this always seemed to consider Jack his
particular friend. In a few days it became perfectly tame,
and showed no inclination to run away. I shall have more to
say about Chico by-and-by.

The canoe was progressing. The boys had cut their spars
in a very creditable way, and now commenced chopping out
boards of sufficient width for the paddles. They had, however, ample time for exploring the neighbourhood. The
morning after the capture of Chico they had gone out at an
early hour, when, just as we were beginning breakfast, we
heard their shouts proceeding from the higher ground up the
stream. We ran out, thinking something was the matter.

"We have seen two huge baboons," exclaimed Leo. "If
we had had a gun, we should have killed one of them, at all
events."

David and I accompanied the boys along the banks of the
river for some distance, when they said we must be near the
spot; and directly afterwards we saw two creatures, one seated
on a fallen trunk on the top of the cliff, gazing out over the
stream. I examined them with my glass, which I then handed
to David.

"Those are baboons," he said. "Their faces more resemble
those of dogs than of monkeys; and hideous-looking monsters
they are. It was fortunate you boys did not encounter them.
You must take care and not go unarmed so far from our
Castle."

"I should say they were nearly as large as gorillas."

"Now the sun is shining on them, I can see their huge
black faces. That big fellow on the trunk has a hide of reddish
brown colour, though his head is shaded with light red, and
his limbs are of a fawn colour. He is, I suspect, the *Cyno-
cephalus anerbis*. See! he is sitting down, scowling round

him maliciously, as if in search of an enemy, or meditating on
his own bad deeds.   They always move over the ground on
all fours, and often descend in numbers on a plantation, and
carry off all the fruits they can lay hands on.   We must take
care to keep them at a distance, for from what I have heard
they are as daring as the gorilla, and, though not so powerful,
more mischievous."

"Let us see if we cannot frighten them," said Leo ; and
before we could stop him, he rushed out, clapping his hands
and shrieking loudly.

The baboons gazed at us with looks of astonishment, when
several others, scrambling out from the neighbouring rocks,
assembled in a body.   They seemed to be consulting together
whether they should advance, when Leo and Natty again
shouted.   This seemed to decide them ; and they began,
instead of running away, to approach us in a menacing attitude.   I now saw it was time to fire.   I took aim, and hit the
leader.   He stopped for an instant, and, giving forth a loud
cry between a bark and a roar, turned round, and with his
companions made off into the rugged country up the river.   I
must say I was very glad thus to be rid of them ; for although
I had often seen baboons in captivity, when I thought them
disgusting-looking creatures, in their wild state as they had
just exhibited themselves they looked ferocious and terrible in
the extreme.

David told us they often go hunting in packs like wolves,
and on those occasions do not hesitate to attack the largest
wild animals.   Sometimes they will assault even elephants,
while they without hesitation encounter the leopard and
hyena.   The leopard, however, retaliates, and when he finds
one alone springs on it, and seldom fails to come off the victor.

The mandrills are another species of baboon who inhabit
this region.   They are remarkable for the brilliancy and

variety of their colour. Often their cheeks are striped with violet, scarlet, blue, and purple, which looks not unlike artificial tattooing; the nose is blood-red; the loins, which are almost bare, are of a violet-blue colour, gradually verging into a bright blood-red; the tail is short, and carried erect. Though very fierce in their wild state, they are more easily tamed than the other baboons. I had seen one in a London menagerie, who went by the name of Jelly, and who really knew how to behave himself, as he could sit upon a chair, and drink out of a pewter can, and smoke a pipe as if he enjoyed it.

Every day we met with various small monkeys in whole troops, skipping about the trees, and looking down upon us wherever we went. Kate was much alarmed when she heard of the boys' encounter with the baboons, and entreated them in future not to go from the Castle without a third person well armed.

"But," said Leo, " give me a gun or Stanley's pistols, and I will fight as well as anybody."

"And I will back him up," said Natty.

"Yes; but Leo might miss the wild beast, and you might hit Leo, and so I am afraid you would have a very unsatisfactory account to give of yourself when you got home," said Stanley.

"By which observation, Captain Hyslop, I conclude you are descended from an Irishman," observed Senhor Silva; " for if Natty was to kill Leo, and a wild beast was to carry off Natty, I do not see how they could come and give an account of themselves."

"Had poor Terence O'Brien uttered the expression, I should not have been surprised," said Kate, laughing at her brother. " But I hope such a dreadful event will not occur, and that Leo and Natty will be content not to make use of firearms till they are a little more accustomed to them."

" There I have you, sister," said Stanley.   " How are they to be accustomed to them unless they use them ?   Well, as we are brother and sister, it is not surprising that you should make such a remark ; and I believe our dear mother comes from Ireland, which I suppose will fully account for the same. However, in my opinion, the sooner the boys learn to use fire-arms, under the circumstances in which we are placed, the better.   It is very important that boys should learn to swim, ride, and row, if they are to go out into the world.   I must give them regular shooting lessons.   They will then be able to use their guns to advantage when called upon to do so."

As soon as breakfast was over we hurried down to the canoe. The outside was now completed, and there was ample work for all hands in cutting out the inside.   We commenced with axes, clearing away as much of the wood as we could.   When this was done, we lighted a fire.   We had some pieces of bar iron : these were made red hot, and we were thus able to smooth away the parts the axe could not so well reach.

CHAPTER VII.

E were working away at the canoe : the boys keeping the fire up ; the rest of us heating the irons and burning out the inside ; Jack amusing himself and us by singing a sea-song to the tune of " Come, cheer up, my lads ; " while Chickango was indulging himself in shouting a native ditty of which we could neither make out the words nor very clearly the tune,—it had reference, I fancy, to our canoe-building, to which he was wishing all manner of success. Suddenly a loud, trumpeting sound saluted our ears ; and looking round to ascertain whence it came, we saw far away in the forest a huge elephant, which we naturally concluded had been attracted by our voices. He stood whisking his ears and holding his trunk out in a somewhat threatening manner. Our guns stood against the trees at some little distance. Chickango gave a warning shout.

" Hide behind the trees, or climb up the nearest you can reach," said Senhor Silva.

Stanley seemed in no way disposed to follow this advice, but rushed to his gun.

" Come, boys ! come with me," cried Jack, " and we will be up a tree."

Timbo followed Jack and the two boys. Jack sprang up to a low bough of a tree, and then, stooping down, with Timbo's aid helped up the boys. David had some time before this gone back to the Castle to remain with his sisters. Senhor Silva also seized his gun, and ran off to a distance. Chickango rushed behind a stout tree; whilst I, seizing my weapon, stood by Stanley's side. Just then I recollected that it was only loaded with small shot, which, of course, would not have been of the slightest avail against the monstrous animal. Again the elephant sent forth a loud trumpeting, and rushed towards us. Stanley took aim and fired, but whether the animal was struck or not we could not tell : at all events, it came rushing on furiously towards us.

"Fly behind a tree!" cried my cousin, suiting the action to the word; but the elephant had his eye fixed on me, and scarcely had I reached the trunk of the one I had selected when he was close to me on the other side. Confused by the fearful noise he made, I knew not which way to turn. He seemed in no way disposed to quit me. He kept dodging round and round the tree. "Run to the next tree!" I heard Stanley cry out. "You may get up it and escape him!" Glancing over my shoulder, I saw that the boughs were low down, and being a good climber, I had fair hopes of success. It seemed my only prospect of escaping a fearful death. I watched my opportunity, and when the elephant was on the other side of the big tree, I ran to the next, and springing to a bough, caught it, and soon swung myself up to the next. He caught sight of me, however, shrieking and trumpeting with rage. Even now I did not feel that I was out of his reach. I had just time to scramble up to the boughs above my head, when he was close under the tree. As I did so my cap fell off. I knew that the animal could reach to a great height with his trunk, and did not feel secure till I had clambered up to one still higher. I was then able to look down on my assailant, when I saw that he had seized my cap as it fell; and that probably saved my leg being seized by his trunk, for he could without difficulty have reached it. I could only hope that my cousin would in the meantime be able to reload and kill the creature.

I looked round, but could not see him. Presently I observed two young elephants coming out of the bush, when the old one, giving a glance up at me to see that I was safe, ran towards them and began fondling them with his trunk; the infant monsters (for they were as big as cart horses) returning his caresses with their own proboscises. What had become of my friends I could not tell. Every instant I expected to hear the sound of Stanley's rifle, but it was silent. I now began to fear that some accident had happened to him, for I was very sure that he would not have deserted me. Jack and the boys were, I hoped, safe; and I knew that the blacks could very well take care of themselves. At all events, I thought, Senhor Silva will return—though, to be sure, he is no sportsman, and may not wish to encounter the elephant. By degrees the elephant's anger seemed to abate, and he stood whisking his tail and flapping his ears, playing with his young charges. Still I knew very well that should I venture down, his anger would revive, and he would rush at me as before. I determined, therefore, to wait patiently till my friends could come to my rescue.

As I was looking round, trying to discover where Jack and the boys had got to, I saw the head of a negro moving among the brushwood. I thought it must be that of Chickango; but presently I caught sight of another and then another creeping along like serpents, now moving slowly, now more rapidly. I concluded that their leader had his eye on the elephant, as whenever he stopped they stopped, so that they could scarcely be distinguished on the ground. Each man carried a large

spear in his hand. They increased in numbers, approaching
from all quarters. It was evident that they had tracked the
elephant, and at length closed in on him. They were fierce-
looking warriors, and I could only hope that they might prove
friendly, as from their numbers they would have been awkward
people to deal with as enemies. I expected every moment that
the elephant would discover them, for I thought, by the way
he stopped and looked about, that he was aware that danger
was near. The hunters, however, remained perfectly still, and
I could scarcely have believed that some fifty or more human
beings were close to me, ready in an instant to spring up into
active exertion. I anxiously watched their proceedings, keeping my eye on the man I supposed to be their leader. Once
more I saw him stealthily moving on ; then suddenly springing up from behind a tree, he darted his spear with tremendous
force right into the elephant's neck, and before the creature
could look round, had again disappeared behind the tree. The
young elephants had caught sight of their assailants, but instead of flying, rushed up again to their guardian, one of them
intertwining its trunk round his neck. And now from every
side the hunters started up. Spear after spear was darted into
the elephant's back, till it literally bristled with shafts.
Whenever the creature turned round, he was met by new
assailants, while the young ones were meantime untouched.
The hunters probably knew that they would fall easy victims
when their guardian was killed. The poor creature turned
round and round, apparently to defend its young charges from
the spears of its assailants. Its life blood was now flowing
fast. When the blacks observed this, they continued to plunge
more and more weapons into his body. The young ones
turned round, and, I thought, cast reproachful, if not angry,
glances at the assailants of their guardian. Fresh hunters kept
coming up, and discharging their weapons, again retreated

under shelter. At length the poor elephant could no longer move. Its huge head and trunk fell slowly down, then sinking on his knees, fell gradually over on one side. The poor young elephants were quickly despatched; and the hunters came rushing up, shouting and singing over their prizes. The chief then stuck a club into the ground, with a hideous-looking figure carved on the top of it. On this they all joined hands and began dancing and shouting more furiously than before, going round and round their prey. The chief and others then brought pieces of the meat, which they placed before the idol. This I now knew to be a *fetich,* as all idols as well as charms are called throughout Negro-land. I was afraid every instant that I should be seen. Hitherto there had been little risk of that, as they were all so eagerly engaged in their assault on the elephant. I supposed that my friends must have seen the hunters; at all events their loud shouts would now make their presence known.

Presently one of the hunters looking up caught sight of me in the tree. I thought he would have fallen back with astonishment. The spear he was holding in his hand dropped to the ground, his hand sank by his side, and he kept gazing up, rolling the large white balls of his eyes round and round as if he was going off in a fit. His exclamation drew the attention of his companions to me. Many of them seemed as much astonished, exhibiting their surprise in various grotesque ways. What they took me for I could not tell; probably some strange animal or a spirit of the woods. The latter was, I believe, the case. I made signs to them that I wished to be their friend, putting my hand out as if to shake theirs, and then began slowly to descend the tree. Still they did not seem to have made up their minds whether they would wish for a nearer acquaintance. I tried to explain to them, however, that I had got up the tree to avoid the elephant, and as they had killed it, for which I was much obliged to them, I could now come down without fear. Whether they understood my signs I could not tell, but by degrees I saw that their fears were somewhat quelled. Stopping at each branch to make signs, I at length reached the lowest, when once more stretching out my hand I dropped to the ground, and, in as composed a way as I could, walked into their midst. They possibly had never seen a white man before, and looked at me with as much astonishment as an European who had never heard of the existence of negroes would have looked at them. They now crowded round me, and began to examine my dress. Some put their hands on my face and rubbed it, as if expecting the white colour to come off. Others examined my hands, while one fierce-looking fellow poked his fingers through my hair.

Familiarity, it is said, breeds contempt; and when I found that they were making these advances, I feared that, instead of looking upon me as some superior being as they at first did they might at length ill-treat me. One of them found my cap, which the elephant had thrown to the ground. After examining it and putting it on my head, he instantly pulled it off again and clapped it on his own woolly pate. The chief hunter next seemed disposed to take possession of my jacket. I knew it would not do to show any signs of fear, so rushing at the man who had taken my cap, I seized it from his head and held it tightly in one hand, while I resisted the efforts of the chief to draw off my coat. Having satisfied their curiosity for the present, they at length left me alone, and returned to the carcases of the elephants.

Their first care was to cut off the tusks of the old elephant, which were of great size. This done, they gave vent to their exultation in loud shouts, and then set to work to cut up the flesh. They seemed to prefer the young ones, the flesh of which was divided among the whole party. They next assailed the larger beast, when each man was loaded with as much as he could carry. This work occupied some time.

Considering that it would be wise to take my departure, I was on the point of doing so, when several of the blacks seized me by the arms, and making me understand that I was to accompany them, pointed to a huge piece of elephant flesh, which they signified it would be my duty to carry. This was more than I had bargained for, and I positively refused to go with them. They now began to move off, and two or three attempted to drag me along. I shouted at the top of my voice, resisting with all my power, " Help!—help!—Jack!—Timbo! —Chickango!" I had got some way, and was afraid I should be carried off as a captive, when I heard a shout at a little distance, and presently saw Timbo and Chickango running towards us. They were followed by Jack, Stanley, and David, the two boys bringing up the rear. The hunters, when they saw my friends advancing, faced about, looking at them with glances of astonishment. I heard them all talking together, seemingly asking each other who these strangers could be. They had recovered their spears, which, still red with the blood of the elephants, they held in their hands, ready to dart at Timbo and Chickango. Seeing this, my friends halted, and placing their muskets on the ground, held up their hands as a sign of peace, addressing their countrymen, who quickly replied, turning their glances every now and then at me. Again Timbo spoke to them, and this time with greater effect than at first. The expressions of anger and fear which I had observed in their countenances gradually wore off, and they looked with a more kindly expression towards me. Presently they turned the points of their spears to the ground, when my two friends advancing, took the chief by the hand, and immediately those who held me brought me to the front. Chickango, who had taken up the thread of the discourse, went on speaking very vehemently, and advancing, led me out of the throng. Timbo immediately seized my hand. "Go away—quick now, Massa Andrew. Perhaps dey change deir mind. See I here come de captain and Senhor Silva, and de t'ree young gentlemen. Dese niggers t'ink you white spirit, and no dare hurt you."

By this time the rest of our party had come up, and great was the surprise of the hunters when Senhor Silva addressed them in a language they could understand. I do not know exactly what account he gave of us, but the result was that we were all in a short time shaking hands, and apparently the best of friends. They even begged that we would accept of some of the flesh of the elephants they had killed—to be sure it was part of what they themselves could not carry off. Our new friends now invited us to visit them at their village, which was situated on the summit of a hill about four or five miles off, but so surrounded by woods that we had not seen it.

From their wild looks and manners, we were not sorry when at length they took their departure. Timbo called them Bakĕlĕs, and gave no very flattering description of them. We were thankful that they had not caught sight of our canoe. They might prove friendly, but should they, as was possible, attempt to molest us, it might be advisable to leave their neighbourhood, when we should certainly have a better chance of escaping by water than by land.

"Then we had better have another craft built without delay," observed Jack, who heard Stanley make a remark to this effect. "And now that we know how to set about it, we should get another built much quicker than the last."

On reaching the Castle we found the young ladies greatly alarmed at our absence. Senhor Silva, it appeared, had rushed back and called out David, telling him to come without delay to our assistance. Some of us were up trees, besieged by an elephant, and he could not tell what had become of the captain and the rest of us. Stanley had, it appeared, lost his bag of bullets, and had made off to where we were working at the canoe to find them. This had prevented him from firing at the elephant; and not being able to find his ammunition, he also had gone back to the Castle, from which he found David issuing with a rifle and some bullets. Soon afterwards they were met by Timbo and Chickango, who also had observed the approach of the hunters, and had advised them not to show themselves till the elephant was killed and they were in good humour after their victory. Jack and the boys had in the meantime remained up the tree, and, like me, had been watching all that was taking place. When they saw that I was made prisoner, they had slipped down, and, unperceived, had hastened to the Castle for assistance. Kate, on hearing the account given of the savages, strongly urged us to commence our journey without delay to the south.

"But you see, miss," observed Jack, " it will take us some time to build another boat, and it may be that we shall become good friends with these people before then. Timbo says that if we know how to manage them, we shall be able to get on very well, and maybe we shall do them a good turn, and they will help us."

Our first canoe was now completed, and we lost no time in commencing a second.

"It would be as well, I think," said Senhor Silva, " after we have cut down another tree, to take the bull by the horns, and visit these people at once. If we show confidence in them they are less likely to injure us, and, at all events, we can be on our guard against any treachery they may meditate. I know these native tribes well. If we show that we do not fear them and are prepared to resist aggression, they will seldom venture on an attack."

The knowledge that we had a number of natives in our

neighbourhood, who might possibly be evil-affected, greatly changed the sense of security we had hitherto enjoyed. Although, as far as we were aware, they had not found out the Castle, they might do so at any moment, and come and attack us. We agreed, therefore, never to leave it in future without defenders. We accordingly formed ourselves into two parties. While one went out hunting, or exploring, or working at the canoe, the other was to remain in the fortress for its protection.

Stanley, who always considered it best to meet danger in the face, or, as our Portuguese friend had said, " to take the bull by the horns," was anxious forthwith to pay a visit to our neighbours. He begged Senhor Silva to accompany him, and chose myself and Chickango to be of the party; while David, Timbo, and Jack, with the two boys, were left to protect the young ladies. To increase the strength of our fort, we had driven stout poles all round it, and formed what Jack called ports along the walls, through which our muskets could be fired.

" Do not be afraid, Stanley," said Leo, as we were preparing to set off. " If the blacks come, we will render a good account of them. Natty and I can now fire a musket as well as any of you, and we have been teaching Kate and Bella. We will beat them off, depend on it."

" I do not think the blacks will come," said Natty; " but if they do, I think we ought to fight. There is no doubt about that."

Natty was always more quiet in his remarks than his friend, but I felt sure there was quite as much mettle in him. With our guns on our shoulders, our friends cheering us, we marched down the hill towards the negro village. Senhor Silva had brought a couple of swords, one of which he wore, and the other Stanley had girded to his side, while Chickango and I carried spears. Stanley had in addition his pistols stuck in his belt. Altogether we presented a tolerably warlike appearance, sufficient, we hoped, to make the savages treat us with respect. After proceeding for some distance we found a native path, which, Chickango said, led to the village. He and I by this time were able to converse pretty well, I having learned some of his language, and he having picked up a good many words of English. We did not always, to be sure, understand what each other said, but we made out our meaning by signs when words failed us. An open space at the foot of the hill, where plantains were growing, showed us that we were near the village, though it was so completely concealed by trees that not till we were actually at the gates did we discover it. It consisted of a long street of huts, the doors facing each other, with the blank walls on the outside; very similar indeed to those we had constructed, though ours was on a much smaller scale. At each end were gates, which were now left open. Several men came rushing out, with their spears poised, as we approached, but on Chickango addressing them, they lowered their weapons, and gave us a friendly greeting. Their skins were somewhat lighter than the coast natives. They were a tolerably good-looking race for Africans. Their only dress was a piece of matting worn round the loins, and their ornaments, necklaces formed of the teeth of wild animals, and rings round their arms and legs. The women whom we saw had a number of these rings, while their hair was dressed in various ways with no little care. Nearly all the people had, slung over their shoulders, a grass-cloth bag or purse, very neatly made, in which they carried various articles. The chief had neat grass-cloth mats spread for us, and taking his seat on one, he begged us to sit down on the others. Senhor Silva then presented him with some tobacco, greatly to his delight, and he instantly produced some well-carved pipes, when, fire being brought, he commenced smoking with evident satisfaction. It is curious that savages in both the eastern and western hemispheres should so delight in the much-abused weed. As we sat smoking the calumet of peace—for such we hoped it would prove—the chief informed us that he had been residing at that spot about a couple of years, but added: "I fear we shall soon have to move towards the coast, for already we hear that the fierce Pangwes are advancing in this direction; and unless you white men will help us, we cannot hope to oppose them." He described the Pangwes as a terrible people, and great warriors. "It is said that they eat up all the enemies they kill," he added, shuddering as he spoke. "Such may be our fate; for as they come not only in hundreds but in thousands, we cannot hope to withstand them."

Senhor Silva replied, that although we, as visitors to their country, could not interfere in their quarrels, yet we should be glad to negotiate with their enemies should they make their appearance.

The chief laughed at the notion. "You might as well attempt to turn the torrent of yonder river," he answered, "as to try and induce the savage Pangwes to turn aside from any undertaking on which they have resolved. There is only one thing they understand, and that is the argument which your muskets can hold. If you wish to aid us, you must come with them and plenty of ammunition, and you may then make the Pangwes turn aside to some other district."

I need not further describe our interview, but it ended in a most satisfactory manner, the Bakèlès promising to be our friends, and to help us should we require their aid. Having concluded our visit, we took our leave, and commenced our return homewards. As we made our way through the forest we saw vast numbers of apes playing about the trees, and kept a bright look-out on either side lest we should come suddenly upon a lion or leopard—an animal still more to be dreaded, on account of the distance it can spring. We trusted to the guidance of Chickango, for alone I doubt whether we should have been able to find our way. As we were moving along, suddenly, from among the leaves of a palm, I caught sight of an odd-looking face peering out at us, and apparently examining us with much curiosity. The nose was white, and a thick fringe of white hair surrounded the cheeks. The face was black, and the body was of a dark colour, stunted hair covering the top of the head. I could not help bursting out into a fit of laughter at the odd look of the creature, for it reminded me of an old woman. It was unlike any ape I had before seen. Hearing the noise, it turned up its eyes with a look of astonishment, and then springing to the nearest branch, ran off, twirling its long tail, into the depths of the forest. As we agreed not to kill any animal uselessly, we let it go without firing a shot; for we were too far off from home to carry it with us. It would have been useless to collect objects of natural history, except very small ones, as we could not have conveyed them with us on the long journey we expected soon to make.

We were slowly making our way through the forest, keeping our guns ready to fire at a moment's notice, when, in the thick shade of some closely growing trees, we caught sight of another huge ape, with a young one by its side. At the first glance I fancied it was a gorilla, but the second showed me that it was a differently formed animal, and not nearly so large as that monster of the woods. The mother and infant were gambolling together; the young one tumbling head over heels, then leaping on its mother's shoulders, then rolling about the ground, while she turned it over and over, apparently to the little creature's great delight. We all stopped, and concealing ourselves behind some thick brushwood, watched the creatures.

"There you see a very curious animal," said Senhor Silva to me. "That is what you call in Europe a chimpanzee."

The old animal had an intensely black face, while that of its young one was yellow. It was between four and five feet in height. Suddenly it stood up on its legs and walked a few paces, stooping somewhat like an old man. The position it assumed enabled it to look at us, when, with a sudden cry, it seized its young one in its arms, and sprang up the nearest tree, exhibiting a wonderful agility. I should have had no heart to shoot the creature, and I was glad to see Stanley, though he instinctively lifted his piece to his shoulder, drop it again.

"No," he said; "we must let the mother and child live. David must go without the specimen. We could not carry even the skin home, and one young monkey as a pet is enough So Master Chico shall have no rival."

We had not gone far when we met with two paths leading through the thickest part of the forest, both of which we concluded would conduct us homeward.

"If you and Chickango will take one, Senhor Silva and I will take the other," said Stanley. "We shall have better prospect of sport, and two guns are sufficient to contend with either lions or panthers."

We accordingly separated, I taking the path to the right.

When passing through an African forest, it is necessary for a man to keep his eyes about him in every direction. The path I was following led to an open space, which had been used as a plantation by the natives, I guessed, by finding a few plantains growing on it. Passing across it, we discovered another path, which led further into the forest. "Dis no man path," observed Chickango. "Elephant make it." I had no doubt he was right from the appearance of the opening, the boughs on either side being broken down, and the ground being trampled by the feet of large animals. Though I might have been somewhat proud of "bagging" an elephant, as Stanley would have called it, I had no great wish to encounter one. Still, as from the direction I judged the path to take, by observing the way the sun's rays penetrated the thick foliage, I thought it would lead us homeward, I did not like to turn back. We therefore proceeded along it. Elephants are tolerable road-makers, as wherever they can get through an army may follow.

We went on for about an hour, till we were in a denser forest than I had yet seen. Creepers innumerable hung down from the boughs, twisting round them, and forming a complete network in all directions; while huge fern-leaved plants covered the ground, waving gracefully above our heads. We were in a complete labyrinth of shrubs, plants, and creepers, out of which alone I could certainly never have extricated myself. "No fear, me find a way," said Chickango, "while sun up," and he pointed to a small opening above our heads, through which the sky could be seen. We went on a little way; but it appeared to me that we were getting more and more involved in the mazes of the forest. I looked at Chickango. He had always been faithful. I could not suppose that he now intended treachery; and yet could he have had any private communication with the natives we had been visiting, and agreed to deliver the white men dead or alive into their hands? I was following close behind him, for often there was not room for us to walk two abreast, I should say to creep rather, in and among the underwood. Suddenly he turned round and touched me on the arm, making a significant gesture to be silent. Then he crept slowly on, crouching down close to the ground. I followed his example as well as I could, though it was difficult to get on in that attitude. Presently he stopped, as did I, behind the crooked trunk of a half-dead tree, and listening, I heard a

loud flapping noise as if some machinery were at work. Then
rising a little from my cover, I observed a high brown back,
looking like some vast mound among the foliage. It soon began
to move, and the head and ears of an enormous elephant came
into view. It appeared to me that the creature's eye was
turned towards us. If so, I could not but expect that he would
quickly come to ascertain who were the pigmy intruders into
his domain. Chickango kept crouching down watching the
creature. I fancied I heard a noise on my left side, and
glancing in that direction, I saw another huge head with a trunk
lifted above it. Had I been an experienced hunter, I might
have known how to act. I was afraid of speaking and asking
Chickango what he advised, lest the elephant might hear us.
Even now I could not but suppose that we were perceived.
Presently Chickango sprang to his feet and took aim at the creature's head. I expected to see it roll over, but some white
splinters which flew from an intervening branch told me that
the bullet had been turned aside, though I fancied even then it
must have struck the animal. Instantly its trunk was lifted up,
and, with a tremendous trumpeting which made the forest ring,
and a loud tramping of his heavy feet, he dashed towards us.
I felt that our lives depended upon the accuracy of my aim. It
was my first shot at an elephant, and let any one fancy how the
mighty animal must appear in a rage, and it may be supposed
that my aim was not likely to be very steady. I fired when
he was about a dozen paces off. A practised sportsman
might have waited longer. If my bullet took effect it did
not stop his charge, for on he came directly at us. Chickango sprang on one side. I attempted to follow his example
by springing on the other. As I did so my gun caught in a
creeper, which suddenly whisked it from my hand. I dared not
stop to recover it. The creature's huge tusks were aimed at
me, I thought, and I expected the next instant to be pierced

through or crushed to death, or to find myself tossed high in air by his trunk; but, blind with rage, and somewhat puzzled by two enemies, he ran his trunk against the stump of the tree with such force as to pierce the rotten wood through and through. Over it came with a crash; but so firmly fixed, that for a moment he could not shake it off, while the dust from the rotten wood fell into his eyes. His companion, meantime, was trumpeting away furiously, and advancing towards us, but Chickango and I were concealed from him by the thick wood. Other trumpetings were heard at the same time, which showed us that he had other companions besides the one we had seen. I had not forgotten the way I had escaped the previous day, and glancing round, I saw a thick-stemmed tree directly behind me. I darted round it; while Chickango concealed himself behind another. Our assailant, meantime, disconcerted by the piece of the tree still clinging to his tusks, went crashing on through the underwood till he had got to a considerable distance from us. His nearest companion, fortunately a female, followed him. "Load, massa, load!" cried Chickango to me. Alas! I dared not move to recover my gun, and felt that my only chance of safety till I could do so and reload was to keep behind the tree. The male elephant, having cleared his tusks of the rotten wood, lifted up his trunk, and began trumpeting away as a signal of defiance, which was echoed by his other companions in the neighbourhood. Chickango advanced towards him cautiously as before. Once more he lifted his weapon and fired; but, to my horror, I saw the creature again dash forward. The black, though a bad shot, was active of foot, and sprang behind a tree. The other elephant now came rushing forward; while at the same moment I heard another trumpeting sound close behind me, and saw, to my dismay, the vast heads of numerous male elephants, their huge tusks and trunks towering above the underwood. The

various movements I had made brought me back to the neigh
bourhood of the spot where the elephant had made his first charge,
when I caught sight of my gun at a little distance. I dashed
forward and seized it. The attempt was dangerous, for the
elephants were coming towards me. In doing so I lost sight
of Chickango; but I was sure from his previous conduct that he
would not desert me. I again retreated towards the nearest
large trunk I could see, though all the intervening space, it
must be understood, was filled with fallen trees, and creepers,
and saplings, intertwined as I have before described. To stay
there, however, I saw would be more dangerous than to take
to flight, as with so many assailants, sagacious and cunning in
the extreme, I should at once have been surrounded; and as
to climbing a tree, as I had before done, that was impossible,
as no branches were within reach. I had no time either to
reload my piece. I therefore ran, or rather I scrambled,
among the boughs as fast as I could get along. Probably it
was fortunate my gun was not loaded, as I could scarcely have
prevented the lock catching in the creepers through which I
made my way. For some seconds I thought the creatures
were at fault, for I saw them standing still, looking about for
me. It was sharp work. I did not profess to be an elephant
hunter, but I could not help feeling that I was, at all events,
now being hunted by the elephants. Still I persevered, but I
dreaded every moment to find myself caught by a creeper, or
my head in a noose, and then to see the monsters rush towards me.

The big elephant and his companion whom I had at first
met with now again espied me, and, trumpeting and shrieking,
came dashing through the woods towards me, tearing down
the branches, and trampling the young saplings under foot.
Once I stopped and began to load my gun, but again the
elephants advanced, and I feared that I should not have time

to do so before they were upon me. I continued to retreat in the same uncomfortable way as before, scrambling over the fallen trunks, and expecting to see every moment a huge boa, or some venomous snake, dart out from among them. I was thus scrambling on, endeavouring to increase my speed, when one of the dangers I dreaded occurred. I slipped, and my foot catching in a creeper, held me firmly, while I fell forward amid the tangled mass of creepers, out of which I could by no efforts release myself. I struggled in vain. The trumpetings and cries of the elephants sounded loudly in my ears. Just as I had given myself up for lost, a shot whistled over my head. The nearest animal staggered forward till he was within half-a-dozen paces of me. Another and another shot followed. One of them appeared to have been equally successful with the first, for another elephant, turning round as if to move off, sank hopelessly on the ground. Loud shouts followed, and presently I saw my cousin and Senhor Silva forcing their way through the forest, while Chickango darted out from behind a tree where he had taken post, and fired, just at the moment to save my life. The other elephants, frightened by the sound, lifted up their trunks and rushed back into the forest, the crashing sound of falling boughs and the loud tramp of their feet showing the direction they had taken.

Two male elephants with fine tusks, and a cow, were the result of our adventure. The tusks were too valuable to be left, so we immediately set to work to cut them from the heads of the animals.

"Pity Bakĕlés no know of dis," said Chickango. "Dey come and have great feast, and t'ank us."

Stanley and our Portuguese friend told me that they had been directed towards the spot on hearing our first shot, and that then the trumpeting of the elephants had reached their ears. This had made them hurry forward to our assistance.

I was thankful that they had arrived thus opportunely, or I believe that my adventures in Africa would have been terminated. The weight of the tusks was considerable. We slung them, however, on two poles, and carried them between us, though darkness had set in before we reached the Castle, and we felt not a little anxious lest some lurking wild beast might spring out on us — an event very likely to happen in an African forest at night. We, however, reached home in safety, and gave an account of our adventures. Our visit to the Bakělés' village excited great interest; but when I came to describe our adventure with the elephants, I saw Kate and Bella's colour go.

"Oh, Andrew !" exclaimed Kate, "how dreadful it would have been had the elephants reached you ! How providential it was that Stanley and Senhor Silva arrived in time to save your life !"

David and his sister both expressed a wish to visit the Bakělés.

"I am not quite certain that that would be wise," observed Stanley. "At present they look upon us as a party of warriors who may be gone to-morrow; but if they see young people, they will think we have come to settle, and may perhaps be disposed to try and get rid of us."

Kate was very glad to hear that we had let the chimpanzee go.

"I wish, though, that you had brought the young one home; we could very well have taken care of it, and Chico would like to have a playmate," exclaimed Bella.

"Possibly Chico and Chim might have quarrelled instead of played together," observed Senhor Silva; "and I suspect you will find Chico sufficient to look after."

Senhor Silva, though accustomed to the climate, was not so strong as most of us, and the morning after our long expedi-

tion he was unable to rise from his couch.   David said he had
a bad attack of fever; and as the day wore on, he became
delirious, and caused us great anxiety.   He had endeared himself to us by his kind and unpresuming manners; besides which
we knew that he would be very useful in enabling us to travel
through the country—indeed, without his aid the difficulties
of accomplishing the journey would be very great.   Anxious as
we were, we could not all of us remain at home.   David therefore stayed behind with the two girls to attend on our sick
friend, and Stanley begged me to accompany him on a shooting
expedition with Chickango, while Jack, Timbo, and the two
boys continued working on the second canoe.   We were anxious
to shoot some pigeons and small game for our larder; though
I suspect Stanley would have been better pleased to come
across some of the larger animals of the forest.   We had bagged
a good many birds, when a beautiful little gazelle came bounding across our path.   It put me in mind of an Italian grey-
hound, only it had a longer neck and was somewhat larger.
I was quite sorry when Chickango, firing, knocked it over.
It was, however, a welcome addition to our game bag.   He
called it Ncheri.   It was the most elegant little creature I
met with in Africa among the numberless beautiful animals
which abound in the regions we passed through.

We were at the time proceeding along the foot of a hill.
Scarcely had he fired, when a loud trumpeting was heard, and
directly afterwards we saw a negro rushing through the underwood, followed by a huge elephant.   "Up! up the hill!"
cried Chickango, suiting the action to the word.   I followed,
for as we were wishing to kill birds alone, my gun was loaded
only with small shot.   The elephant made towards us.   The
negro stranger came bounding on.   Chickango and I had got
some way up the hill, but Stanley, who stood his ground, was
engaged in ramming home a bullet.   The elephant had all the

time been keeping one eye on the black and one on us. When
I thought he was on the point of seizing my cousin, he suddenly turned on his first assailant. The black darted to a
tree, when the elephant, seizing him with his trunk, threw
him with tremendous force to the ground. This enabled
Stanley to spring up after us ; and the hill being very steep,
with rolling stones, we hoped that we were there safe from the
attacks of the now infuriated beast. It cast a glance at the
unfortunate black, who was endeavouring to crawl away along
the ground. Again the elephant was about to seize him with
his trunk, and in an instant would have crushed him to death,
when Stanley, raising his gun, fired, and struck the creature
in the most vulnerable part—behind the ear. The ball must
have entered the brain, for, sinking down instantly, it rolled
over, and, we thought, must have killed the black by its weight.
We hurried down, hoping that there might yet be time to
save the poor fellow's life, regardless at the moment of our
victory, which, with hunters in general, would have been a
cause of triumph. As we got round, we found that the black
had narrowly escaped being crushed to death ; indeed, as it
was, his legs appeared to lie almost under the monster's back.
We drew him out, however, and to our satisfaction found
that he was still breathing. Chickango said that he belonged
to the Bakëlés, and was probably a chief hunter among them.
As, however, we were much nearer our own abode than their
village, Stanley and I agreed to carry him with us, somewhat,
I fancied, to Chickango's astonishment. "Oh ! he black
fellow, he die ; what use carry ?" he remarked. Of course
we kept to our own opinion, hoping that with David's skill
the poor man might recover. He was unable to speak, and
was indeed apparently unconscious.

"Had my rifle been loaded with ball, I should have saved
that poor fellow the last fearful crush ; and in future we must

not go without one or two of our fowling-pieces loaded with ball," observed Stanley, ramming down a bullet into his rifle.

Chickango and I did the same. We then constructed a rough litter, on which we placed the injured negro. We bore him along, my cousin and Chickango carrying the head and I the feet part of the litter. We found the weight considerable, especially over the rough ground we had to traverse, but the life of a fellow-creature depended upon our perseverance. Chickango carefully noted the spot where the elephant lay, that we might return as soon as possible for some of the meat and the tusks, which were very large. We reached the spot where our friends were cutting out the canoe just as they were about to leave it, and we were thankful to have their assistance in carrying the stranger. Kate got a great fright seeing us coming, thinking that one of our party had been killed. David instantly applied himself to examining the hurts of the negro. He found that his left arm had been broken, and the ribs on the same side severely crushed. " The injuries might be serious for a European," he observed; " but the blood of an African, unheated by the climate, escapes inflammation, and I have hopes that he may recover." Senhor Silva had recovered his senses, though still very weak, and when I went to see him, he expressed his gratitude for the attention with which David and the young ladies had treated him. Chickango was very eager to set out immediately, in order to bring in the elephant's tusks and some meat, but Stanley considered that it was too late in the day, and put off the expedition till the following morning.

We were somewhat later in starting than we had intended— Jack and Timbo accompanying Stanley, Chickango, and I. We carried baskets and ropes, to bring with us the ivory and a supply of meat. On reaching the spot, however, where the huge monster lay, we found that others had been before us.

The tusks were gone, and a portion of the flesh. Innumerable birds of prey, also, were tearing away at it, or seated on the surrounding trees devouring the pieces they had carried off, while several hyenas, already gorged, crept sulkily away, doubting whether they should attack us or not. The spectacle was almost ghastly, and it showed how soon a mountain of flesh might disappear in that region. Chickango was greatly disappointed, as not a particle of flesh which he could touch remained, while, of course, we regretted the loss of the valuable tusks. On our way back, we caught sight of a number of beautiful little monkeys skipping about in the trees. Chick-ango called them oshingui. They were the smallest I ever saw. Below the trees where they had their abode ran a small stream; and Chickango told me they were very fond of water, and were never found at a distance from it. On the same trees, and playing with them, were numerous birds, called monkey-birds from their apparent attachment to those creatures. We saw another very beautiful little bird, with an extremely long flowing tail of pure milk-white. It had a crest on its head of a greenish black, and its breast was of the same colour, while lower down the feathers were of an ashy brown. Snow-white feathers on the back rose up, like those of the birds of para-dise, to which it had a strong resemblance. Soon after this I saw some creatures on the ground, and catching hold of one of them, I found that it was an enormous ant of a greenish white colour, with a head of a reddish black. The fangs were so powerful that when I put my fingers to them, they literally tore a piece of flesh out.

"Why, these creatures would eat us all up, if we were to encounter them as we did those the other day," I remarked.

"No fear, massa," answered Timbo. "Dey no come in same way. Dey no go into house, no climb tree, and only just a few hundred or t'ousand march together."

It was satisfactory to hear this, for really I felt that should an army invade us, we might have more reason to dread them than the blacks themselves. I was not sorry to miss the elephant flesh, for I had not forgotten the tough morsels we had placed between our teeth when presented to us by the friendly blacks soon after we landed.

CHAPTER VIII.

UR first canoe had been ready to launch for some days, and we were eager to try it. We had, however, to cut a road through the brushwood down to the river's bank before we could do so. This task accomplished, placing it on rollers, the boys assisting, we easily dragged it down to the water.

"There, Master Leo, I told you she would not be lopsided," exclaimed Jack. "Not she; see! she sits on the water like a duck; and them paddles will send her pretty briskly through it, depend on that."

We all jumped in, and eagerly paddled about, well pleased with the success of our undertaking. Though capacious, however, it was evident that she would not carry the whole of our party and luggage, and I was glad therefore that our second canoe was nearly completed.

"We will have races!" cried Leo; "Natty shall steer one, and I the other. Won't it be fun!"

The boys, taking the paddles, showed by the way they handled them that they would soon be able to manage her. They wished, indeed, to start at once down the river, but as it was already getting late, we were compelled to return to the shore. We found a secure place where we could conceal the canoe under some bushes, and having done so, returned home-wards. Senhor Silva was somewhat better, and the strange negro had sufficiently recovered to speak. He told Chickango that he belonged, as we supposed, to the village we had visited, that his name was Igubo, and that he had the reputation of being one of the best hunters of the tribe. " And so I am," he added; " but had it not been for my white friend there, I should have been slain at last by my huge enemy, of whose brothers I have killed so many." Though he could have had but a glance at Stanley he recognized him at once, and begged Chickango to thank him for saving his life.

The next day the boys were very eager to go out in the canoe. " No, no, young gentlemen, time enough by-and-by," said Jack. " You come and help Timbo and I to finish off the other, and we will get on with it while the Captain and Mr. Crawford take a cruise."

" But, I say, we have not settled what they are to be called," exclaimed Leo, as we walked along.

" I have been thinking about that," said Natty. " What do you say to calling one the *Panther*, and the other the *Leopard*? They are proper names for this part of the world."

" And so would be the *Crocodile* and *Hippopotamus*," said Leo; " and as they are water animals, those names would be more suitable."

" But they are not pretty names," argued Natty. " Would not the *Giraffe* and *Gazelle* be better?"

" We ought to have got Kate and Bella down to name them," exclaimed Leo.

" Come, what do you say, Mr. Crawford?" said Natty. " Do not you consider the *Giraffe* and *Gazelle* are two pretty names?"

" They are prettier than the others," I replied, " though they are not quite so appropriate perhaps; but as all sorts of

names are given to vessels, I do not know why our canoes
should not have the prettiest names we can find."

At last Leo came round to Natty's opinion, and it was agreed
that our two canoes should be called after the names he proposed, the first launched being called the *Giraffe*. The boys,
I saw, were very anxious to accompany us, but still they
went away with a good grace with Jack and Timbo. We
hoped to obtain a good supply of wild fowl, and perhaps to
shoot some larger game from the banks. Though I had my
gun with me, I assisted Chickango in paddling the canoe,
while Stanley sat with his gun ready to shoot whatever might
appear. We had knocked over a good many wild fowl, which
made us wish that we had a dog with us to bring them out,
as we had a good deal of trouble in rowing after them. At
length Stanley shot a beautiful flamingo, which went away
paddling down the stream at a great rate. We pursued. We
were not far from the banks, when suddenly I felt so tremendous a shock, that I thought we must have run on a rock,
and immediately afterwards a huge head appeared above the
water and dashed towards us. The hippopotamus, for such it
was, and a very large one, seized the boat by the gunwale, and
threatened to overturn her. At the same moment several
other monsters rose with their snouts above the water. I felt
that we should have a poor chance of escaping if the canoe was
upset, for I thought that the monsters would immediately make
at us and tear us to pieces, or swallow us whole, for their
mouths seemed large enough to take any one of us down at a
gulp. I seized my gun, as did my cousin, who sprang to his
feet, and levelled his piece at the monster's head. " Fire!
massa, fire! or he upset boat and kill all we," cried Chickango,
leaping up to the bow of the boat, and holding up his hands
with a look of horror. I heard the wood crunching under the
creature's teeth. Stanley, who never lost his presence of

mind, balancing himself in the bow of the boat, took aim, and at the moment I expected to find the boat dragged under, and probably we ourselves attacked by the other monsters, he fired. The bullet struck the creature in its most vital part, near the ear, and penetrated the brain. It opened its huge jaws and sank back into the water, beneath which it disappeared, while its companions, alarmed by the report, swam off, leaving us unmolested.

There we were, floating calmly on the stream, and I could scarcely believe that an instant before we were engaged in a fearful encounter. The canoe, however, gave evidence of the power of the creature's teeth, for part of the gunwale, though it was of considerable thickness, was literally crunched up. Several holes were made in the bottom, through which the water was running. We soon had out our knives and set to work to plug the latter, which we quickly did, before much water had rushed in, and that was soon bailed out with our hats. Our canoe had received too much damage to allow us to continue our voyage, and we therefore paddled back, hoping that we might never again be engaged in a similar adventure.

" You see, young gentlemen, it's just as well you did not go in the canoe," observed Jack, when he saw what had happened. " Why, that creature would have bitten you in two if he had caught you in his jaws just as easily as you would crack a nut. It will take us a pretty time to repair this damage. However, it is as well matters are no worse. Take my advice, in future we will go cruising in company, and if a beast like that munches up one canoe, we shall at all events have the other to get home in."

As most of the next day was spent in repairing the canoe, we did not go off in her. The young ladies I found had become very anxious for a change. Bella complained much of not being allowed to run about outside the Castle by herself.

" Could not you find me some pretty animal to ride upon ? ' she said. " I have seen many passing along in the distance, and if you could catch a couple you could soon tame them, and then Kate and I could ride about with you wherever you go."

" What were the animals like ? " asked Stanley.

" Something like horses, or perhaps large donkeys, but they galloped along so fast that I could not very well distinguish them," she answered.

" They must have been zebras or quaggas," said David; " though, if Bella has seen them, I do not know how we could have missed them."

" Because we have been up on the height and can look over the country, while you have been either busy inside or down in the valley," answered Bella. " Is not that a good reason?"

" I am afraid, however, that even if we were to catch a quagga for you, we should have a hard task to tame it," said David; " but we will try what we can do; perhaps, however, we shall find some other animal which will answer the purpose. What do you think of an ox ? They are used more to the south, and make very good steeds, though a little difficult to guide perhaps."

" I will tell you what ! " exclaimed Leo. " If the rest will not go to the south, what do you say to starting off with Natty and I, and we will have an independent expedition, and take Chico with us. Natty and I will paddle and you shall steer, and Chico can sit in the bows and keep a look-out ahead. What do you say to that, old fellow ? "

The ape had at that moment entered the room, and walked up to Leo, whom he looked upon as his especial playmate, though he seemed to consider Jack his chief protector.

I was glad to find that Senhor Silva.was improving. Our negro guest was also much better, and seemed anxious to return to his people. His wives and children would be looking for him, and he thought he could very well make his way through the forest to his home. David, however, persuaded him to stay a few days longer, till his arm and ribs were properly set.

Two weeks passed away without any unusual occurrence. The other canoe was now finished and ready for launching, but the heat of the weather prevented us from willingly making any exertion, and had it not been for the necessity of procuring food, on many days we should not have left the house. We discovered at a little distance the remains of a deserted village, and outside it grew a number of plantains, as well as pumpkins, and other fruit, which, although not so good as those carefully cultivated, were very valuable. We also found many wild fruits growing in the forest; pine-apples, especially, were very fine, and there were nuts of various sorts. Chickango discovered a quantity of ground or pea-nuts, which, though bitter, and somewhat unpalatable, were very nutritious, and he and Timbo ate them readily.

At length our guest was well enough to take his departure. His two countrymen accompanied him for some distance, and Senhor Silva had generously given him several articles which he valued highly—a few yards of cotton, a knife, and some tobacco were among them. He begged Timbo and Chickango to express his gratitude, and I really believe, from the expression of his countenance, that he felt it.

Two days after this, early in the morning, we were surprised to see him approaching the Castle. I went out to meet him. He took my hands, and looked into my face with an imploring glance, which showed that he was much distressed, and then accompanied me into the Castle. The moment he saw David ne ran up to him, and then pointed in the direction of his own home. Then he ran to Leo and Natty, and stroked their heads, as if he was weeping over them. Timbo, who had been in the cook-house, now came out, and having exchanged a few words, Timbo said, "Igubo got home, found children bery ill; want doctor come cure them."

This was plain enough. "Tell him I will go gladly," said David; "but either you or Chickango must accompany me to interpret."

"I will bear you company also," I said. "I feel sure we can trust to him, but his people may not be so well disposed, and if we all three go armed we may make them respect us."

Directly breakfast was over we set out, greatly to Igubo's satisfaction. He hurried along, leading us through elephant tracks, till we reached a path formed by the natives which led to the village. Igubo conducted us immediately to his house, round which a number of people were collected, and inside was a man with his face painted and his hair dressed out with strange ornaments, performing all sorts of antics.

"Dat de fetich man," said Timbo. "He do no good. He t'ink he enchant de sick children. He one 'postor."

"Little doubt about that," I observed; "but we must take care not to offend him. But you tell them that white man's doctor has come, and that if he will go and carry on his incantations outside we will go inside and try ours, and there can be no doubt that the two working together will produce more effect than one alone."

"You no t'ink dat, Massa Andrew," said Timbo, looking up in my face. "No, I only tell dem he go out, we go in. White man know how to cure children better dan de black."

We found two fine boys about twelve and fourteen years old, both in a raging fever. David, I should have said, had come provided with a few medicines, which he thought most likely to be of use, and he now sent all the people out of the house except the mother of the boys and our friend. "Tell nim," he said to Timbo, "that he must get me some pure water." This was easily procured from a stream which came rushing down the side of the mountain at no great distance. David gave each of the boys a cooling draught, and made their parents understand that they were to take no food except such as he ordered. He watched by the children till they at length fell into a profound sleep, charging Igubo not to allow any-body to enter the house. David then proposed that we should take a turn through the village, of which we had not seen much on our previous visit. I need not again describe the village. We had not got far when we met several slaves bringing us a number of fowls, some bunches of plantains, and baskets of cassava. These they placed at our feet with a message from the chief to say that we were welcome, for he had heard of our brave deeds. We of course received them, and they were carried to a sort of verandah in front of Igubo's house, while through Timbo we returned our thanks to the chief. He himself soon afterwards made his appearance, fol-lowed by several attendants. Unless by his anklets and necklace, and the rich tattooing on his breast, he was not to be distin-guished from the rest of the people. His only clothing was a piece of fine matting, worn round the waist in the form of a kilt.

David was unwilling to leave the boys, and we therefore con-sented to remain till the following day. They were then somewhat better, but when we proposed going their father entreated that we would remain. David explained that he was wanted at home, that one of our party was sick, and that if Igubo would follow his directions the boys would probably recover.

" Dat's de bery t'ing dey will not do," said Timbo. " He say, if you go, de boys go too. We make carriage and take dem."

" The best thing, probably, that can be done," said David ; and we accordingly agreed to let the boys be brought with us. The litters were soon constructed, and were by David's

advice covered over thickly with branches of trees, so as com
pletely to shade them from the heat of the sun. Eight stout
fellows undertook to carry them, and all things being ready,
we bade farewell to the chief, who, however, seemed rather
angry at our departure.

"He no good man," said Timbo, as we came away. "Better
go dan stay. I find out he take elephant's tusks and de meat
de oder day, but he no tell us, lest we ask to have dem again."

We considered it wise not to say anything about the elephant's tusks, and, glad to get out of the village, we proceeded
homewards.

"Whom have you brought?" exclaimed Leo, when he saw
us arrive.

When we told him, he and Natty expressed themselves well
pleased at having some companions. "We will look after
them," said Leo.

"And I will teach them to read," exclaimed Natty. "I
hope they will not want to be going away, though. We must
nurse them in the meantime, and try and get them well."

"Poor little fellows," said the ever kind Kate, when she
saw them. "We will do all we can for them, though they
look very ill."

The eyes and cheeks of the young negroes were sadly sunk,
for fever makes the same ravages in their frames as it does in
those of white people. The father, though he saw his boys in
safe keeping, still seemed unwilling to leave them. He had
done what was quite contrary to the customs of his people,
and he told Timbo he was afraid, if he was long absent,
that the rest of his family might be ill-treated. He accordingly, after looking affectionately at them, and expressing his
thanks to us all, but to David especially, took his departure.
I should have said that we brought away the presents made to
us, which proved a welcome addition to our bill of fare.

CHAPTER IX.

"WHEN are we to see the *Giraffe* and *Gazelle* launched, and to have our promised excursion on the river?" asked Kate, the evening after Igubo had left us.

"Oh do, Stanley!" cried Bella. "It is cruel to keep us so long shut up like captive princesses in your Castle, and as the natives are friendly and you can avoid the hippopotami, there can be no danger."

"The *Gazelle* is not yet launched," answered Stanley; "but as soon as she is in the water you can come and see her."

"Oh, but we should like to see her at once, and help you to launch her," said Kate. "If you will start to-morrow morning as soon as it is daylight and the air is still cool, we will accompany you."

The young ladies gained their object, and we were all on foot even before the sun had risen, ready to set out. They would not wait for breakfast, but insisted on carrying provisions and a kettle to boil our tea. David wished to remain to look after his patients, and Senhor Silva was not yet sufficiently strong to bear us company.

"Remember we are to paddle your canoe, girls," cried Leo; "and Andrew will steer for us; and if Timbo will come with a musket or spear, to do battle with any hippopotami or other river monsters, we will allow him to go also."

As we had the rollers with which we had launched the other canoe, and the road had already been cut, the labour of dragging the *Gazelle* to the water was much less than it had been in the former case. We all cheered as she was launched into the water.

"May you bound over the waters of the river as your namesake does over the prairie," exclaimed Bella; "and carry us safely to the south, there to end your existence in a respected old age!"

"Bravo, Bella!" cried Leo, clapping his hands. "You have uttered my speech to perfection, and now you shall have the pleasure of the first paddle our new craft has made. Come, Andrew, come, Timbo, we will lose no time; we can get back for breakfast."

The *Gazelle* floated even more gracefully than her sister canoe. The boys jumped in with their paddles, and Timbo and I holding her to the bank while the ladies stepped in, we followed them, the black taking his place in the bow with another paddle, and I sitting in the stern and steering with a fourth. Chickango and Jack were in the other canoe, and were soon after us.

"Come, let us have a race; we will beat you!" cried Leo, flourishing his paddle; and Natty seconded him, though he saw very well that Timbo and I were really doing most of the work.

We pulled rapidly down the stream, startling numerous birds, some with beautiful plumage, greatly to the delight of Bella. We had not gone far, when a huge head appeared near the bank.

"Oh, what a monster!" exclaimed Bella, shrieking with alarm. "That must be one of those dreadful river-horses which so nearly ate you all up the other day."

"Oh no ; he only nearly bit the boat in two," said Natty; "and we will not let him come near you now."

"We will keep out of his way, at all events," I observed, turning the canoe round.

Stanley just then fired at a water fowl, and immediately several dark heads rose above the water to see what was the matter, and a huge monster, not hitherto perceived, came rolling off the bank ; but he, as well as his companions, quickly disappeared beneath the surface. Remembering what had be-fore occurred, I could not help dreading that one of them might rise up and strike the bottom of our canoe.

"Don't you think we had better go on shore?" said Bella, looking back on the spot where the river-horses had appeared. "Kate, you will want to be there some time before Stanley, to get the breakfast ready."

Little Bella's courage had evidently oozed away. However, as I knew it was possible that one of the hippopotami might strike us, we paddled up the stream as fast as we could go. Soon afterwards I caught sight of another creature resting on a sandbank, with a hideous long snout and a scaly tail and short thick legs. It was a monstrous crocodile.

"Oh do, Andrew, make haste and get on shore !" exclaimed Bella. "What a horrible creature ! I did not expect to meet with such monsters."

I tried to comfort her by assuring her that the crocodile would not attack us, and would more likely swim away than follow us. On landing, we hauled up the canoe, and then commenced collecting sticks for a fire. Kate's kettle was soon hissing merrily, suspended by a high tripod over the fire, and by the time the provisions were spread, Stanley and his companions had arrived. While we were so engaged, we saw, approaching among the trees, a black man, with a shield on one arm and a spear held in the other hand. His arms and part of his body were tattooed in curious lines. Round his neck he wore a necklace of alligator's teeth, while his hair was so dressed as to form a long tail behind, and his beard was twisted into two curious horns, which stuck out from his chin. Round his loins was the skin of a wild beast, and at his side a broad short sword in a sheath; a sort of cross-bow hung at his back, with a quiver full of small arrows. Altogether, with the shield and spear I have mentioned, he looked a formidable warrior to those who were not possessed of fire-arms. The shield, though capable of turning the darts and spears of his equally savage foes, would have availed him little against a modern rifle ball.

Bella eyed the warrior with a glance of terror.

" Do not be afraid," said Natty, placing himself before her. " Leo and I will fight for you."

"Yes, even though there were an hundred such fellows." said Leo. " He looks very different from our friend Igubo. I wonder what he has come for."

Chickango advancing, a conversation ensued which lasted some minutes. The countenance of the warrior fell. We saw him glancing now over one shoulder, now over the other. Then suddenly he turned, and without uttering another word, ran off as fast as his legs could carry him through the forest. Chickango, who had his rifle in his hand, raised it. Stanley shouted to him not to fire, and while he hesitated, the stranger had darted behind the trees. The black returned, uttering words which, though incomprehensible to us, showed that he was very angry. At length, when somewhat calmed, Timbo, who had been unable himself to understand what was said, learned from him that the stranger was one of a band of Pangwes who were advancing towards the territory of the Bakélés. He had come, apparently unaware that there were inhabitants so near. He had first begun to threaten us with the vengeance of his people should we oppose their progress; but on Chickango telling him that a large number of Bakĕlĕs were in the neighbourhood, and that, should his people venture to come that way, they would speedily be driven back and destroyed, he had become alarmed, and so, in spite of his boasting, afraid of being captured, had taken to flight. Still the account which Chickango gave of these Pangwes made us very anxious. The people of his tribe, he said, had for long been at war with them, and had frequently been defeated. They had come from a long way off in the interior, and year after year had been advancing towards the coast. They were not only fierce and cruel warriors, but cannibals, and capable of committing every atrocity.

"What do you think about it all?" said Stanley to Timbo, who had been interpreting Chickango's account.

"Dog dat bark not always bite, massa," answered the black. "Me t'ink dat dey see our rifles and run away."

"I am of Timbo's opinion," I could not help observing "However, we must send and let our friends at the village know of the approach of their enemies; but unless we are attacked, we must on every account avoid fighting. The sooner we can embark and proceed on our voyage the better."

"I believe you are right, Andrew," observed Stanley; "but still I do not like the thought of running away; besides, we cannot leave those two black boys to the mercy of the savages, though if we carry them with us, their father will not know what has become of them."

"I tell you what I do, massa," answered Timbo; "I go and tell Igubo that he come and fetch dem, and den we send out scout to know what de Pangwes are doing."

Our further boating for the day was, of course, put an end to; and having concealed the canoes in the thick brushwood which grew down to the river's bank, we proceeded homewards, with the exception of Timbo, who hastened off to the Bakĕlés' village.

Senhor Silva looked very grave when he heard what had occurred. " Those Pangwes are fierce fellows," he said, " from what I know of them ; and though they may not venture to come within range of our firearms, yet they may surround us and starve us out. We shall act wisely if we at once prepare for our voyage, and commence it as soon as Timbo returns."

" But about these two boys, what shall we do with them ?" asked David.

" I am afraid their fate must be a sad one," was the answer, " whether their father comes for them or not. If he takes them away, they will probably fall into the hands of their enemies ; or if they are left here, they are too likely to perish from hunger."

" Oh, then let us take them with us," said Kate, and little Bella echoed her words. " Surely the canoes are large enough to carry them, and it would be terrible to leave them to die."

" They shall have part of my share of food," said Bella.

" And mine and mine," added Leo and Natty.

" I would rather leave Chico behind," said Leo, " though I am afraid those dreadful savages would eat him."

" Oh, we must carry him too," said Natty; " for I am sure when we stop at night he will be able to forage for himself; he will find out roots and fruit when very often we are not able to discover them."

We did not spend much more time in talking. It was arranged that we should start immediately on the return of Timbo. We therefore at once set to work to pack up our goods and to collect all the provisions we had in store to carry with us. As we could not tell into what regions the river might carry us it was important to kill some game and to collect as many plantains as we could carry off from the deserted village

Chickango and the two boys undertook to set off for the latter object, while Stanley and I went out with our guns into the woods. We were unusually successful, and in an hour had bagged as many pigeons and other birds as we could carry. We found, as we neared the Castle, Natty and Leo staggering on under a load of plantains.

"We shall have no fear of starving now, at all events!" cried Leo, "for Chickango has got as many more. As we came along, however, he started off to the top of the hill, where we understood him to say he could get a sight of the Bakělés' village, and I suppose that he will be soon with us."

We were disappointed on our arrival at finding that Timbo had not returned.

"I am afraid that some accident has happened to the poor fellow," said Stanley; "or he may have been incautious, and fallen into the hands of the savages."

David and Jack had been so well employed, that, with the assistance of the young ladies, everything was prepared for a start.

"I wish that we could be off," said David; "but we must not leave our faithful Timbo behind."

"Well, if you will all go down to the boats, I will remain here and bring him up as soon as he comes," said Natty. "We shall thus gain time."

"No, no; I cannot let you do that," I said. "I will remain, and you must go."

Natty, however, positively refused, and Stanley had to exert his authority, as our leader, to make him accompany them. Very unwillingly, he at length consented to do so, provided I promised, should Timbo not appear in the course of an hour, to follow them. The matter was arranged, and our party were taking up the loads they proposed to carry, when Chickango made his appearance among us. His countenance

expressed alarm, and he was too much agitated to explain him·
self.   At length Senhor Silva understood him to say that, on
looking towards the Bakĕlĕs' village, he had seen smoke ascending—that it grew thicker and thicker, and then flames burst
forth, and he was convinced the whole village was on fire.

"Depend upon it, the Pangwes have done this," he observed;
"and, flushed with their victory, they will very soon march to
attack us.   We must either prepare ourselves to stand a siege,
or lose no time in escaping."

"Then let us at once commence our march," said the
captain; "but, Andrew, I do not like that you should run the
risk of falling into the hands of these savages, which you will
do if you remain behind."

"I know my way down to the river so well," I answered,
"that I can easily join you should I see them approaching,
and I will, meantime, keep a look-out from the height above
the fort.   Depend upon it, they have too much respect for our
firearms to venture an attack, unless with their whole body
At all events, some time must elapse before they can be here.
My only anxiety is about Timbo, should he have fallen into
their hands."

"You will promise, Andrew, not to remain more than an
hour?" said Kate, as she and Bella, each carrying a load pro-
portionate to their strength, went out of the fort.   "We shall
be very anxious till you join us."

I watched the party as they descended the hill.   I did not
think the young ladies had much cause to regret leaving the
place; but still they turned a glance behind them, as if they
were quitting it with sorrow.   Though difficulties and dangers
might be before them, still I hoped that they were on their
way to a more civilized and healthy climate.   In the hurry of
departure Chico had been forgotten, for he was quietly snooz-
ing in his usual corner of Jack's hut.   Leo and Natty had

already left the fort, when they discovered that he was not with them. "Chico, Chico!" they both cried out, and hearing his name called, he ran out, and sprang up upon Jack's shoulder, who had already got as much as he could well carry. Nothing, however, would shake Master Chico off. I could not help thinking even at that moment of Sinbad the Sailor, and the Old Man of the Sea. "Well, I suppose if you will not walk, I must carry you," exclaimed Jack; and away he went after the rest, Chico glancing about him with a look of surprise at the sudden exodus of his friends.

As soon as they were gone, I closed the gates and climbed out of a window in the back of the fort. This I did, that should the Pangwes arrive, they might not discover the flight of our party, and might spend some time in making preparations for the attack. I then ascended the hill, with my telescope, which I had retained, but could see no one moving in any of the open places I could command. In the distance, however, I observed dense clouds of smoke and bright flames ascending above the forest, which I was sure must proceed from the village we had visited. What was the fate of the unfortunate inhabitants? I knew too well the way that negro warriors are accustomed to treat those they conquer, and I could not help picturing to myself the horrid spectacle of women and children murdered, and those who had escaped slaughter carried off to be sold as slaves to the cruel dealers in human flesh, and, more than that, in the hearts and souls of their fellow-creatures. I looked at my watch. I calculated how long it would take my friends to reach the canoes. I was thankful when I felt sure they must have had time to get on board, and thus to be in comparative safety. Time went slowly on. I kept looking at my watch, but still Timbo did not appear. The hour had nearly passed.

At length, with great regret, I descended the rock, and

took my way towards the river.   I had just passed the
Castle, when I caught sight of two figures moving towards
me among the trees below.   They might be scouts of
the enemy.   I hesitated what to do.   Concealing myself behind some brushwood, I lifted my glass to examine them.
Great was my satisfaction when I saw that one of them was
Timbo; the other was a negro whom he was assisting along,
and who appeared to be wounded.   I hurried down to meet
them.   Timbo, when he saw me, made a sign to me not to
shout, pointing behind him to make me understand that he
was pursued.   As I approached, I saw the negro was Igubo.
He recognized me, and it seemed to revive his strength.
Without stopping to inquire what had occurred, I took his
arm, and assisted Timbo in hurrying him on towards the
river's bank.   When he found this, he made a significant
gesture towards the Castle.   " He ask for his sons," whispered
Timbo.   " Tell him they are both safe, I hope, in the canoes."
A gleam of satisfaction passed over the countenance of the
wounded man, and he made fresh efforts to struggle on.

We had great reason to hurry, for ever and anon I could
hear the shouts of the savages in the woods behind us, though
still they appeared to be at some distance.   Blood was flowing
from Igubo's side.   I fortunately had a handkerchief, and in
spite of the necessity for haste, I insisted on stopping to bind
up the wound.   I was afraid that otherwise he would bleed
to death.   We gained by it, indeed, for he was afterwards
able to move more rapidly, and the flow of blood appeared
almost staunched.   As we approached the river I caught sight
of two figures among the bushes and tall reeds which lined
the bank.   Could our enemies have got ahead of us?
Presently we saw one of the figures dart out from their con-
cealment, and then, to my satisfaction, I recognized Leo.   He
and Natty soon came running towards us.   They had been on

the watch, it appeared, having grown anxious at my non-appearance. The rest of the party were seated in the canoes. We assisted the wounded man into the one in which David was, with the two young ladies and Jack. A place had been left for me there also. Igubo, not seeing his boys in it, uttered an expression of disappointment. We lifted him up, however, and showed that they were in the other canoe. When satisfied, he submitted to have his wounds more completely and scientifically bound up than I had been able to do. Meantime Jack had taken the steering-oar, while Timbo and I seized the paddles. A few hurried words from Timbo explained to Stanley what had happened, and without further delay we shoved off from the bank, and began to make the best of our way down the stream. Natty had come into my canoe, while Stanley called Leo into his. Mine was the *Gazelle*. It was the best of the two, the other having been injured by the hippopotamus. Stanley had placed his sisters in it, trusting rather to Jack's seamanship than to his own. His canoe being the lightest, he took the lead, that he might give timely notice to us should any sandbanks be encountered in our course, and, what were perhaps more to be dreaded, any wild rapids, down which it might be dangerous to proceed. Chico had seated himself in the bow of the canoe, as if he had been placed there to keep a look-out. Natty had taken a paddle, and Kate begged that she might use another till her brother had finished attending to poor Igubo's wounds.

Not till we had got a little way down could I ask Timbo what had occurred. " Oh, Massa Andrew," he answered, " me no like talk about it. De Pangwes come, and stay hid in de night close to de village, and just before de sun get up, —de sun dat is so bright and good, make de trees grow, and cheer de heart of man,—dey steal out wid de sharp sword and de spear, and de moment de Bakĕlĕs open de gate, rush in

and kill all de women, children, and old men; and some stay
outside and kill dose dat run away, and catch de young men
and knock dem down, and tie deir hands, and take away to de
slave-dealers. Igubo jump over de wall, and kill two or
t'ree who came after him; and dough dey stuck de spear in
his side, he get away. As I got near de village I hear de
cries, and know too well what dey mean; so I hide, for I fear
if I run dey see me and follow; but when I found Igubo drop
down just near where I was, I rushed out and lift him up
and bring him along; and de Pangwes just den no see us,
because some young men who had got swords and bows and
arrows 'tack dem, and fight bravely; but dey all killed, and
den de Pangwes set fire to de village, and you know de rest."

Timbo had scarcely finished his account when he shouted
out, " See, see! Dere dey are! Dey come dis way!"

We had all been so busy in paddling the canoe and watching our leader that we had not looked either to the right
hand or the left. Stanley, for the same reason, had not seen
what was taking place on shore. We now saw a large body
of black warriors shaking their spears, and beating them
against their shields, as they rushed on towards the bank of
the river. They had evidently the intention of stopping us.

" On, on!" cried Stanley. " Put your best strength into
your strokes; the river is broader a little way down, and we
may escape their arrows and spears if they attack us."

" Don't you think, sir, we had better get a broadside ready
and give it them?" exclaimed Jack. " They are more
likely to treat us with respect if we show that we are well
armed."

" I would advise you not to fire unless hard pressed," said
Senhor Silva. " We will show our muskets, but they are
fierce warriors, and even should a few be killed, the rest would
not be daunted, and would probably pursue us till a more

narrow part of the river is reached, when they might over-whelm us with their spears and poisoned arrows."

" Let me now take the paddle, Kate," said David, who had placed Igubo at the bottom of the canoe, resting his head on a bundle. " My arm is stronger than yours, my sister, and in case the savages attack us, you and Bella must lie down at the bottom of the canoe."

The canoes glided rapidly down the stream, making the water hiss and bubble under their bows. Had we not had the two helpless girls to protect, the adventure would have been an exciting one, which few of us would have objected to go through. The Pangwes, shouting and shrieking, and shaking their spears and shields, had now reached the banks of the river. It seemed scarcely possible that we could escape them. Not, however, till David had again and again pressed them, would his sisters consent to place themselves in greater safety at the bottom of the canoe. The crew of Stanley's canoe plied their paddles vigorously, and kept just ahead of us. We needed no exhortation from him to follow their example.

We had now got almost abreast of where the savages were standing. Every instant I expected to see them draw their bows, with those deadly poisoned shafts ; or hurl their spears, which I knew too well could reach to a great distance. I saw Timbo eyeing them very calmly.

" If we were to fire a broadside into them now, it would soon put them to flight," cried Jack.

We, however, kept on without apparently noticing them. As we approached, they increased their shouts. Some of their chiefs seemed to be going among them, urging them to rush into the stream. Happily the river was here much wider than above us, and continued so for some distance down. A sand-bank appeared in the middle. We trusted that a channel might be found on the right side of it, away from where the savages stood. We now saw several men with swords in their hands, urged by their chiefs, rush into the stream.

"See, see!" cried Timbo; "what are those creatures on the sandbank?"

I looked ahead, and there observed eight or ten large alligators and several small ones basking on the sandbank. Our approach somewhat startled them; and off they slid into the water, swimming towards the bank where the Pangwes were collected. They apparently caught sight of them at the same time. One of the leading swimmers at that instant threw up his arms, and, uttering a shriek, was drawn down under the water. The others, seeing the fate of their companion, turned round, and, in spite of the shouts and exhortations of their chiefs, swam back to the shore. The alligators pursued them, and two others were carried down before they could reach the banks. So eager were the monsters that we saw their snouts rising above the water even at the very bank, when hundreds of spears were darted at them. Aimed in a hurry, the missiles probably glided off their scaly sides. We could not discover whether any were killed.

Now the Pangwes, finding that their attempt to cut us off had failed, began hurling their spears at us, and sending showers of light arrows, many of which fell fearfully close to the canoe. Some stuck in the sandbank, inside of which we were making our way. It showed us the danger of having to pass our enemies where the river became narrower. The only advantage we should there possess would be the greater rapidity of the current. We continued to ply our paddles with might and main. Now we had passed the sandbank, and a wide extent of water lay between us and the negro army. They, however, appeared to have discovered that should we get far ahead we might escape them altogether; and we saw a large body moving away to the southward. We could not help fearing that there might be some bend in the river, or narrow passage, where they might still hope to cut us off. Our utmost efforts must be exerted, therefore, to gain the place before they could reach it. There was still another danger. We might ground on one of the sandbanks, or some point might project from the western side and compel us to go round nearer to the eastern bank. I, of course, kept these thoughts to myself, and did my utmost to send the canoe along, and to keep up the spirits of my companions.

"If we get within reach of them," sung out Stanley from his canoe, as he saw them moving along the bank of the river, "we must instantly take to our arms and give them a volley. It will not do to let any of their arrows come near us."

"Ay, ay," I answered. "Our muskets are, I believe, all loaded."

"All right, sir," said Jack. "I loaded them before I placed them in the canoe, and I do not think those black fellows will stand a taste of our bullets."

Poor little Bella looked very much frightened when she heard us talk of firing.

"They will not fire unless there is absolute necessity for it," I heard Kate say to her. "You know, Bella, it will only be done if we have to defend ourselves."

The current was so strong and our canoes moved so swiftly that we were quickly leaving the main body of Pangwes. We heard their shouts of rage and disappointment as they saw us escaping them. Horrid as were those shrieks and cries, they of course only made us paddle the harder; but still I felt anxious lest the smaller body I have spoken of might outstrip us.

"Suppose the Pangwes try to cut us off at another place, could we not haul our canoes up and make our escape overland?" exclaimed Natty, showing that he had understood the reason of the movement we had observed.

"We might escape them, certainly, for the moment," I answered ; "but we could not proceed on our journey without our canoes."

"But we might return and get them, or drag them overland," he observed.

"That would be a task, I fear, too great for our strength," I said. "But your suggestion, Natty, is worthy of consideration, if we are hard pressed."

I told Stanley what Natty had said.

"I hope we shall not be obliged to do that," he answered. "Paddle away, lads ; we shall soon, I hope, see the last of them."

On we went, the river now making its way through a thick forest, the trees coming down to the very water's edge ; now again it opened, and low prairie land was seen on the eastern side. The level appearance of the country made me fear that the river might make some bend such as I supposed our enemies were attempting to reach. The banks were, however, too high to enable us to see to a distance. At any moment they might appear on the shore. At length the banks became somewhat lower, and, standing up, I caught sight of a body of men hurrying across the prairie. They were, however, at a considerable distance behind us ; and now it evidently depended on whether we should reach the supposed narrow place before them or not. I had often read of heroines ; but as I looked at the calm countenance of Kate, showing that she was resolved to go through all danger without flinching, I could not help thinking that she deserved especially to be ranked as one.

I could see as I gazed over the plain, besides the negro army, numerous animals scampering across it, put to flight by their appearance — herds of quaggas, zebras, buffaloes, and various sorts of deer, the lofty heads of a troop of giraffes appearing above them all. Innumerable birds flew amid the boughs of the trees, and wild fowl rose from the sedgy shores, or gazed at us from the mud-banks as we shot by. Here and there a huge hippopotamus raised his head, and gazed with his ferocious eyes, wondering what new creatures had invaded his territory; while scaly alligators lay basking in the sun, or swam about seeking some creature to devour.

"If we get clear of the savages we shall have no fear of starving," observed Natty, as he saw the herds of wild animals I have described.

"You are right, Natty," said Jack; "and as to getting clear of them, there is no doubt about that."

"I have been praying that we may escape them," said Natty; "and that makes me think we shall."

"Right again, Massa Natty," observed Timbo. "It great t'ing to know dat we have got One to take care of us when we can no take care of ourselves. He hear de little boy prayer just as much as de big man."

Had Timbo joined us at an earlier hour, we might have escaped the dangers to which we were exposed; but still I was thankful that we had got him with us. As I looked ahead I saw that the river was making a bend towards the east. It was what I had dreaded; but the danger—if danger there was—must be run. Again I asked Stanley whether he thought it would be wise to haul up the canoes, and try to escape overland, should the river be too narrow to enable us to keep out of the range of the poisoned arrows of our enemies.

"That must be our last resource," he answered. "We must first try the effect of our firearms. Their blood be upon their own heads, if we kill any. I have no wish to injure any of them, even though they may be seeking our lives, if we can by any possibility avoid it."

I felt much as Stanley did. To desert the canoes would be

to expose the young ladies to fearful fatigue and danger, and
was to be avoided by every means.

We now entered into the reach I had expected to find.   It
was, however, as broad as the part we had lately passed through.
We took the centre of the stream rather than cut off the angle,
lest our enemies might be concealed on the bank.   And now,
going along it for some distance, we rounded another point
projecting from the west, and found ourselves in a still broader
part.   It was somewhat shallow, we judged by the numerous
little islands and banks which rose above its surface.

"Hark!" said Natty, suddenly; "don't you hear the roar
of water?"

I listened, and felt convinced that some waterfall or rapid
was near us.   I shouted to Stanley.   We ceased paddling for
an instant.

"It may be a cataract," he answered; "but I have hopes
that it is simply the sound of rapids.   If so, we may pass
through them."

"A dangerous experiment!" observed David.

"It depends upon their character," answered his brother,
from the other canoe.

"But, without a pilot, would it be possible?"

"We must land and survey them first," shouted Stanley.
"We shall have no difficulty in doing that; and if we cannot
pass them, we must try and drag the canoes over the land.
That, at all events, can be done."

We found as we proceeded that the roar of waters increased;
and there could be no doubt, from the way the river ran, that
a rapid was before us.   We went on till the water was already
beginning to bubble and hiss.   The bank on our right afforded tolerably easy landing; so, running the canoes to it, we
secured them to some trees which grew close down to the water.
Stanley sprang out, and called to Timbo to accompany him.

"We shall be able to judge whether we can safely pass through them," he said. "I will be back quickly. Yes, we will take our rifles ; we may find them necessary."

He said this as Senhor Silva handed them out of the canoe. They were soon out of sight among the thick under-wood which grew near the banks. It is very different, I should say, from the underwood in England ; composed rather of huge leaves, reeds of enormous height, and other plants of the Tropics. The opposite side was also covered with wood, so that we were unable to ascertain whether the Pangwes were in the neighbourhood or not. We were, however, so much concealed by the foliage among which our canoes were moored, that an enemy might have passed on the opposite bank without per-ceiving us. We waited anxiously for the return of Stanley and Timbo. At length they appeared.

"We can do it," Stanley exclaimed. "The water is rapid but clear, and we may easily steer our way clear of the huge boulders through which it passes."

Once more we shoved off. Each man screwed up his nerves for the trial ; for no slight trial it would prove—of that I was certain.

"Stanley is so cool and calm," observed Kate, "I have no fear."

His canoe led. In a few minutes we were in the strength of the current. On we glided, like arrows from a bow. We had little else to do than to guide our canoes. Still we kept paddling, so that we might the more easily, if it were possible, turn aside from any danger ahead. Now a huge boulder rose up on one side ; now we darted through a passage which only afforded room for the canoes to pass. Now the water ran smoothly without a bubble ; now it hissed and foamed as it passed over a shallower bed. There was an excite-ment in the scene which made our spirits rise. I felt almost inclined to shout at times as we dashed on. Yet an instant's carelessness might have proved our destruction. We appeared to be decending a steep hill of water at times; now wavelets rose on either side, and threatened to leap into the boat.

Our eyes were fixed on our leader's canoe, and his on the water ahead, through which he was to guide us. For one moment I cast my eyes on the eastern shore, and was sorry that I had done so, for there I saw a number of dark forms collected just below the rapids. What they were about I had not time to observe. I said nothing; it would be time enough when we had shot the rapids. On, on we went. We were in a sea of foam, the water roaring, bubbling, and hissing. I feared that Stanley's skill could scarcely carry the canoe through; but he had noted the point, and his experience told him that there was sufficient depth. Now a wave washed aboard on one side, now on the other, now came hissing over our bows; but we dashed through them, and I saw before us a calm and lake-like expanse. In another instant we were free of the rapids, and floating calmly on the lower portion of the river.

Once more I cast my eyes to the spot where I had seen the blacks. They were our enemies; of that I had no doubt. I pointed them out to Stanley.

"What can they be about?" he asked.

Timbo looked at them. "Building rafts," he answered. "Dey are shoving off even now. Dey knew we must come dis way, and hoped to cut us off. But hurrah! hurrah! we got down sooner dan dey!"

Several rafts of reeds, such as I before described, were shoved off from the bank. We did not stop to examine them; but plying our paddles with might and main, we continued our course towards the point where we believed the river made its exit out of the lake.

CHAPTER X.

THE savages on the raft, which had already got some way out into the lake, saluted us with showers of arrows; but, happily, we were too far off for them to reach us. Already our arms ached with our long paddle, but it was no time to rest. We knew not whether, vindictive as they appeared, they would attempt to pursue us, or whether others might not have gone further down along the margin of the lake, with the hope of even yet intercepting us at the narrow part which we saw. Evening was approaching, and the difficulties of the navigation, should the night prove dark, would be greater.

"I see some objects on the left bank," cried Natty.

"Never fear, we will slip by them," said Jack. "To my eyes they have got four legs, and will not hurt us."

We speedily neared the point where the lake-like expanse narrowed into the proportions of a river. The creatures seen by Natty were now discovered to be a herd of zebras, which had come down to the river's bank to drink. They gazed at us as we passed with a look of astonishment; but, though they kept moving here and there, as if asking each other what we could be, they did not take to flight, but continued scampering round and round as horses do in a field, stopping every now and then to take another look at us. They quickly, however, returned to the water, for they probably knew that unless they made haste they would be interrupted by some of their re- morseless foes—lions, panthers, or hyenas—which might come down to the same spot to quench their thirst before setting forth on their nightly rambles in search of prey. They were beautiful and graceful creatures, very unlike the poor patient ass with which we are acquainted in England, and accustomed to associate with everything that is stupid and obstinate. Yet the zebra and the ass are nearly related; indeed, the former is classed by naturalists as an ass. I shall have more to say about them by-and-by.

Evening was rapidly drawing to a close. Still, although the alarm which the zebras had caused us when first indis- tinctly seen had subsided, we thought it possible that some of our savage foes might still be on the watch for us further down the stream, or, should we land and rest, that they might over- take us before we again got under weigh. " It's wisest, accord- ing to my notions, to keep well ahead of an enemy if you have to run from him, and as close as you can to his heels if you have to chase him, till he hauls down his flag!" exclaimed Jack, vigor- ously plying his paddle. " What do you say, Mr. Crawford?"

I heartily agreed with him. The thought of what would be the fate of my young relatives would have nerved my arm for even greater exertions than we were called on to make. We still, therefore, continued paddling, in spite of our fatigue, with might and main, anxious to put as many leagues as pos- sible between ourselves and our enemies before we stopped. The sun set in a glorious glow of ruddy light on our right, shedding a hue over the tops of some lofty hills which appeared on the opposite bank. The stream increased in rapidity: but still, as far as we could see, was free from danger. There was yet sufficient light from the sky, though it could not be called twilight, to enable us to continue our course.

"If the navigation is as open as at present, we will continue on for another hour," shouted Stanley. "We shall then be safe from the savages, and may have a quiet rest, I hope, after our day's work."

"Ay, ay, sir," answered Jack from our canoe. "We have not worn our arms off yet; though, if you don't mind stopping, maybe the ladies would like a bit of pigeon and a bite of plaintain."

"Oh, no, no," exclaimed Kate. "Do not stop for our sakes, if you are not tired. We feel no hunger, and would rather not delay a moment till you think it safe."

We accordingly paddled on. By degrees the glow faded from the sky, and darkness settled down over the landscape. Still Stanley continued leading. Presently I saw on our left a silvery arch rising over the hills. It increased rapidly, and soon the full moon rose in the sky, shedding its light over the waters.

"We do not get sight of such a moon as that in old England," cried out Leo from the other canoe. "It is often there more like a patch of red putty stuck on to a wall; but see! this looks like a mighty globe of pure fire floating in the heavens." So indeed it did.

"Do not be disparaging our good old English moons," cried out Natty. "You forget the harvest moon; and, though it is not quite like this, it is a very beautiful object to gaze at, and useful to those who have to carry home the full-loaded waggons of corn."

Our spirits were rising as we felt we were escaping from the danger we had encountered. I hoped, too, our hearts were grateful. The bright light of the moon now enabled us to proceed with almost as much ease as during the day. As we

sped on, however, we saw numerous animals on the banks
coming down to drink; but we passed them too rapidly to
ascertain what they were. I think we must have continued
paddling on two hours longer, rather than one. Stanley
seemed unwilling, so long as we could move our arms, to stop;
indeed, the cool air of night renewed our strength; and, for
my part, I felt that I could have gone on till daylight, if
necessary, for the sake of securing the safety of the young girls
depending on us for protection.

At length the ground on our right rose considerably above
the plain. "I think I see an island ahead," cried Stanley. "If
so, it may suit us for a bivouac, and may be more secure than
the mainland." As we went on we found that he was right.
The island appeared to be about four or five hundred yards
in circumference, with numerous trees growing on it, which
would afford us the means of forming huts, and give us wood
for our fires besides. Fortunately, we had no need of provisions, as we had an abundance in the canoes. We took the
passage on the west side, and, going to the further end of the
island, found a small bay, into which we steered the canoes.

"We must act the part of invaders and drive out any previous occupants," observed Stanley as he stepped on shore.
"Kate and Bella and the two boys, with the wounded black
and his sons, must remain in the canoes till we can find a safe
place for encamping. David will stay behind for your protection. Now, my friends, we will advance into the interior."

At the word we all stepped on shore. There was a small
extent of open ground extending a few yards from the water's
edge. This would, at all events, afford us space for our encampment. Had it been a dark night, we should have run a
considerable risk if any savage animals existed on the island; but
during moonlight neither lions nor panthers will assail a man,
unless hard pressed by hunger. We had our axes in our belts.

and were thus able to clear our way over the rocky ground among the underwood and trees, mostly growing wide apart. As we advanced, we shouted to each other, now one now another firing his gun and stopping to reload. Suddenly a loud splash told us that some animal had leaped into the water. Now another was heard, and in a short time we reached the northern end of the island, having completely passed over it. We were satisfied that whatever creatures had been there had taken their departure, and we now returned to prepare for our encampment. In the meantime, we found that David and the boys had been landing the provisions. We had all become pretty expert in cutting down trees; and, as many of those in our neighbourhood were small, we soon had a sufficient number to make a small hut for Kate and Bella. This was erected with a rapidity which would have astonished people at home. As there was no fear of rain, we were not very particular as to the roof; and the abundance of vines enabled us quickly to weave a network round it, through which no panther, nor even a lion, could force its way. Less substantial structures were erected for the rest of the party. The boys were busy in collecting dry wood for the fires; and in scarcely more than half an hour we had formed a village which might have served us for many weeks if necessary, provided the weather remained dry. The two young blacks had, in the meantime, under the superintendence of Kate, been preparing our supper. She insisted that she was in no degree tired, and would not be idle. Igubo sat up, with his back supported against a bale, giving directions to his sons. A number of birds were forthwith roasting before the fire, while an ample supply of plantains were being baked on the ashes. Our cookery was of necessity somewhat rough, but we were grateful to those who prepared our food, and I could not help fancying it tasted better done by their hands. A sufficient amount of wood had been collected to keep up four good fires during the night. One was placed on the river side, to scare any animals which might approach from the water; one at either end of the camp; and one on the forest side, though we hoped that we had driven off all enemies from our island. As soon as supper was over, Stanley recommended all hands to retire to rest.

"But, massa," said Timbo, "we escape great danger; sure we t'ank Him who preserved us."

"Indeed we ought to do so," said Kate; "and we are thankful to you, Timbo, for reminding us."

"I am sure my father would," I heard Natty say to Leo.

Stanley took a pace or two up and down, and then turning to Timbo, said, "You are right, old friend; but it would be somewhat out of my way, I am sorry to say. David, I must ask you to take the lead."

The young doctor, though full of talent, felt, I saw, a diffidence under the circumstances; but, mustering courage, he undertook to lead us in prayer; and with expressions which came, I am sure, from his heart, he returned thanks to the God of mercy for our preservation from the great dangers we had passed, and implored protection for the future. I heard Natty, who was kneeling near me, repeat his words with deep earnestness; and I was sure also that Kate and little Bella were pouring out their hearts in prayer. Though Timbo was the only African who could join us, the others were, I believe, greatly impressed with the scene, which, I had reason to know, was never forgotten by them.

Chickango and I had been appointed to keep the first watch, while Senhor Silva and Jack were to relieve us. In a short time the rest of our party were fast asleep, with the exception of David, who, as soon as his sisters had entered their hut, drove some stakes round the entrance, so that even a snake could not find its way in. After pacing up and down for some time with my gun in my hand, I told Chickango I would try and make my way to the other side of the island, as a full moon shining down among the trees enabled me to do without much difficulty. Its beams shed a silvery light on the water, which flowed calmly by. I soon reached a spot whence I could see the opposite shore, across a channel which divided the island from the mainland. As I stood there, I fancied I saw creatures moving along the banks, then I discovered five or six elephants approaching the water. They came to the edge, and, dipping in their trunks, poured the cool liquid down their throats. Presently a herd of giraffes came with a swinging trot across the ground, their heads moving about from side to side as they swung forward their long legs. They appeared, however, rather cautious of approaching till their more powerful companions had quenched their thirst. Just then, from a point a little on one side, several smaller animals made their way down to the bank; and, as they drew nearer, I discovered them to be a male and female lion, with their whelps. They stood watching the elephants, now and then uttering a low angry sound, yet never breaking into a roar. I stood rivetted to the spot, thankful that we had chosen the island for our encampment; for had we been on the mainland, we must have found our post untenable. They were, however, not the only visitors to the water. A huge rhinoceros, which I recognized by the horn on his nose, advanced with a heavy tread; and several buffaloes, and other animals which I took to be wild boars, joined the assemblage. The elephants, it appeared to me, kept the other animals in awe, for all stood at a distance from each other, slaking their thirst after the burning heat of the day. Many, probably, had come from a distance to seek for water. The giraffes were the only ones which continued in motion, they evidently being unwilling to approach while their savage enemies the lions were in the neighbourhood. For-

tunately for them, I was not possessed with the instincts of a
hunter, or I should probably have shot one of the lions; the
female especially, as she kept looking at the elephant, with her
cubs by her side, offering me a mark which I could not well
have missed; but, in the first place, I should have disturbed
my friends, and then I thought to myself, " Why should I kill
one of these creatures, which are but following their natural
instincts? and, as they are not likely to attack us, no good can
be attained."

At length I thought that Chickango would fancy some accident had happened, and might be induced to leave his post
to search for me. I therefore returned to the camp. I had
nearly reached it when I fancied I heard a sound behind me.
I turned round, an indefinite feeling of horror suddenly
seizing me. I called to Chickango: he sprang forward. At
that instant I saw a huge creature creeping along through
the underwood. Chickango was by my side. He raised his
gun, and gave a loud shout. The animal sprang up a tree.
He fired, and a large panther fell to the ground. The rest of
the party, starting from their beds, came hurrying up. The
creature was not quite dead, but a blow from the negro's axe
quickly finished it. My friends congratulated me on my
narrow escape; and indeed I was thankful that I had been
again preserved. The creature must have remained on the
island. Probably the moonlight prevented it springing on me
at once, as it might easily have done.

It was some time before quiet was restored to the camp.
David hurried back to assure his sisters that there was no
danger; for they had naturally been alarmed by the shot, and
the cries of the party as they sprang up from their sleep. The
adventure made us increase the number of our fires on that
side of the camp; while Stanley, declaring he had had sleep
enough, joined us on the watch. As may be supposed, I felt

no inclination to make another trip about the island by myself, lest a companion of the animal we had killed might take a fancy to spring upon me. I must confess I was very glad when Timbo and Senhor Silva came to relieve me; but nothing could induce Stanley again to lie down.

No sooner had I placed my head upon the bale of goods which served me as a pillow, than I was fast asleep. I was aroused by Natty's voice,—

"Oh see, Mr. Crawford!—it is worth looking at. The sun has just risen."

I sprang to my feet, and found all the camp already up. The sun at that instant was showing its upper edge above the mountains, looking like an arch of fire, tinting the distant mountains with a soft tinge of the same hue, and casting a ruddy glow over the broad stream which flowed at our feet; while the whole sky was covered with a rich orange glow, deepening towards the horizon into the brightest vermilion.

"We will lose no time in proceeding on our voyage," said Stanley; "so the sooner we can get through our breakfast the better."

As the fire was ready, the water was soon boiling, and we contented ourselves with the cold meat and plantain which had been cooked on the previous night. The canoes were immediately reladen; and quickly embarking, we once more commenced our voyage down the stream. As we opened the wider part, we looked northward along the banks, but could discover no signs of our enemies; and we hoped, therefore, that we had completely distanced them. The number of animals which we saw on the banks showed us that we were not likely to meet with many inhabitants. This was satisfactory, as we could not tell how they might be disposed towards us. Although the heat was great, our spirits felt lighter in the belief that we should meet with no enemies: and we continued paddling along, Chico standing as before in the bows of the canoe; the boys, as usual, joking with each other; while Jack every now and then burst into one of his sea-songs, an entertainment with which he had not indulged us since we were engaged in building the canoes. The *Giraffe*, as before, took the lead. We paddled more leisurely than on the previous day, as we should soon have worn ourselves out had we continued the exertion we had then gone through. Thus, in spite of the heat, we were able to continue on for some hours.

We landed at noon on the western bank, where a group of trees afforded us shade, which we greatly needed; indeed, the heat of the sun had become so great that we could scarcely have continued longer exposed to its rays. We as before beat the bushes in the neighbourhood to ascertain that no animal lurked among them, and then lighted a fire to cook our dinner. As may be supposed, the birds that had been killed on the previous morning were no longer fit for English palates; but our black friends, without ceremony, consumed them. We had therefore to wait until we had killed some fresh game.

Stanley, Senhor Silva, Timbo, and I took up our guns to proceed inland. The scenery on the banks was very beautiful, the trees not growing in dense masses, but scattered in groups, like those in a gentleman's park in England. Beautiful flowers covered the open spaces. Among some of the groups of trees we observed the orchilla weed hanging from the branches. This is one of the exports of Africa, and is used as a dye stuff. There was a beautiful little shrub which Chickango called the *mullalo*. It bore a yellow fruit. He gathered several—which he said were good to eat—and we found them full of seeds, like a custard-apple, with a sweet taste. A larger tree was covered with white blossoms, their fragrance reminding me of the hawthorn at home; but the flowers of these were as large as dog-roses and the fruit the size of big marbles. Chickango pointed to the flowers ; not so much to admire them, as because numerous bees were sucking their sweets. " Dere ! dere !"—and he pointed out several hollows in the neighbouring trees. " Me come back, and get for eat," he said. From another shrub—which our companion called the *mogametsa*— he picked a quantity of fruit, which had the appearance of a bean with pulp round it.

" Why," cried Leo, as he tasted the *mogametsa*, " it is just like sponge-cake—capital stuff ! We must take a quantity to the camp."

Another very nice fruit was the *mawa*, which grew abundantly on low bushes. Indeed, we found a number of edible bulbs and bushes. Among them I must not forget to mention the *mamosho* and *milo*. The latter is a sort of medlar, which all hands pronounced delicious. Indeed, there was no fear of our starving in this region. There were great numbers of birds also ; but I will describe them by-and-by.

Troops of animals passed us, among which the giraffe was conspicuous. We were just emerging from the wood, when we saw a single giraffe following a large herd at a distance, having from some cause been separated from his companions. On he went, swinging his tall head from side to side to keep time with the motion of his legs, which put me in mind of the way spiders move. He was passing a clump of trees, when a terrific roar reached our ears. The poor animal attempted to increase his pace; but before he could do so, a huge lion sprang from a thicket, and with one bound alighted on the giraffe's back.

" It is too far off for a bullet to reach him," observed Stanley, "or I would try to rescue the giraffe by killing the lion."

" It would be useless, for I suspect the giraffe's fate is sealed," said Senhor Silva. " The grip with which the lion seized his neck is sufficient to end his days. In spite of the giraffe's strength, the king of the forest will soon have him down."

The giraffe continued his course, going away from us, so that our chance of shooting the lion decreased. Still we pushed on, hoping that the terrified animal might turn, and bring his murderer closer to us. On he went, however, uttering cries of terror, the rest of the herd scampering off at full speed, which soon carried them away from their unfortunate companion. The life-blood was flowing fast from the giraffe's neck; but he struggled on in spite of the immense weight of the creature on his back and the agony he must have been suffering. In vain he reared up—in vain he struggled. Presently we saw him sink to the ground, when the savage beast flew at his neck, and soon finished his sufferings.

"Take care," said Senhor Silva; "we must not approach too near, for if we attempt to dispute his prey with the lion, it will make him more savage than ever."

"Our guns will settle that question," answered Stanley, still hurrying on.

I kept by his side, and the boys followed. Not till we were within fifty paces did the lion perceive us. He was then standing over his prey, which he had already begun to rend. Raising his head, with his claws on the carcase, he eyed us fiercely, sending forth terrific growls of anger. Still he did not move · and Stanley had now an opportunity of taking steady aim. Still we advanced nearer. The lion perceiving this, with a roar which even now rings in my ears, gave a bound towards us. I raised my rifle and fired; but my arm must have trembled, and I confess I felt little able to take steady aim: the ball only grazed the lion's head. He was now within a dozen paces of us. Leo and David were standing a little on one side. Stanley raised his gun. He fired; but, to my horror, no explosion followed.

" Now ! now !" he cried out.

The boys saw what had happened, and both, levelling their pieces, fired. The lion gave a bound in the air, and fell backward.

" Hurrah ! hurrah !" shouted Leo and Natty ; " we have killed the lion !"

" No ; it was my shot did it," cried Leo.

" It was mine," exclaimed Natty ; " I am sure."

" You both had the honour," exclaimed Stanley, as he knelt over the monster's head. " Here are two shot holes, and either would have killed him."

As may be supposed, the boys' triumph was very great. Chickango, however, was better pleased with the giraffe.

" Here meat enough for one week," he exclaimed, as he began to cut away into the giraffe's flesh.

As we had no prejudice in taking an animal killed immediately before our eyes, though we might have objected to it had we found it dead, we all assisted Chickango in cutting up the animal, each of us taking as much as we could possibly carry.

" You stay here," he said. " Take care no oder lion come. I go call oders ;" and loading himself with twice as much as we could have attempted to carry, he hurried back to. the camp.

The rest of the party soon arrived ; and we had now an ample supply of food for several days, if it would keep so long. Not delaying to kill any birds, as the rest of the party were waiting for their dinner, we hurried back to the camp. We found that Timbo had not been idle, and had caught several fish, which were of good size, and pronounced wholesome. We found Igubo's sons—the eldest of whom was called Mango and the other Paulo—creeping along the banks at a little distance down the river.

" They are after something," observed Jack, " for they have been making a couple of harpoons; and they seem to know pretty well what they are about."

Presently we saw a creature which at a distance looked like a young crocodile leap off the shore into the water. Mango's harpoon was rapidly darted at it; and he was now seen hauling up the creature, which was struggling to escape. He and his brother soon despatched it with blows on the head, and, leaving it on the bank, crept on a little further. Presently another creature was harpooned in the same way by Paulo; and they now came back looking highly pleased, and dragging the reptiles after them. They were about three feet long, with a high ridge running along their backs, and with hideous heads.

" Bery good eat," exclaimed Chickango when he saw the little monsters.

" What!" cried out Leo; " you do not mean to say you would eat those hideous creatures?"

" I suspect we shall have no objection to do so," said David. " They are varanians, a species of water-lizard, very similar to the iguanas of the New World, which are considered great delicacies. Ugly as they look, they are perfectly harmless."

The fires were already lighted, and without loss of time young Mango and Paulo set to work to skin their prizes. Chickango stewed a portion of them in our big pan. The flesh looked remarkably white and nice. First I took a piece; David followed; then Leo put in his wooden fork.

" Why, it is capital!" he exclaimed. " Kate, you must have some. Bella, I am sure you will like it."

In fact, in a short time we were all partaking of the varanian meat, which we preferred to that of the giraffe. We had a dessert of great variety, if not to be compared to some of our English fruits; but we were very thankful to get such nice and wholesome food. The fruits, indeed, were particularly cooling and pleasant to the palate. Chickango, who had disappeared, soon came back with a quantity of honey, which he had taken from the hollows in the trees we had seen on our shooting expedition. It was, as may be supposed, a welcome addition to our repast.

We were still seated at our meal, when a low rumbling noise reached our ears. It continued for some time, and looking out towards the east, whence it appeared to come, we saw dark clouds collecting. Presently vivid flashes of lightning darted forth, and reiterated roars came pealing through the air. "We must get shelter up immediately," cried Senhor Silva, " or the young ladies will be wet through; and our goods may suffer too." The canoes had been well secured to trunks of trees, though not unladen. We immediately got out the axes, and commenced cutting down the smaller saplings and straight branches of trees as rapidly as we could. These we placed on the side of the bank, covering our rude hut over with large leaves and heavy boughs on the top, which we secured by ratans to prevent their being blown away. Everything that could be injured by rain was immediately brought up, leaving room for the young ladies and poor Igubo in the centre.

" Oh, we can perch ourselves on the top of the baggage," cried Leo. " There will be room then for all hands inside."

While we were working away the clouds came rushing on over the sky, the flashes of lightning becoming every instant more vivid and frequent. I had hitherto seen nothing like it on shore. The most vivid flashes of forked lightning darted from the clouds, apparently playing round the summits of the taller trees, and then descending, went zigzagging along over the ground. Others were seen traversing the river in all directions. It was a grand but terrific scene. The blacks looked alarmed, and poor Chico chattered as if he would shake his teeth out, and clung to Jack's neck for protection. The thunder roared and rattled louder and louder, till we could scarcely hear each other speak : while sometimes the whole atmosphere seemed filled with flame. Presently huge drops began to fall. They came thicker and thicker, till they splashed down upon the river, throwing up miniature waterspouts all over it. The roar of the splashing and pattering was quite deafening. The wind, too, howled through the trees, which threatened to come down upon our heads, though we had placed our hut as far from them as possible. In a few minutes the water, which had been perfectly clear, became thick and muddy, and branches of trees and logs of wood were seen floating down the stream.

"We should be thankful that we are safe on land," said David. "Will this last long, Senhor Silva?"

"Sometimes such storms are over in half an hour," was the answer; "but they may last for a couple of days. Should this do so, we may congratulate ourselves on having the canoes to escape in, for the river may speedily swell, and cover the very spot where we are sitting."

This was not satisfactory news; at the same time, it was better to know of the probability of such an occurrence, that we might be prepared for it. The river was rising—that was evident—and now flowed down in waves which would have been almost sufficient to swamp our canoes; while torrents of water came rushing down the banks, and threatening every instant to sweep away our hut. Happily we had formed it on a little elevation on the bank, so that the stream turned on either side, and the risk was therefore lessened. Fiercer and fiercer raged the storm. The waters increased rapidly. It seemed as if the very clouds were emptying themselves upon the earth.

"I hope we are not going to have another deluge," exclaimed Leo.

"Of course not," answered Natty. "Don't you know that one is never to occur again? To be sure, this river may overflow its banks, but we have our canoes to get away in if it does."

I was afraid little Bella would be alarmed, but she kept gazing up at her sister, and seeing her countenance calm and tranquil, sat contented by her side, without speaking, however. In spite of the rain, I every now and then put my head out to ascertain that the canoes were safe; for as the waters rushed down, I was afraid lest the stumps to which they were fastened might be carried away. So thick was the rain that we could scarcely see across to the other side. Suddenly, as if by word of command, it ceased; and though the thunder continued to rattle towards the west, and flashes still issued from the clouds in the east, all quickly became serene. The sun burst forth again upon our heads, and the leaves of the trees and shrubs glittered for a few minutes as if covered with diamonds, though the sun rapidly dried up the moisture. The hut had become very hot, and I was just going out of it, when I saw the head of an animal crawling out from a neighbouring bush. At first I thought it was some creature, till I saw a long body following. It was a huge serpent. It came wriggling over the ground directly towards the hut. "Ondara!" shouted Chickango; "shoot! shoot!" Stanley sprang down from his seat, and aimed at the monster's head. I did the same. The creature, after convulsively twisting and turning itself into huge coils, lay still. We hurried down to examine it. On measuring it, we found that it was upwards of fifteen feet in length. David examined the head, and pronounced it to be venomous.

"Yes, indeed," said Senhor Silva. "It is the largest of all venomous serpents, and if the stories told of it are true, so virulent is the poison that it causes almost instantaneous death.'

We had reason to be thankful that we had escaped the two
dangers. As we were anxious to proceed on our voyage, having
now an ample supply of provisions, we once more embarked.
I was afraid, from the thickness of the water, that we should
have difficulty in avoiding any banks in our course; but it very
soon cleared, and we proceeded as before.

As we were paddling along a sudden sickness seized me
Whether it was from over-exerting myself, or from the heat of
the sun, I could not tell. Still I tried to go on. At length I
felt my paddle slip from my hand. Natty had just time to
catch it, and to save me from falling forward on my face. I
was placed in the bottom of the canoe, alongside poor Igubo,
and knew no more.

*   *   *   *   *

For days and days I lay in an unconscious state, utterly unable to move or speak or think.

Some time after this I had a dreamy consciousness of existence, but often for hours together I knew nothing of what was
occurring. I felt myself now and then lifted out of the canoe.
I knew that David was attending me, and at other times a
sweet face bending over me, and fair hands holding a fan and
driving away the flies. Once I heard Natty whispering, " Oh,
he will die! he will die!"

"I pray Heaven he may not," was the answer; "and David
thinks he will get through it. But he is very ill."

Then again I fell off into a dreamy state. Now and then
I knew I was on shore, and once more on the water. I was
conscious of the movement of the canoe, but what was happening round me I could not tell. I heard shots fired, and then
strange voices shouting and shrieking, but I could not utter a
word, nor could I understand what was said to me. After a
time the power of thought came back, and I knew when it was
day and when it was night, and I was able to discover that

many days and nights had passed away. Still I could not ask questions. An awning had been placed over the stern of the canoe, under which I lay. I remember seeing Igubo paddling away, as strong as the rest of the party, and though there was the mark of the wound in his side, it was perfectly healed. This showed me that a considerable time must have elapsed since I had been attacked. I discovered also that we were ascending a stream, but even then I could not speak. Shortly after this I felt myself lifted up and placed on a sort of palanquin, and carried along over the ground. I knew that I was remaining for some time, and that my little cousin Bella was sitting by my side fanning my face, and now and then moistening my lips, or giving me a slight portion of food. After that, I was once more lifted into the canoe. The river must have been far narrower than any we had passed through, for even as I lay in the bottom of the canoe I could see the trees on either side.

I had a relapse. I knew nothing more till one day I opened my eyes, and saw my cousin Kate seated near me, and Bella on a low stool at my side, with a book before her. Kate was working away most assiduously, as was her wont. Not far off in a corner sat Chico, as busily, though not so usefully, employed in cracking nuts. We were in a large airy hut, formed, as far as I could see, very much after the fashion of those we had before constructed. I was so placed as to be in the shade, and at the same time to obtain as much air as possible. I heard the voices of Leo and Natty at a little distance. They were engaged in some work, I concluded, and were laughing and talking merrily. I tried to speak, and I must have uttered a sound, for instantly Bella sprang up, and, casting her bright eyes on me, ran to her sister. "Oh, he is awake, and looks as if he knew me!" she exclaimed. Kate cautiously approached, and I saw her looking down upon me with an eye of pity and interest.

" Are you better, Andrew ?" she whispered.

" Yes, thank you," I could just utter in a low voice, "much better."   I wanted to say more, but could not.

" Leo ! Leo !" she cried out, "call David ! he will be so glad to hear that Andrew has returned to consciousness."

I could just catch sight of the boys running past the hut.

" Where are we ? what has happened ?" I asked.

" Oh, that would take too long to tell you," answered Kate. " You have been very ill for several weeks, and we have all been mercifully preserved from many dangers. You shall know all about it by-and-by.   We are safe now, I hope, and Stanley has sent for assistance ; but I must not talk more now."

CHAPTER XI.

HANKS to David's skill, and the preservation of the medicine-chest, under God's providence, I gradually recovered my strength. Several days passed, however, after the one I have mentioned when I returned to consciousness, before I could converse, or David would allow me to listen to a narrative of the events which had occurred since I was taken ill. My friends were employed in building huts and a stockade on a high hill which they had selected as a location to remain at till means of proceeding to the south could be procured. It was some hundred miles to the north of Walfish Bay, the nearest point where Europeans were located.

The first day I could sit up (I remember it well), Kate was by my side. A fresh breeze blew in at the open door of our hut, cooling my fevered brow. How beautiful all nature looked. We could gaze over a wide expanse of country, with blue hills on the left, and thick forests gradually breaking into scattered clumps of trees, and an open prairie reaching to the horizon towards the south. Below us I saw an extensive lake with a river flowing into it.

"There," said Kate, "is the stream down which we came to this spot. How thankful I was when we reached it, for

David said he had no hope of your recovery till we could find
a resting-place, with pure air and a more bracing climate than
we were passing through. It was dreadful to have you exposed so long to the damp night air, and the miasmas which
arose from the river; but we are in safety now, and I try to
forget all the dangers and anxiety we endured. It may be
many weeks or months before we can again set out; but by
that time, David says he hopes you will be thoroughly restored to health, and we shall be able to journey on with light
hearts, and, I hope, find friends to welcome us at the end."

"Oh, yes, dear Andrew," exclaimed Bella. "You have no
idea how frightened we often were; for we thought if the
savages had stopped us or taken us away from you, that you
would certainly have died. Sometimes we thought you were
dead, you were so quiet and pale; but when you are well
again, we shall not mind anything."

"Hush, hush!" said Kate, "we must not talk to Andrew
of what has passed. All is well now. Stanley is delighted
with the place. There is an extraordinary abundance of game
of all sorts, as he calls the wild animals which rove over those
plains. Sometimes we can see from here herds of buffaloes,
and cameleopards, zebras, and all sorts of deer and quaggas;
and there are savage animals too—lions, rhinoceroses, and
leopards, and elephants; indeed, he will not allow the boys to
go far by themselves lest they should be attacked."

"No, indeed," said Bella; "for though Stanley does not
always tell us his adventures, I suspect he has some narrow
escapes. In the river and lake, too, there is an immense
number of hippopotami and crocodiles. The boys went down
to bathe soon after we arrived, and had a fright, which will
prevent them ever doing it again. They were both in the
water when a huge crocodile darted across towards them, and
they had just time to scramble out and run away, leaving

their clothes behind them, when Jack and Timbo, who were fortunately near, rushed down and drove the creature off."

"It was indeed a mercy they were not seized," said Kate. "But we must not talk more to you now, Andrew. Stanley says he could not have wished to go to a finer spot, and it is only necessary to be cautious to avoid danger from any of them."

"Ah, here come the boys, and they have got a beautiful little animal between them. What can it be?" exclaimed Bella. "See, it has got small horns, and looks a graceful creature."

"It must be an antelope of some sort," said Kate; "but they will tell us."

The boys, who were coming up the hill, soon reached the hut. "We have got a koodoo! It is for you, Bella," they exclaimed in the same breath. "Chickango and Igubo caught it this morning, and have given it to us; but we are to take great care of it. See, it is already almost tame, but if we were to let it go it would soon be off." Kate made a sign to them. They both stopped and looked eagerly at me.

"O Andrew, how glad I am to see you sit up," cried Natty, on discovering that I knew them. "We were very unhappy about you; but now you will soon be yourself again, and till you are well enough to go about, our koodoo will give you plenty of employment, for Chickango says he requires careful nursing, just like one baby. We are to feed him with milk, and in a little time he will become as tame as Chico, though he will not play so many funny tricks, perhaps."

The little koodoo, when brought up to Kate and Bella, allowed itself to be stroked, and put out its tongue and licked their hands, though I saw from its startled eye and the tremor in its slender legs that it was as yet far from happy in its captivity. In a short time David came in, and after he had congratulated me on my improved looks, examined the little animal.

"Yes, indeed, it is a pretty creature," he observed; "but the full-grown one is still more beautiful. I saw several two mornings ago, which had taken shelter during the night in a thick wood which clothes the side of the hill at a short distance from this, and as they did not perceive me, I was able to observe them at leisure. The female is without horns, but the male has magnificent spiral ones upwards of three feet in length, which rise erect from his exquisitely-formed head, and give him an air of nobility and independence. The animal is about four feet high at the shoulder, and the general colour is a reddish gray, marked with white bars over the neck and croop. When walking slowly its action is very graceful. While watching the beautiful creatures I caught sight of a leopard lurking in the neighbourhood. I fired just in time to save the life of one towards which he was stealing. I missed the leopard, for I was at a considerable distance; but the report frightened the koodoos, and away they went, leaping over bushes, stones, and all impediments at a rapid rate, while the savage beast stole off, vowing vengeance, probably, against me for having disappointed him of his morning meal. The koodoo lives chiefly on buds and leaves and the young shoots of trees and bushes, and it is said that he is capable of going a long time without water. He is of a very timid disposition, but I am told, however, that when hotly pressed or wounded, he will sometimes face about and attack his pursuer. But we must now see about getting food for our young captive. We were, fortunately, on our way here, able to purchase half-a-dozen goats from some natives who had brought them from the south, and we must devote the milk of one of them to him."

"But how can you make him drink it?" asked Bella.

"Just as we give it to babies," said David, laughing. "I

will make a sucking-bottle for him. It can very easily be done. See! that small gourd hanging up will answer the purpose. I will fasten a piece of linen and a small quill in the mouth, and we will try the little creature."

"I will go and milk the goat," cried Leo, rushing out. "You come and help me, Natty, though."

Meantime David prepared the bottle, and in a few minutes Leo returned with a calabash full of milk.

"It is lucky I went," said Natty, "for the goat had refused to be milked at this hour, and had knocked Leo over."

"Yes, and she would have knocked you over, too, if I had not held her legs," said Leo. "However, we managed it."

"Why, how did you do that?" asked David.

"Oh, we tied her hind-legs to a post on one side and her fore-legs to another, and I held the head while Natty milked," said Leo

"Poor goat!" observed Kate. "I suspect she will not allow you to play that trick again."

The bottle was filled, and no sooner was it put to the little koodoo's lips than the creature began pulling away in a very satisfactory manner, every now and then giving a butt at it as it might have done when obtaining milk from its mother. It satisfied us, however, that there would be but little difficulty in bringing up the creature. Chico had eagerly watched the operation from his corner in the hut, though he did not approach the new comer. As soon as the deer had done with the bottle, David hung it up, when the monkey, fancying himself unobserved, instantly made for it, and, greatly to our amusement, applied it to his own lips, and began sucking away till he had drained it dry. He then quietly attempted to hang it up again, though in this he failed, and the bottle fell to the ground.

"We cannot afford to give you milk, Master Chico," said

David; " but I will soon cure you of that trick." Saying this, he went to his medicine-chest, which stood near, and having filled the bottle with water, put in a little powder, which he shook up. He then returned the bottle to its usual place.

" Now, take care, Master Chico, what you are about," he observed. " You are not to touch that bottle, recollect."

Chico looked at the bottle with longing eyes for an instant, then turned away, as if it was a matter of perfect indifference to him. In a short time he came down, and began to examine the little stranger, who seemed, however, in no way pleased with his presence.

" Oh, we will soon make you good friends," said Natty. " I hope we shall have a happy family before long. Do you know, Andrew, we have already got several creatures, and have managed to tame many of them, so that they feed on the hill-side in view of the hut, and come back at night regularly, for fear of wild beasts."

" Now, boys," said David, " we have talked with Andrew long enough, and I think we must leave him to Kate's care again. Your chattering is too exciting, and he has not got strong yet."

" Oh, but we will be very quiet, and merely listen to him, if he is inclined to talk," said Leo.

" That is the very thing I do not wish him to do," observed David.

" I feel quite strong," I said. " Pray do not send the boys away unless they wish to go."

However, the doctor was inexorable. While we were speaking, Chico had stolen back to his corner. Presently I saw Leo eyeing him, and hiding his face for fear of laughing.

Chico by degrees made his way up to the bottle, and slily unhooking it, put the spout to his lips and began tugging away with might and main. Presently casting it from him, with a loud chattering he rushed back to his corner spluttering and spitting vehemently. Leo now gave way to his laughter, in which all the party joined. Even Kate could not resist laughing, nor could I, though my merriment was somewhat faint, I suspect. Chico looked indignantly at us, as if he did not at all like being made fun of.

"I told you," said David, holding up his finger, "if you would drink from that bottle you would repent it."

He now took the bottle, and offered the contents to Leo and Natty, which they naturally refusing, he emptied it, and washed it out thoroughly. "It is quite clean now for Master Koodoo," he observed.

"Now, boys, take off your new pet, and try how quickly, by gentle treatment, you can tame it."

"I must ask Chickango and Igubo to get me one," exclaimed Bella. "I should like to have a beautiful creature like that for a pet, and I am sure I could soon make it love me."

"That must depend on whether one happens to jump into a pit," said Leo. "That was the way this one was caught. The mother managed to scramble out, but was shot while attempting to help her young one."

"Yes, and it seemed very cruel to kill the creature at such a moment. I should not like to have done it," observed Natty.

"That I am sure of," whispered Bella. "Natty would never wish to hurt any creature."

The boys now led off the little koodoo. Stanley soon afterwards arrived, followed by Jack, with some beautiful birds and several rock rabbits which they had shot. They congratulated me warmly on being so much better. I caught sight also of Timbo, Igubo, and his two sons.

"What has become of Chickango?" I asked, afraid, from not seeing him, that some accident had happened.

"The faithful fellow has gone to Walfish Bay with Senhor Silva," said Stanley. "We attempted in vain to find a native who would carry our message, and at last our Portuguese friend, though knowing the fearful risks he will run, undertook the journey, when Chickango insisted on accompanying him."

"Well, Mr. Crawford, I am main glad you are getting well again," exclaimed Jack, when the rest of the party had retired. "I would have given my right hand for your sake, and often when I thought you were going to slip your cable, I was ready to burst out a-crying; but, as Timbo says, God is very merciful, and now I hope you will come round pretty quickly, since you have weathered the worst point, where, so to speak, there were most rocks and shallows, and are now in smooth water."

I saw Timbo watching at a distance, and as soon as Jack had gone, he too came up.

"Oh, Massa Crawford, it do my heart good to see your eye bright again, and colour come back to de cheek. Me now no fear. You soon all right. I pray God night and day dat you get well, dat I do, and I go on praying still, for God hear de prayer of de black fellow, just as he hear de white man. Oh, Massa Crawford, it a great t'ing to be able to pray. If I no do dat I t'ink my heart sink down to the bottom of de river where de crocodiles crawl about; but when I pray it rise up just like a bird wid de big wings, and fly up, up, up into de blue sky."

I thanked Timbo warmly for his regard, but still more for the prayers he had offered up; and I felt as sure as he did that they had not been disregarded. My father's exhortation, I am glad to say, often came back to my mind. It was very delightful lying there in the shade, with the beautiful landscape and its countless numbers of inhabitants, and listening to Kate reading the Bible, in which we often came to passages, some peculiarly applicable to our position—so it appeared to me—others describing the wonders of God's works which we saw displayed before us, and his love and mercy to man.

In a few days I had so much recovered that my friends insisted on carrying me down to take an excursion on the lake. The day was cool, for a fresh breeze played over the water. Leo and Natty begged to have the pleasure of paddling me.

" And we will go too, shall we not? " cried Bella to her sister. I was glad to find that Kate consented.

" And I must go to look after you," said David, " and Timbo will stay at home to take care of the house."

" Very well, if I go as captain," said Jack; " but I cannot let you go and run your noses into the mouth of a hippopotamus or alligator, either of which, I have a notion, you would be likely to do."

Stanley and the two black boys had gone off in the *Giraffe*, as he wished to shoot. I wished to walk down, but found, on attempting it, that I could not; indeed, I had become so thin that I was no great weight for my friends to carry. As soon as we had taken our places in the canoe, we shoved off. I was able to sit up and enjoy the scenery. To the west rose the lofty hills on the side of which our village was placed, for so I think I must call it, while on the left were woods with fine trees, and here and there a break through which the broad prairie could be seen extending as far as the eye could reach towards the south. We got glimpses of numerous animals moving in and about the woods, and some scampering over the plain. It was already late in the day when we embarked. As the weather was fine and the lake perfectly calm, we paddled down the centre to enjoy the greater purity of the air, away from the banks. The trip was so enjoyable that we were tempted to go further, perhaps, than was prudent. At length, unwillingly, David begged Jack to turn the canoe's head home-wards. As we were paddling along, we caught sight of Stanley's canoe entering a creek out of the lake.

"Oh, see, see!" cried Bella, "what thousands of animals! I never saw so many collected together."

Such indeed was the case. On the point nearest the lake some twenty or more huge buffaloes were standing drinking at the stream. Further on a whole herd of quaggas had come down, while through the woods could be seen the graceful horns of a troop of koodoos and other deer, though it was difficult to distinguish them among the trees. But we were more immediately interested with the numerous birds we were passing. It would be difficult to describe them all; but David, who was a good ornithologist, told us their names. Amongst them was one which seemed to run about on the surface employed in catching insects. It had long thin legs, and extremely long toes, which enabled it to stand on the floating lotus leaves and other aquatic plants invisible to our eyes. A lotus leaf, not six inches in diameter, was sufficient to support its spread-out toes, just as snow-shoes enable a heavy man to get over the soft snow. It was the *Parra Africana*. Then there were numbers of the pretty little wader, which looked exactly as if it was standing on stilts, from the length of its legs, while its bill appeared to be bent upwards, instead of downwards, as Leo declared it ought to be. David called it an *avoset*. "See," said David, "the use of its bill!" It was wading in a shallow; and the form of its beak enabled it to dig up insects out of the soft sand far more easily than if it had been straight. We saw vast numbers of the large black goose walking about slowly and feeding. It had a strong black spur on the shoulder, with which it can defend its young. David told us that it forms its nests in ant-hills, and, of course, eats up the inhabitants. Among the several varieties of geese was the Egyptian or *Chenalopex Ægyptiaca*. It flew along over the surface, but appeared unable to rise. It would have been impossible to count the ducks which sat on the banks. Stanley fired among them, and almost filled his canoe with a few shots, as he afterwards told us. He had killed in one shot nearly twenty ducks and a couple of geese. But they were only some of the smaller birds. Further up were spoonbills with nearly white plumage; a tribe of stately flamingoes, such as I have before described; numbers of the *demoiselle* — an extremely graceful and elegant-looking bird—and a light blue crane, and another crane with light blue and white neck. We must have counted fifty or more specimens of the *Ibis religiosa*, and vast flocks of the large white pelican, which came following each other in a long-extended line, rising and falling as they flew. David cried out that they looked as if they were all fastened together like a thick rope made to move like a serpent. There were also innumerable plovers, snipes, curlews, herons, and other smaller birds. A number of those strange birds, the scissorbills, were flying about near a sandbank on one side. They had snow-white breasts, black coats, and red beaks. We observed the hollows in which their nests were placed in the sandbanks, for they made no attempt to conceal them. "What brave little chaps they are!" exclaimed Leo. "See!" Some crows had approached as he spoke, when the scissorbills flew after them and drove them off. As we drew near, however, the crows took to flight, when the little scissorbills hung down one of their wings, and limped off, pretending to be lame. This trick did not, however, save the life of one of them, at which David fired for the sake of examining it. On getting the bird into the canoe, we found the lower mandible almost as thin as a carving-knife. The bird places it on the surface of the water as it skims along, and scoops up any minute insects which it meets with in its course. Its wings being very long, and kept above the level of its body, it can continue thus flying on for a considerable time, till it has supplied itself with an ample meal.

"By feeding at night, it probably escapes being snapped up by some hungry crocodile, which it would be if it fed thus close to the water in the day-time," observed David. "The scissorbill has great affection for its young, as indeed have most water-birds."

On another bank we saw a number of pretty little bee-eaters congregated together. The bank was perforated with hundreds of holes conducting to their nests. As we passed by they flew out in clouds, darting about our heads. Then there were speckled kingfishers, and also beautiful little blue and orange kingfishers, which we saw dash down like shots into the water searching for their prey. There were sand-martins something like those seen in England; and from the trees also, as we passed under the banks, rose flocks of green pigeons. I must, however, bring my account of the feathered tribes we encountered in our trip to an end. Stanley's gun soon created dismay and astonishment among them, and often the air, as he fired, seemed literally filled with birds. The zebras and quaggas started off and took shelter in the woods; but the buffaloes more firmly stood their ground, eyeing us with aston-ishment, and evidently not understanding the effect which a bullet would produce should it hit one of them. Suddenly too, from out of the water rose several huge heads of hippopo-tami, which made Bella cry out with dismay, for though we were by this time well accustomed to them, she had never got over her alarm at seeing the monsters.

"Oh, let us paddle away from those dreadful creatures!" she exclaimed. "I am sure they are going to swim after us.

See, see! Oh, how horrible if they should seize Stanley's boat! They are between him and us. He will never be able to come back."

"Do not be afraid, Miss Bella," said Jack. "The captain will give a good account of them. A bullet would soon send any one of them to the bottom."

Jack, however, shouted out to Stanley, and pointed to the hippopotami. He had by this time got his canoe so full of birds that he could scarcely carry more, and he now came paddling after us, utterly regardless of the monsters. As he passed by, though they gazed at him with their savage eyes, and mouths half open, they did not attempt to approach; and the blacks continued to shout and shriek to keep them at a respectful distance. Stanley, having put specimens of the birds he had shot into our canoe till we could scarcely receive more, went back to knock over, as he said, a further supply, while we paddled homewards. David had now plenty of occupation in examining our prizes, while the boys paddled slowly onwards, assisted by Jack, who not only paddled, but steered also. We found Timbo waiting for us at the landing-place with the litter to carry me. He had a gun over his shoulder, and appeared to be keeping a bright look-out on every side, shouting every now and then at the top of his voice.

"What is it, Timbo?" asked David.

"Me see big lion!" he answered. "He mean mischief. Just now roar and roar again. He would like carry off Massa Andrew, but we no let him."

"Oh, never fear," cried Jack. "We will keep the biggest lion at bay if he should come near us, and will give him a shot which will make him wish he had kept away."

"The lion is not likely to come near us when he sees so many people," said David; "but we will be on our guard against his approach."

I was immediately lifted on to the palanquin, and Jack and Timbo carried me up towards the house. All hands loaded themselves at the same time with birds, and Kate and Bella fastened as many at their backs as they could carry. Even then they were obliged to leave many behind for a second trip. David and Leo walked by the side of Bella, while Natty led the way. We had got half-way up the hill, when, from a thicket at some distance, a loud roar proceeded, and we saw the head of an enormous lion appearing from among the bushes.

"Roar away, old fellow," cried Jack. "It will be the worse for you if you come here."

"Shall I fire? I might kill him," said David.

"No, massa, no," answered Timbo. "If you hit him he come on in great rage. He now only angry because he dare not come near. Each time he roar we roar back, and dat keep him away;" and Timbo setting the example, the whole party set up a loud shout, with the exception of Kate. Little Bella, however, made her shrill voice distinctly heard. For my own part, I could not have attempted to shout. It showed me how prostrate I had been, for even now I had difficulty in slightly raising my voice.

Our shouting brought Chico to the door. As soon as he saw us he came hopping down the hill; but the next time the lion roared he gave a spring backwards, and turning round, rushed back into the hut.

"We must go down and warn the captain," said Jack; "for if he does not know that the lion is in the neighbourhood, the beast may surprise him; and, at all events, he will want assistance in bringing up the birds."

"We will go, then," said Leo and Natty; and they set off together.

David, in the meantime, secured our cattle-pen, which

probably had attracted the lion to the spot. At each side of
the entrance a circular hut had been built, answering the pur·
pose of the gateway towers of a castle. Igubo and his two boys
occupied one of them, and Jack and Timbo the other. They
were built of reeds closely bound together, and the doors were
of the same material. These were strong enough to resist the
attack of any wild beast, and were always kept closely shut at
night. I felt somewhat tired after my day's excursion; but
some supper my kind cousins soon prepared restored my
strength. They had got ready a more substantial meal for
Stanley and his attendants, who now arrived.

"What do you think, Mr. David?" I heard Jack exclaim
"If a big alligator has not got into the canoe and eaten up all
the birds while we were away! It is fortunate we brought up
as many as we did. However, the captain has got enough and
to spare."

"We will be even wid him," said Timbo. "Igubo say he
kill alligator. If he find him he get dem all back to-night."

"Tell him he had better not make the attempt," said Jack,
"or maybe the lion will pick him up on his way to the river.
We must give a good account of the brute to-morrow, or he
will be doing us mischief."

There was ample work that evening in plucking the birds and
in salting down the larger number. I should have mentioned
that a salt spring had been found on the side of the mountain;
without it, indeed, I doubt if we should have been able to
remain at the place, for we had already finished our supply of
that necessary article.

There was no necessity to warn the rest to secure their doors
at night. One man, it was agreed, should keep watch, as it
was very likely the lion would attempt to get into the cattle-
pen. As I lay asleep in my hut the roar of the lion entered
into my dreams. Sometimes I thought he was flying at Kate,

and I was in vain endeavouring to defend her. Once he had carried off Natty; and I saw Leo, his namesake, seated on his back and digging a spear into him. At last I started up, and was sure the sounds I heard were real, and no mere fancies of the brain. The whole of the inmates of our camp were on foot, and I heard them calling to each other. Presently there was a shot, followed by another tremendous roar.

" Can you see him?" I heard Stanley cry out.

"No, sir; he has made off," answered Jack.

" I thought I hit him," exclaimed Stanley.

" T'ink not," said Timbo. " He no like sound of gun."

After a time they all went back to the huts. I think I said I slept in David's, for he acted as my nurse throughout my illness, and no one could have been more gentle and kind. Next morning Stanley and the boys hurried out to see if there were any marks of blood; but none were discovered, and it was therefore plain that the lion could not have been hit.

My companions had not been idle, I found, for they had cultivated a considerable piece of ground, and enclosed it, on one side of the cattle-pen. People in England have little notion how rapidly fruits come to perfection in the Tropics, where the account of Jonah's gourd is completely realized. Thus, in time, we had all sorts of vegetables, which contributed greatly to keep my companions in health, and to restore my strength. Stanley's gun also supplied us amply with animal food of the greatest variety, so that we were never on short allowance. Igubo and his sons were expert fishermen, and caught as many fish as we required. There were often more than we could eat fresh; the remainder were sun or smoke-dried, and, hung up, kept for a considerable time. The fishermen had to be careful not to fall into the jaws of croco-diles, who were constantly on the watch; and thus they often had to beat a rapid retreat to escape from the monsters.

Up to the time I am speaking of we had received no visits from the inhabitants, but Stanley, in his more extensive shooting excursions, had fallen in with a few, though the nearest village was about four miles off. It was situated in a valley to the north of us. The people appeared peaceably disposed. They seldom or never ventured far from their homes, having the means of supporting life and abundance of game round them. They also cultivated the soil sufficiently to obtain enough vegetables for their wants. Stanley had won their friendship by making them presents of birds and some animals, and in return they begged him to accept a supply of manioc, which Mango and Paulo brought to us. They look upon it as their staff of life, and as it is produced with very little labour, it well suits their habits. Stanley described the plantation which surrounded the village. The plants, he told me, grow to the height of six feet, and the leaves are often cooked as a vegetable; indeed, every part is useful. The roots are about four inches in diameter and eighteen long. To cultivate it the earth is formed into beds about three feet broad and one in height, and into these pieces of the stalk are placed about four feet apart. In about eight months, or sometimes rather more, the roots are fit to eat. There are two sorts, I ought to say. One is sweet and wholesome, and fit to eat when dried, and can at once be beaten into flour for making bread or cakes; the other is bitter, and contains poison, but is more quickly fit for food than the sweet sort. To get rid of the poison it is placed for four days in water, when it becomes partly decomposed. It is then taken out, stripped of the skin, and exposed to the sun. When thus dried it is easily pounded into a fine white meal. It is then prepared for food as ordinary porridge is made, by having boiling water poured upon it by one person, while another stirs it round till it is thoroughly mixed. Our black companions were very fond of it; but while

we could obtain more substantial food, few of our party would
condescend to eat it, except now and then as a change. The
poison is of so volatile a nature that it is quickly got rid of by
heat. Timbo made the meal into thin cakes, which, when
baked on an iron plate, were pronounced very good. David
told us that it was called cassava, as well as manioc, and that
its scientific name was *Jatropha manihot*. After a few trials
he contrived to manufacture a kind of starch, which I had often
seen in England under the name of tapioca. He was delighted
when he succeeded in producing it, and Kate at once made
some very nice puddings from it, by mixing it with honey
to give it flavour.

We obtained also from the village some yam roots, which
had greatly the taste of potatoes, though of a closer texture.
They also were placed in the sun to dry before being cooked,
and we found by putting them in dry sand that they would
keep well for a considerable time. The yam is the root of a
climbing plant which David called the *Dioscoreo-sativa*. It
had tender stems, eighteen to twenty feet in length, and sharp-
pointed leaves on long foot stalks. From the base of the roots
are spikes of small flowers. The roots are black and palmated,
and about a foot in breadth. Within they are white, but externally of a very dark brown colour. Besides this another sort
was brought to us a little time afterwards, called the *Dioscoreo-
alata*, very much larger than the former. Some, indeed, were
fully three feet long, and weighed nearly thirty pounds.

" How it would delight an Irishman's heart to see a potato
as big as this root!" exclaimed Leo. "It would be a hard
matter, however, to find a pot big enough to boil it in, or to
steam it afterwards, to make it mealy."

CHAPTER XII.

THE boys were continually asking Timbo and Igubo when they were going to catch them another pet. They were with me one day when the two men arrived loaded with the flesh of an animal which Stanley had shot.

" What is that ? " I asked.

" He bery good eat," was the answer; " like a little horse."

" But what is it called in England ? " I inquired.

" Him zebra," he answered; " mark over back. We cooky for supper."

" I wish Stanley had caught him alive," said Leo. " Now, Timbo, cannot you manage to get a young one for us, or a couple, and then we could break them in, and make them carry us."

" Him no carry no one," answered Timbo. " He wild. Kick off, even dough you stick on like Chico."

" But we could soon teach Chico to ride it. I suspect that it would puzzle even a zebra to kick him off."

" We will try," said Timbo. " We go and make many pit-falls; but take care, Massa Natty, you no tumble in when tiger or leopard dere."

I found that the men had already dug some pit-falls, though

hitherto, excepting a koodoo, nothing had been caught in them.

Next morning they set off to visit the pits, accompanied by the boys. In rather more than an hour they came back, Leo and Natty dragging a beautiful little animal between them, while the two men brought the head and skin and a quantity of meat of another. David, who was with me, ran out to meet them.

"They have got a gemsbok!" he exclaimed; "one of the most interesting of the antelope tribe. It is known also as the oryx."

"How did you catch it?" he asked.

"We found it in the pit!" exclaimed the boys at once; "the mother and the young one. Poor little creature. The mother fought so furiously that the men were obliged to kill her, and not till then could we get the young one out. But it will make a capital playmate for the koodoo."

"It is very hungry," said David. "We will try if it will take some milk."

While Leo and Natty ran off to milk a goat, the men held the little animal, which, though it trembled, made no attempt to escape. David examined the head of the larger one. It had beautiful horns, nearly three feet in length, slightly curving backwards, and of a shiny black colour, and very slender. The mane and tail were very like those of a horse, while the shape of the head and the colour were those of an ass, the legs and feet, however, showing it to be an antelope. Both the horns were so exactly equal that I could fancy a person taking a side view of the animal might imagine them to be one and the same; and David said that the gemsbok has often therefore been supposed, by those who have seen it at a distance only, to be the unicorn which the ancients believed to exist. The little calf was of a reddish cream colour, and was so small that the horns had scarcely yet appeared. Timbo told us that the gemsboks were generally seen in small herds. Probably this one and its calf had been separated from their companions, as no others had been taken. It is one of the swiftest quadrupeds of Africa, indeed its speed is almost equal to that of the horse. Herds of them are generally found in districts devoid of water, as they can go a long time without drinking, having receptacles in their inside somewhat like those of a camel, though much smaller, for retaining fluid.

As soon as the milk was brought David tried to feed the little creature with a spoon instead of the bottle, and after a few attempts it willingly swallowed the milk. He then applied the bottle to its mouth, and as soon as it found out its contents it sucked it eagerly; he had hopes, therefore, of being able to bring it up. Kate and Bella, summoned by the boys, now came in to inspect their new pet. It allowed Bella to stroke it and pet it without evincing any fear, and when she fastened a handkerchief round its neck it followed her willingly.

" What a dear little creature ! " she exclaimed. " We must give it a name, though. I do not think gemsbok is pretty. I like oryx better."

" I am afraid, however, when he gets his horns he will prove rather a dangerous companion," observed David, looking at the head of the larger animal which lay on the ground outside the hut. " It will fearlessly encounter the most savage animals of the desert, and instances have occurred where it has succeeded in killing even a lion or tiger which had incautiously sprung on its dagger-like weapon of defence."

" Oh, it will be a long time before those grow ! " said Leo. " We can pad them, or cut off their tips, and then it can hurt no one, even in play."

Stanley and the two black boys were out hunting, and in

the evening they appeared, the boys carrying slung on two
poles an animal which we saw at once was alive.

"Why, it is a little horse!" cried Leo, running down the hill

"The zebra we have been so longing for!" exclaimed
Natty.

They soon arrived at the encampment, Stanley highly delighted with his prize. It was curious that the two animals
they had been so long wishing for should have been taken on
the same day.

" He has given us a great deal of trouble," said Stanley, as
the little creature was brought up to be inspected by the girls
and I. " I doubt if we shall ever make him reconciled to
captivity. He struggled and kicked so much that we had
great difficulty in getting hold of him, and as to making him
come along with us, that was impossible. The harder we
pulled one way the more determined he was to go another.
and at length Mango suggested that we should sling him as
you see, and he could then no longer help himself. But it
was no easy matter to get him into the slings."

The little zebra was somewhat more clumsily shaped than
a pony's colt, and about the size of one three or four weeks
old. A pen had been built for the koodoo, and into this the
two animals were now introduced. Koodoo gazed at them
with looks of astonishment, but in a short time ran up to the
little oryx and seemed to welcome it.

" I do not know whether they understand each other's language," said Bella, " but it strikes me koodoo is telling little
oryx that he is very well treated, and recommending him to
be reconciled to his fate."

The zebra, however, would not go near them, and whenever they approached ran off round and round the pen. In
a short time it became hungry, and David, accompanied by
Timbo, with a calabash of milk, went in to try and feed it.

Timbo had some difficulty in catching it, for whenever he drew near it kicked out viciously, and then scampered off. It was, however, at length caught, and though at first when David tried to put the milk into its mouth it kept its teeth closed just as as a child does when medicine is offered it, it at length allowed some to be poured down its throat.

I was now sufficiently recovered to walk about the camp with the aid of a stick. Sometimes Kate and Bella assisted to support me, and when Leo and Natty were within they were always ready to offer me their arms. We never ventured to leave the camp without a guard; for since the first visit of the lion I have described to our neighbourhood we had frequently heard his roar, although he had not ventured to come nearer. Our life, indeed, was not altogether free from anxiety, for we could not hide from ourselves the danger which the hunters especially ran from wild beasts, nor could we be certain either that the natives in the neighbourhood might not some day prove treacherous. Stanley, grown bold by immunity, increased the length of his expeditions, and frequently did not return till after nightfall. One day he went out accompanied by Igubo and his two sons, leaving the rest of us to work in the garden and to keep watch over the camp.

"How long shall you remain away?" asked Kate. "It makes us feel so anxious when you are absent."

"You can dispense with our protection for a couple of nights, I hope, at all events," he answered. "We have no enemies to fear; and, in truth, two nights spent in the wilds at a time are sufficient to satisfy even my love of sport. If we had waggons to carry our provisions, and horses to ride, the case might be different; but even if we get the game we cannot bring it back, so you may rely on our reappearance at the time I propose."

I did not see them in the morning, as they went away before

I had risen. Stanley had been absent two days, when, as the weather was cool, the boys begged me and their sisters to come down and take a paddle on the lake. I was able, I thought, to walk down and back again with their assistance, and as David thought I should benefit by the amusement, he advised me to go, Timbo remaining, while Jack went as captain. Chico, as usual, accompanied us, and hopping into the canoe, took his seat in the bows. As we paddled along we had abundance of matter to interest us, in the numerous birds which skimmed along the water or sat perched on the trees. Bella pointed out some beautiful turtle-doves, which were sitting happily on their nests above the water gently uttering their low coos to each other. Not far off we espied an ibis perched on the stump of a tree, shattered probably by lightning.

"I should like to bring her down for her impertinence," cried Leo. "Listen to her loud "Wa—wa—wa." She is trying to drown the voices of your favourites, Bella."

Though we passed close by, the ibis seemed in no way disposed to move, but continued shouting "Wa—wa—wa." However, she was not allowed to cry alone, for near her sat three fish-hawks piping away in the same fashion. Leo was about to stop and take a shot at one of them, but Kate intreated him to let the bird alone, and we rowed on, leaving him and his companions piping away to their hearts' content. Presently we saw a moderately-sized bird, like a plover, darting here and there, and uttering a peculiar sound. "Tinc—tinc—tinc," cried Leo; "what is that you say?" Presently a white-necked raven, which was sitting on a stump some way down, flew off, shrieking with fear, as the plover pursued it.

"Well, that is a coward," said Leo. "He is running away from a bird half his size."

"Very wise," observed Jack. "Timbo, when he was out with me the other day, told me they call him the 'hammering iron,' on account of his 'Tinc—tinc—tinc' cry. But it is not his cry which makes the raven fly off. He has got a sharp spur on his shoulder, just like that on the heel of a cock, and he could dig it into the raven, and soon draw its life-blood."

On went the plover to a bank a little way ahead, where it pitched on what we thought at that distance was a log of wood. As we paddled up the seeming log turned into a huge crocodile basking in the sun.

"Stop paddling," I cried to the boys. "Let us see what the plover is about."

It ran along the back of the reptile, but stopped on the top of its snout, and then with perfect fearlessness actually flew down into its gaping mouth. I then recollected an account I had read of a bird on the Nile of that description, which is known by the name of siksak—the trochilus. It is stated by two or three credible witnesses that it performs the part of tooth-picker to the monster. Whether it was so occupied or not we could not tell, but presently the crocodile appeared to rouse itself up and to crawl towards the water, into which he plunged, diving down out of sight.

"There goes Master Tinc—tinc—tinc flying away. I suppose he will go and warn his other friends," said Jack. "That is his business, so Timbo says; and when these birds are about you can never get a shot at a crocodile."

As we continued paddling on we were convinced that they had been warned of our approach, for they all betook themselves to the water long before we got near them. Proceeding we reached a part of the river where the banks were steep and composed of sand. Presently we saw a creature crawling out of the water, and making its way up the bank.

"What creature can that be?" asked Natty.

"A water-turtle!" I exclaimed; for I recognized it from the descriptions I had seen of it.

Presently it came to a steep part of the bank, and as it was climbing up it fell, and lay helpless on its back.

"We will make prize of him," cried Jack. "Paddle away, boys."

We were soon up to the bank, when Jack sprang out of the canoe, and before the turtle could recover itself he had seized it in his arms and placed it in the bottom of the canoe. There the creature lay utterly helpless. While the canoe's bows were on the shore, Chico, who had got tired of sitting so long in one position, made a spring on to the land to pick some fruit which grew on a low bush at no great distance. The boys were so interested in watching the turtle that, without seeing that Chico was absent, they shoved off, and had already got to some little distance when they discovered that we had left one of our company behind. Chico, having filled his paws with fruit, ran down the bank.

"Hillo, old fellow!" exclaimed Jack, "we will come in for you."

The current, however, just then took the canoe's head, and we drifted some way down before we could turn back. At that instant we saw a ripple in the water, and presently the huge head of a crocodile was projected above it. The monster darted forward; and poor Chico, before he was aware of his danger, was seized by its huge jaws. In vain we cried out and shrieked at the top of our voices. The crocodile had got hold of its prey. Chico struggled, but he was as helpless as a mouse in the fangs of a cat. "Oh, save him, save him!" shrieked out Bella; but it was too late. Though the boys paddled with might and main, before they reached the shore the crocodile sank beneath the surface, dragging the poor ape with him. A little circle alone marked the spot where it had gone down.

"There is one who will pay you off for that," cried Jack,

looking into the water as if in search of the crocodile. " When Igubo hears of it he will be after you, depend on it."

We all felt sad at the loss of our pet, and much as we had enjoyed the early part of the trip, it certainly spoiled the pleasure of the remainder.

" Poor Chico ! " exclaimed Natty every now and then. " I little thought you would come to so untimely an end."

Bella cried outright, and Kate could scarcely restrain her feelings. We now proceeded back to the landing-place, and Jack and the boys having drawn up the canoe to the spot where she usually lay concealed, we commenced our return home. My young cousins and Natty assisted me up the hill. We had got to about half the distance, when a loud roar came from the thicket I have before mentioned. " Roar away !" cried Jack, " you will not frighten us." Bella, and even Kate, could not, however, help trembling at the sound ; indeed, there is something peculiarly terrific in the cry of the lion in his native wilds. I trusted that he would confine himself to roaring, and not attempt to approach nearer. The boys and Jack looked to their guns.

" We will be ready for him if he dares to show his face," cried Jack. " Now, you young gentlemen fire first, if he looks as if he was going to attack us. I will keep my fire in case you miss."

The lion, however, allowed us to gain our home, where we found David and Timbo looking out for us, and ready to fire at the beast should he approach.

" But where Chico ?" cried Timbo when he saw us. Jack told him what had happened. " When I tell Igubo, he soon punish crocodile," he said. " Igubo great crocodile hunter."

" But what have you there, Jack ?" asked David, as he saw the turtle which Jack had brought up on his back. " Well, you have indeed a prize, for the turtle will be a pleasant addition to our bill of fare.

When the girls went to their hut, we examined the water-turtle, which Timbo and Jack at once prepared for cooking. Opening it, we found that it had upwards of thirty eggs in its body. The shells were flexible, and the same size at both ends, like those of the crocodile.

"Dis make one bery fine dish," said Timbo, "and de liber is first-rate. We hab it ready for when de captain come back."

"We must leave the charge of cooking it to you," said David, "for I doubt whether my sisters will understand the art so well."

Part of the turtle was cooked, and supper made ready, but still our friends did not appear. Night drew on, and we became somewhat anxious. At last David advised his sisters to take supper and to return to their hut, while we sat up waiting for the party. Hour after hour passed by, and still they did not appear. At last David insisted on the boys and I going to bed, while he and Jack and Timbo kept watch. Every now and then we could hear the roar of the lion in the distance, replied to by their loud shouts to scare him away. I could only hope that my cousins were asleep: as for myself, I could not close my eyes. Not a breath of wind was stirring, not a sound was heard except that ominous roar which occasionally broke the silence of night. At length David came in, pretty well tired out, and lay down, saying that Jack had undertaken to keep the morning watch. I also, in spite of my anxiety, at last fell asleep. I awoke suddenly with the sound of the lion's roar in my ears. It seemed far louder and nearer than before. Could it be fancy? The morning light was streaming in through an opening over the door, which we had left to admit air. Again I heard that fearful roar. I started up, for it seemed to be in the very midst of our camp. I thought of my young cousins and the boys, who were likely enough to have gone out early. I sprang to the opening, and there I saw, in the very midst of the cattle-yard, an enormous lion, his head lifted up proudly, while his huge paws were placed on one of the animals he had struck down. Never had I seen so magnificent a creature—his vast mane covering his neck and shoulders, while his tail waved to and fro as a signal of defiance, looking up as if he saw an enemy approaching. The other animals, terror-stricken, were trying to force their way out of the yard. I could see no one. What had become of Jack and Timbo I could not tell. They could not have deserted their posts, for both had given too many proofs of courage to make me suppose so. Calling to David, who was yet sleeping soundly, I seized my gun; but when I returned. the lion had gone, with the animal he had struck down. David and I rushed out of the hut. At that moment there were several shots. Looking out in the direction from which the roars had previously come, I saw the lion bounding away along the hill, still apparently unwounded.

"Has he gone? has he gone?" I heard Leo and Natty shouting out. "Yes, yes! and he has carried off our little gemsbok!"

"But where are Jack and Timbo?" I asked. "How was it they let the creature come in?"

"They heard some shots in the distance, and thinking that they were fired as signals by Stanley and his party, they were just setting off to meet them, and the lion must have taken that opportunity of coming into our camp. They had not got far, and must have caught sight of the lion as he was making his escape. It is a mercy the girls are safe!"

As they were speaking, Jack and Timbo came back. "Well, I never did think he was going to play us so scurvy a trick," exclaimed Jack, "or we would not have left the camp. But what do you fancy those shots can mean, Mr. Crawford?"

Both David and I agreed, however, that they were
probably fired by Stanley or his companions, either at some
animals, or as a signal to give us notice of their return; and
we therefore begged Jack and Timbo to proceed as they had
purposed, while we remained on the watch for the lion, should
he venture to come back.   Kate and Bella now came out of
their hut, and great was their grief at hearing of the loss of
one of their pets—the most promising, indeed, of all, for in a
few days it had become so thoroughly tame, that it would
follow them about like a lamb.   They, like us, had been kept
awake the greater part of the night, and, owing to this, had
not been aroused by the sound of the lion's voice; indeed, the
events I have described occupied less time than I have taken
to write them.   The boys now employed themselves in collecting the trembling animals, who had not yet recovered from
the fright which the appearance of their dread enemy had
given them.   The little koodoo and zebra had, however, been
safe in their pen, or they would probably have run off, and we
should have seen no more of them.

" I did not fancy that a lion could have leaped so high a
palisade," said David; " but I see we must take other measures
to secure our camp for the future.   I believe that even a lion
cannot break through an enclosure of prickly pear, and I propose that as soon as Stanley comes back we all set to work to
surround our camp with a thick line of it; and if we fasten a
fringe of its sharp leaves to the top of our fence, we shall be
able to bid defiance to either lion or leopard.   I doubt, indeed,
if elephants, or even human beings, would willingly assail such
a fortification as ours will then be."

I fully agreed with David, and we settled that we would
immediately set to work and collect the cactus plants which
grew in abundance on the hill-side.

" They will be very hungry when they do come back," said

Kate, " and therefore, Bella, you and I will prepare breakfast for them forthwith."

David and I assisted them in getting the repast ready, and our anxiety was shortly relieved by Leo running in exclaiming that Stanley and all hands were coming up the hill. " They have got no end of game," he added—" birds and beasts enough to feed us all for a month to come, if we were in Siberia and could freeze the meat and keep it till we want it, instead of being in the middle of Africa. Unless we can salt it pretty quickly, however, it will not be fit for much by to-morrow morning."

Having thus delivered himself, he ran out again, and down the hill. In a short time our friends arrived, pretty well tired out, however, for they had been to a long distance, and found it impossible to reach home the night they had intended. They had therefore encamped a few miles off, and started again at dawn. The shots we had heard had, as we supposed, been fired by them to give us notice of their approach. They had pushed on without stopping for breakfast, and were duly grateful to Kate and Bella for enabling them thus speedily to satisfy their hunger.

Stanley said he had a long account to give us of their adventures. They had fallen in with a native village, the inhabitants of which appeared to be inclined to be friendly, and had invited them to join in a grand hunting expedition. " I will tell you all about it as soon as I have eaten something," said Stanley. " But what is this I hear of a visit from a lion? Did the brute actually dare to leap into the midst of our camp and carry off one of its inmates? It shall not be the fault of my rifle if he does not pay dearly for his freak before another sun rises."

CHAPTER XIII.

E were, of course, eager to hear Stanley's adven-tures.

"Finding the day tolerably cool, though I doubt if our friends in England would have called it so, we pushed on further south than we have ever gone before," said Stanley. "The country, though wooded in parts, was generally open, and we had little diffi-culty in making our way across the prairie. I have never seen such large herds of buffaloes, zebras, gnus, rhinoceroses, and giraffes. Had we been mounted, we should have had no diffi-culty in coming up with them, but on foot it was a very dif-ferent matter. Often, as we got up to them, almost within range of our rifles, they were off again, leaving us standing alone, without a hope of overtaking them. As the sun rose higher and grew hotter, the buffaloes and rhinoceroses retired to their coverts, as did many of the other animals, the zebras and giraffes alone defying the sun's rays. I now hoped, by keeping under shelter of the woods, we might the more easily surprise some of the animals we were in search of. Before proceeding further, however, I proposed that we should open our wallets and dine; and having selected a shady spot under tree at a little distance from the forest, where there was probability of our being surprised by any prowling leopard or hungry lion, we formed our noonday camp. We had not sat long, when Mango came in and told us that he had seen the head of a buffalo projecting from the forest at some little distance, and that he was sure there must be several there. I had been so annoyed at not killing anything, that, without finishing my dinner, I set off with Mango to try and reach the spot unobserved by our expected prey. We at once got under shelter of the wood, and worked our way along through the borders of the forest, hoping to get up to the spot without disturbing the herd. Mango at length made me understand by signs that we had now reached the place where he had seen the buffalo. I can tell you they are very different animals from those we met with further to the north. These are pictures of brute strength and ferocity, their horns, short and curling, but pointed like daggers, meeting at the roots, where they form a thick mass, serving as a helmet to the animal. I was afraid of coming suddenly upon them, for I knew that if startled they would be off before I could obtain a shot. Mango was positive that we were near them. He suggested at last that we should climb a tree, whence we might survey the neighbourhood. Finding one, we mounted it, and when I had got a steady footing, I looked round me, hoping to discover the animals. Not a living creature, however, stirred. At last my companion pointed out some dark objects just seen indistinctly through the thick foliage. They were the backs of the buffaloes, I had little doubt. I fired, but nothing moved, and I could not help supposing that I had mistaken some large stone for a living creature. To settle the matter, I again loaded and fired. At the report of the gun, half-a-dozen superb male buffaloes sprang to their feet, and, tossing their heads, sniffed the air for a few seconds, and darted off through the wood. My companion and I immediately descended the tree and I made chase in the hopes

of coming up with them by following their tracks. We proceeded for some little way along the borders of the forest, when
Mango stopping, pointed ahead, and I saw a vast herd of buffaloes—there might have been nearly three hundred of them—
suddenly rushing out of the wood, overthrowing and stamping
down every object they met with in their headlong course. We
rushed back towards the wood, where alone we could hope for
safety. A portion of it projected some way at an angle from
that part whence the buffaloes had issued. They espied us,
however, and came tearing on across the open. We dashed in
among the underwood, but before we had got far they were at
our heels. Two savage brutes led the way. The horns of
the first were almost into poor Mango. A tree with low
branches was near me. It would afford us the only prospect
of safety. Had I stopped for a moment to fire, it would have
been too late, and it might not have served to turn them in
their course. I sprang to the tree, helping up the boy, who
had barely time to get out of the way of the leader's horns,
when the herd rushed by us. I turned round and fired, but
having to cling to the tree, I had great difficulty in taking
aim. The effect of the report was to bring the whole herd to
a halt, and, facing round, they confronted us in one dark and
formidable phalanx, as if they had resolved to besiege us in
our tree. I remembered the way you, Andrew, had been
caught by the elephant, and I fancied that the buffaloes were
about to treat us in the same manner. One or two buffaloes
might have been disposed of, but we had not ammunition
sufficient to kill one half of our assailants, even should each
bullet lay one low. They kept looking at us with savage
glances, as if determined to punish us for our audacity. They
looked, indeed, as if they could very easily have brought the
tree in which we were perched down to the ground; and so
they might, if they had known how to do it. I, however, re-

solved to try the effect of a few shots. I fired one, and felt sure I had hit the animal—a large bull—but he did not move. Again and again I fired, but, strange as it may seem, neither he nor any of the herd moved a foot, though they eyed me and my companion all the time with an ominous look, as if resolving how they should treat us. Every moment I expected them to charge. Suddenly, as I was about to fire for the fourth or fifth time, the whole herd, wheeling about with a curious shriek rather than a bellow, their heads lowered to the ground, and their tails swishing to and fro vehemently over their backs, off they set at a furious pace, which made the very ground tremble under their feet. Mango and I were left to follow them if we chose, or return to camp. We did the latter. I must confess I felt somewhat ashamed of my want of success when I resumed my seat by the fire. I consoled myself, however, with a couple of pigeons which Igubo had in the meantime roasted. Though we saw vast quantities of game of all sorts, we were equally unsuccessful, and at length I proposed to return, when Igubo pointed out some smoke rising over a belt of forest which appeared before us. He said that he was sure it arose from a native village, and as I was anxious to make the acquaintance of our neighbours, I resolved to push forward and visit them. I sent Igubo on ahead to win the confidence of the people by showing them that he was unarmed. He soon made a signal to us to come on, and I found him and the chief man apparently on the most friendly terms. The chief, a remarkably stout black, wore a scanty petticoat, with a fillet of crocodile's teeth round his head, a similar ornament on his neck, and bracelets on his arms. He was attended by a drummer, who, as I approached, beat with might and main to do me honour. His followers were armed with shields made of reeds, very cleverly woven, sufficiently long to protect the whole body and legs, and about three feet broad. At their backs hung quivers of iron-headed arrows, and two short broadswords were slung to their sides. The chief invited us into his hut. It was of good size, with a verandah in front. In a short time his wife and her attendants brought a large mess of manioc flour and some pieces of cooked meat, but what it was I did not at first inquire. After eating some, Igubo told me that it was zebra's flesh. In a hut opposite the chief's house, I observed the figure of an animal. On examining it I found that it was formed of grass, plastered over with soft clay. The eyes consisted of two cowry shells; and a number of bristles, which appeared to be taken from elephants' tails, formed a sort of frill round the neck. It was more like a crocodile than any other animal; but Igubo inquiring, was told that it was a lion, though certainly it was very little like the king of beasts. On further inquiries, I found that it was the principal idol, or fetich, of the inhabitants, and that when the chief or any of the people are ill, their fetich men, or priests, assemble before it, and pray and beat drums, either to propitiate it or to arouse its attention, that it may drive away the evil spirits which they believe are the cause of the malady."

"Poor people, dey know no better," observed Timbo; for, with the privilege of an old servant, he did not scruple to join in our conversation at all times. "I go and talk to dem and tell dem better t'ings. I tell dem dat dere is one God who lubs dem, and when dey are ill dat dey pray to him. Dat he hear dem, when de fetich hab no ears to hear, and no way to do dem good."

"Oh yes, Timbo," said Natty, "I should like to go with you to those poor savages. It is sad to think that they should be so ignorant. I am sure it is our duty to try to tell them the truth."

"Yes, Massa Natty, we will go, please God," cried Timbo, looking at Natty with a glance of approbation.

" Timbo and I must beg your pardon for interrupting you, Captain Hyslop," said Natty. " Pray go on."

" Unfortunately, I could not understand their language sufficiently well to enter into such matters," observed Stanley. "I was going to say that their village was surrounded by palisades, very similar to those we have seen. The people were clothed in even more scanty garments than usual. On finding that we came without any hostile intentions, and were more likely to give than receive of them, they cordially welcomed us. They were in a state of commotion, nearly the whole village being prepared to turn out on a grand hunt. When they understood that we also were hunters, they invited us to accompany them. They had been forming for some time past a huge trap, called a *hopo*, about three or four miles away, near a stream in the neighbourhood, at which large numbers of game were accustomed to assemble. As the narrow end was toward the village, we were able to examine it on our way. The hopo consists of two hedges formed of stakes and boughs driven into the ground at a considerable distance from each other, toward the end opening into the wild part of the country where animals are likely to be found, and closing in toward each other till they almost approach. They then form a narrow passage, some sixty yards long, at the end of which a pit is dug, eight or ten feet deep, and fifteen or more in length and breadth. We found that trunks of trees were laid across the two ends, to prevent the animals which leap in from scrambling out again, which they would otherwise very easily do. The pit itself was also surrounded by high palisades, bound together by cross-pieces. Thus it formed a complete trap, from which it seemed almost impossible that any animals which have once entered could escape. The hole was likewise covered over with a sort of matting of green rushes, which concealed the pit below. As I and my dark-skinned companions

proceeded along the hedge, I thought we should never come to
the end of it. I calculated, indeed, that the hedges were upwards
of a mile long, and the same distance apart at their extremities.
The hunters now extended themselves, each man keeping
within sight of the other, forming a circle round the broad
entrance of the hopo of four or five miles in extent, thus surrounding a large area. I could see within it immense numbers
of animals, giraffes, zebras, buffaloes, gnus, pallas, rhinoceroses,
hartbeests, and, indeed, all sorts of deer, large and small. At a
signal from their chief, which was passed along the line, they
began to close in, shouting and shrieking at the top of their
voices. On we went, the semicircle gradually decreasing, till
we were within speaking distance of each other; and every mile
we advanced the animals appeared to grow thicker and thicker,
and I could count a dozen or more creatures of different
species in sight at the same moment. Now a herd of a dozen
buffaloes, now twenty zebras and as many cameleopards might
be seen scampering over the plain, followed by numerous steinboks or koodoos, graceful oryxes or hartbeests leaping and
bounding away before them. Now and then some of the
animals would turn round and charge their pursuers, who fled
on either side, darting their spears and often transfixing them.
The zebras were amongst the most difficult to drive in. They
seemed aware of their danger, and now one, open-mouthed,
would charge at a hunter, who had to defend himself with his
shield; and then a whole herd would break away, and, dashing
through the cordon, gallop back to their native wilds. Still
numbers were driven on. Buffaloes and giraffes were flying
together, all fancying that they were escaping a common danger, while rushing on to destruction. At last the hedges of
the hopo were reached, and on the outer side numerous hunters
were stationed, shouting, and shrieking, and shaking their
spears and shields, still further to increase the confusion of the

terror-stricken animals. When any of them approached the hedge, a well-aimed spear was planted in their sides, the cries of the stricken animals increasing the terror of the rest. On pressed the hunters, driving the game closer and closer together, till, pressed up in one dense mass, even the most wary could no longer attempt to turn and fly. Fearful was the din of the shrieks and shouts which rent the welkin. The leading animals dashed madly forward, thinking to escape from their foes behind. The remainder followed, unable to see over the heads of those in front, but hoping that they had found a way to escape.

"By Igubo's advice, I had gone on the outside; for, in truth, the line of hunters which pressed on through the hopo was exposed to no little danger from the maddened beasts, which even now occasionally turning round, dashed through them, and the greatest activity alone could have saved the men from being trampled on by the terrified animals. Now a huge buffalo would leap into the pit through the slender covering of rushes; now a tall giraffe would go toppling over; an active koodoo or gemsbok would spring over their heads, to fall hopelessly into the same trap. In a short time the whole pit was filled with a living, moving, struggling mass of animals, fearful to look at. The savage hunters, wild with excitement, were spearing with relentless eagerness the poor creatures, those below being borne down by the weight of their hapless fellows who brought up the rear. A beautiful koodoo was among the latter. On it came, leaping away, having escaped the spears of its enemies. It reached the fatal pit. I could not help feeling an interest in the creature. Would it too be added to the victims? It hesitated not a moment, but bounding over the beams, seemed scarcely to touch the animals below, as with a spring it cleared the opposite side. In vain the hunters darted their spears. Off it dashed like the wind, and the satisfaction I felt at its escape made some amends to me for the misery and suffering I had beheld. I literally turned sick with horror, and hope I may never witness such a scene again. The savages, however, seemed to consider it magnificent sport, and stood over the pit plunging their spears into any animal which appeared moving. So far I was thankful, as it put them out of their misery. The hunters did not altogether escape. Some got severe kicks; several had been knocked over and trampled on, in spite of their activity. They had succeeded, however, in driving upwards of forty animals into the pit; for, of course, of those which had been first assembled, a large number had escaped, while a good many had been speared to death before reaching it, and others had escaped into the wilds with spears in their sides, there in most instances to die miserably. Their success put our new friends in excellent humour. They shouted, and shrieked, and danced as they hauled up the animals one by one out of the hopo, and eagerly commenced cutting them up and dividing the flesh. All was meat for their pots—the zebra and giraffe, as well as the buffalo and deer.

"It was nearly evening before the work was over. They pressed us to remain to see another on the following day, but I had had enough of it, and more than enough, indeed. I do not know how the case would have been if I had been very hungry and wanted food. Probably I might have experienced some of the satisfaction which our savage friends did. Igubo and his sons were highly delighted at the number of animals caught, at the same time he acknowledged that the way among his own people of catching game was far less cruel. Further to the north, large nets are spread round the trunks of trees, towards which the animals are driven, much in the manner I have just described. The nets, however, only serve for smaller animals, as large ones would break through them. People are stationed behind the trees to spear any creature of larger size which seems likely to break the nets.

"Our friends pressed on us some of the meat, which, as we had a few articles to give in exchange, we accepted, and parted very excellent friends.

"As I had no wish to spend a night in their huts, we pushed on as far as we could homewards, and did not stop while a ray of sunlight enabled us to see our way. We were pretty well tired with our day's exertions, but it was necessary to light fires, not only to cook our supper, but to guard ourselves against visits from any of the lions or hyenas which might be prowling about. We all therefore set to work to collect wood as fast as we could. While thus employed, I heard young Mango cry out; but on looking round in the direction where I had last seen him, he was nowhere visible. A dread seized me that a lion had carried him off; but again I heard him cry out, and on hurrying forward I was very nearly going head over heels into a deep pit, into which he had fallen. I shouted out to Igubo, who came to my assistance; and with the help of our belts we hauled him up. Mango's chief alarm had arisen from the dread of finding some animal at the bottom. I was very glad, when we drew him up, to discover that, excepting a few slight bruises, he was none the worse for his tumble.

"As may be supposed, we were cautious after this how we moved about, for we well knew that where one pit-fall had been formed, probably many more existed in the neighbourhood. We were glad when at last we had collected a sufficient supply of wood to last us through the night; and I almost fell asleep while putting the meat and cassava bread into my mouth. We had placed our packs by our sides, using some logs of timber for our pillows. Igubo had promised to keep the first watch; and so he did, I have no doubt, to the best of his ability. When, however, I at length awoke, I saw the fire very low, though there was just flame enough to cast its light

on a creature stealthily creeping up towards us. I expected
the next instant to be engaged in deadly combat with a panther or a lion. I sprang to my feet, seizing my rifle and
calling to my companions. The next moment I saw that the
creature was a jackal, and scarcely worthy of a shot. Still
undaunted, he was on the point of seizing one of the packs
nearest to him, when I hove a log of wood at his head. On
this he beat a retreat, uttering a mocking shout of laughter—
so it seemed to me—and quickly disappeared. The alarm he
had caused prevented us wishing again to go to sleep; and
well it was we did not, for directly afterwards the roar of
a lion broke the silence of night. Igubo threw more logs
on the fire, and as the flames burst up we saw two or three
huge monsters stalking round us, but afraid to approach.
Now they came near enough for the light of the fire to shine
on them; but directly afterwards, even before I could get my
rifle ready to shoot, they had disappeared in the dark shades
of the surrounding trees or bushes.

"As soon as it was daylight, we once more commenced our
march. We had not gone far, when the two boys, who were
a little in advance, came rushing back with countenances of
dismay, to let us understand that they had suddenly come
upon some huge beast which was on the point of springing on
them. We advanced, in consequence, cautiously, expecting
every moment to meet the monster. In a short time we caught
sight of a gigantic tiger-wolf, or spotted hyena, sitting under
a bush, and growling fiercely at us. I raised my rifle to fire,
expecting the beast to spring; but it sat without moving.
On getting nearer, what was my horror to see that his fore-
paws and the skin and flesh of the legs had been gnawed away!
Still he showed his savage nature by endeavouring to crawl
towards us. To put an end to his sufferings, I fired at his
head, when he sank to the ground; and Igubo, running up to

him, seized him by the tail, and struck him several times with his knife, though it was not until after repeated blows that an end was put to the creature's existence. How he had been thus mangled, I could not at first understand, till Igubo asserted that it had been done by a lion; that probably they had quarrelled over their prey, and that then the lion had attacked him and mangled him in the dreadful manner I have described. Had we not found him, he would certainly have died miserably in the course of another day or two, and very likely have fallen a victim to an army of soldier-ants.

"We met with several other adventures during the day, and managed somehow or other to lose our way, or we should have reached home before nightfall. Contrary to our intentions, we had therefore to camp out for another night. We had an ample supply of food, but no water could be found, and we had little more than a couple of pints to divide among us, which, though it might have been sufficient to supply an old lady with a cup of tea, was but little to satisfy the thirsty throats of travellers in this burning clime."

When Stanley heard of the attack made by the lion on our camp, he declared that he must set out at once and put a stop to his depredations. After a consultation, however, with Igubo, he agreed to wait till the evening, when they supposed the lion would go down to a spot near the river to drink. It was a small creek, rather, where the banks were sufficiently low and hard to allow the animals to reach the water without difficulty, which they could not do at many places along the borders of the lake on account of the wide fringe of reeds and thick underwood which encircled it.

"Is the gemsbok the only animal we have lost?"

"Oh no, indeed," cried Leo. "Poor Chico is gone!"

"What! did the lion carry him off?" asked Stanley.

"Oh no. A horrid monster of a crocodile," answered Leo. " I wish we could punish the brute."

Igubo seemed to understand what was said. "I do it," he remarked.

"Yes," said Timbo; " he say he kill crocodile; no 'fraid of crocodile!"

How he was going to manage it, however, he did not inform us.

As may be supposed, Stanley dropped to sleep over his breakfast, and was glad directly afterwards to go to bed. Igubo and his boys followed his example; but after a few hours' rest, they again appeared, as fresh as if they had not been undergoing severe exertion for a couple of days under an African sun.

"You come and see Igubo kill de crocodile," I heard Timbo say to Leo and Natty.

Igubo had provided himself with a piece of one of the animals which he had brought home, and which had become no longer eatable. He had fastened it to the end of a long rope, and his sons carried it down to the water. Timbo and Jack, with the two boys, set off after them; and, taking my rifle, I followed to see what would happen.

On reaching the river, Igubo threw in the meat as far as he could, fastening the end of the rope to the trunk of a tree. Then, on his making a sign to us to hide ourselves, we retired behind some bushes. In a short time the rope was violently tugged, and Igubo, throwing off his scanty garments, drew his sharp knife from its sheath, and sprang into the water. I could not refrain from crying out, and entreating him to come back; but he paid no heed to me, and swam on. Presently he disappeared, and I felt horror-struck at the thought that a crocodile had seized him; but directly afterwards the snout of the huge monster appeared above the water, Igubo rising at the same time directly behind it. The creature, instead of attempting to turn, made towards the bank, at a short distance off. Igubo followed; and I saw his hand raised, and his dagger descended into the side of the creature. Still the crocodile did not attempt to turn, but directly afterwards reaching the bank, climbed up it. Igubo followed, and again plunged his knife into the monster's side. Every instant I expected to see him seized by its terrific jaws; but the creature seemed terror-stricken, and made no attempt at defence. Again and again the black plunged in his knife, while the crocodile vainly endeavoured to escape. The next instant Igubo was on its back, and the creature lay without moving. A few minutes only had passed. It opened its vast jaws, each time more languidly than before, till at length it sank down, and, after a few struggles, was evidently dead. Igubo, springing up, flourished his knife over his head in triumph. Jack, running to the canoe, began to launch it. We all jumped in, and paddled off to the bank, Timbo bringing the rope with him We fastened it round the crocodile's neck, and towed the body in triumph to the shore, up which we hauled it.

"Igubo say we find eggs not far off," said Timbo.

Mango and his brother, at a sign from their father, began at once hunting about, and in a short time called us to them. There was a large hole in the bank concealed by overhanging bushes. It was full of eggs, about the size of those of a goose. On counting them we found no less than sixty. The shell was white and partially elastic, both ends being exactly the same size. The nest was about four yards from the water. A pathway led up to it; and Igubo told Timbo, that after the crocodile has deposited her eggs, she covers them up with about four feet of earth, and returns afterwards to clear it away, and to assist the young out of the shells. After this, she leads them to the water, where she leaves them to catch small fish for themselves. At a little distance was another nest, from which the inmates had just been set free; and on a sandbank a little way down we caught sight of a number of the little monsters crawling about. They appeared in no way afraid of us as we approached, and Mango and his brother speared several. They were about ten inches long, with yellow eyes, the pupil being merely a perpendicular slit. They were marked with transverse stripes of pale green and brown, about half an inch in width. Savage little monsters they were, too; for though their teeth were but partly developed, they turned round and bit at the weapon darted at them, uttering at the same time a sharp yelp like that of a small puppy when it first tries to bark. Igubo could not say whether the mother crocodile eats up her young occasionally, though, from the savage character of the creature, I should think it very likely that she does, if pressed by hunger. As is well known, the *Ichneumon* has the repu-tation on the banks of the Nile of killing young crocodiles; but Igubo did not know whether they ever do so in this part of the world. He and his boys collected all the eggs they could find, declaring that they were excellent for eating. They however told us that they should only consume the yoke, as the white of the egg does not coagulate. When it is known what a vast number of eggs a crocodile lays, it may be sup-posed that the simplest way of getting rid of the creatures is to destroy them before they are hatched. It would seem almost hopeless to attempt to exterminate them by killing only the old ones. However, I fancy they have a good many enemies, and that a large number of the young do not grow up.

As we were walking along the bank, we saw, close to the water, a young crocodile just making his way into it; and Mango, leaping down, captured the little creature. Even then it showed its disposition by attempting to bite his fingers. On examining it, we found a portion of yoke, almost the size of a hen's egg, fastened by a membrane to the abdomen ; and when we afterwards carried it up to David, he told us that he had no doubt it was left there as a supply of nourishment, to enable the creature to support existence till it was strong enough to catch fish for itself. Igubo declared that they caught the fish by means of their broad scaly tails. The eggs, I should say, had a strong internal membrane, and a small quantity only of lime in their composition.

We had some difficulty in inducing our friends to believe the account we gave them of Igubo's exploit. He however undertook, if they were not satisfied, to kill a crocodile in the same way another day.

" Oh ! pray tell him not to make the attempt !" exclaimed Kate. " It is far too perilous ; and though he may succeed once or twice, some day another crocodile may come in support of its companion and carry him off."

Igubo only laughed when this was said to him. He had killed crocodiles in that way since he was a boy, and there was no reason why he should not do so as long as he was able to swim.

While speaking of crocodiles, I should observe that the family of huge saurians, to which the monsters belong, is divided into three genera : *Alligator* is peculiar to America · *Crocodilus* is common both to the Old and New World; while a third, *Gavialis*, is found in the Ganges and other rivers on the continent of India. They differ in appearance from each other, but their habits in most respects are similar. The true crocodile, however, frequents occasionally the mouths of large rivers where the salt water enters, and it has been known to swim between different islands at considerable distances from each other. I believe that at the commencement of my journal I have sometimes inadvertently written alligator instead of crocodile, when speaking of the monsters we encountered so frequently.

CHAPTER XIV.

GAIN during the night the roar of the lion was heard. It put Stanley in a perfect fever; but David persuaded him not to go out and attempt to shoot the creature, as he was completely knocked up by the exertion of the previous days. The rest of us employed our time in collecting the prickly pear for fortifying our post, as David had proposed. It was no easy matter, however, to cut the plants down.

"If we were to throw a rope over them, and draw the leaves on one side, we might do it," said Natty.

"A good suggestion," I observed.

We carried it out. While the grown-up members of the party cut down the armed plants, the boys with ropes dragged them in large bundles up to the camp, round which we began to form with them a broad belt. It was hard work; but as there were numerous plants growing about, we had not far to go. We were encouraged to persevere by the assurance that our fortress would thus be almost impregnable to the attacks of wild animals. We yet further secured it by driving in stakes pointed at both ends outside the belt, which thus answered the purpose of a dry ditch, only it was more difficult even than a ditch would have been for unprotected feet to cross over.

At daylight next morning we continued our work, and had made considerable progress before the heat of the sun compelled us for a time to knock off. We had three fires lighted in the centre of our yard, and this probably prevented the lion making another attack, which he might otherwise have done. I was now so far recovered that I was able to accompany David and the boys on short shooting excursions. Although I never took pleasure in slaughtering animals for mere sport, yet it was necessary to kill them for the sake of supplying ourselves with food. The hills above the house swarmed with rock-rabbits, with which we could at all times plentifully supply our table. I had gone out the following morning with the two boys, keeping, of course, a careful look-out, lest a lion might still be in the neighbourhood, when Leo cried out, pointing to a rock above us,—

" See, see! what a curious lump of feathers is up there!"

" What you suppose to be a lump of feathers has, I suspect, a head and wings and claws attached to them," said David. " If I mistake not, that is a *bacha*, a sort of falcon. Probably he is on the look-out for rock-rabbits, and he is hiding his head between his shoulders and crouching down that they may not discover him, but his sharp eyes are watching every movement of his prey. Before long, if we remain quiet, we shall see him pounce down on one of them should they venture out of their holes. The Dutch, I remember, call these rock-rabbits *klipdachs*. Poor creatures, they have good reason to be on their guard against the bacha. While he is there we are not likely to get a shot at one, for, cunning as he is, depend upon it some of the older ones have found out that he is in the neighbourhood."

We watched for some time. Now and then we saw a klipdach pop out of its hole, but presently draw back again, having caught sight of its powerful foe. Now another would

come out, but hide away in its cave very quickly. Still the
bacha remained without moving. He knew that in time the
poor silly little klipdachs would grow careless, and, anxious
for a game at play, would get too far from their homes to skip
back before he could be down upon them. Presently what
David said took place. First one klipdach appeared, and then
another began running about or nibbling the grass close to the
rocks, but it was clear that they were watching the bacha all
the time. Still he did not move, and they began to run
further and further out into the open ground. Then two or
three came out together, and began leaping and frisking about.
Presently the hitherto immovable bacha leaped off the rock,
spreading wide its huge wings, and like a flash of lightning
from a thunder-cloud darted down on a klipdach on which it
had fixed its keen eye. In vain the unfortunate klipdach
attempted to leap away. The bacha had cunningly noted the
road it came. In an instant it was in its claws, the poor little
creature screaming with terror. So rapid was its flight, that
even if we had wished it we could not have killed the bird.
Off it went to the pinnacle of the rock from whence it had
descended, and there began tearing its prey, which, happily,
it soon must have put out of pain. Though we waited some
minutes, not another klipdach appeared, and we had to go on
some considerable way before we again caught sight of any
of the little creatures.

"Well," said David, "I do not know that it can matter
much to the poor klipdachs whether they are shot by us or
caught by the bacha, but at all events we will put them out of
their suffering as soon as possible. Yet I do not think we
ought to throw stones at him. He follows his nature, we
follow ours."

After shooting as many rabbits as we required (by-the-by,
their scientific name, David told me, is *Hyrax capensis*), we

made a circuit, and took our way home along the plain. Leo and Natty were a little in advance, when they came running back saying they had seen a big snake, but before they could shoot it it had got away. Whether venomous or not, of course they could not tell, but Leo declared that, from its appearance, he was nearly certain it was so. It was a somewhat sandy open spot, though a few bushes were near, among which we supposed the snake had hid itself. We of course advanced carefully, when presently in the distance we saw running over the ground a couple of curious-looking birds, with long legs and a remarkable crest, which Leo declared looked like a lawyer's wig. We hid ourselves behind a bush, and the birds, not seeing us, came boldly on. On a nearer approach David pointed out some feathers which seemed to stick out behind the ear.

"They must be secretary birds," he whispered; "known as the *Serpentarius cristatus*. They are determined enemies of serpents, and will attack the most venomous without fear. The secretary bird is so called on account of that crest at the back of his head, which looks something like a pen stuck behind the ear. One might suppose, on account of his long legs, that he should be classed among the cranes and storks, but his curved beak and internal organization show that he belongs to the falcon tribe. His feet are incapable of grasping, and thus he runs along as we see over the sandy ground with a speed which enables him to overtake the most active reptiles.

Presently we saw the birds dart off, and in another instant a large snake rose up before them. One stood still, while the other gave battle to the reptile. The serpent made every attempt to get back to its home, but the bird each time sprang before it with an active leap, and cut off its retreat. Whenever the serpent turned, the bird again placed itself in its front. At length the reptile, as if determined to try what courage would do, raised up its head, which swelled with rage, and displayed its menacing throat and inflamed eyes, hissing fiercely. No human being would have wished at that moment to have encountered it. For an instant the bird stopped, but it was not for want of courage; and spreading out its wings, it covered itself with one of them, while with the other, which was armed with horny protuberances like little clubs, it struck the serpent a blow which knocked it over. Again and again the serpent rose to receive the same treatment, till at length it lay quiet on the grass. The bird instantly flew upon it, and with one stroke of its powerful bill laid open its skull, and then immediately pressing it to the ground with its feet, held it fast. We were unable to see whether it swallowed the head or not, for its companion catching sight of us, they ran off with their prey to devour it at their leisure.

Curiously enough, we were to make the acquaintance of yet another bird before we got home; for, proceeding onwards, we caught sight of a zebra coming towards us. It advanced but slowly, now stopping, now moving on a little way. When it caught sight of us it turned round and attempted to go back. We then saw that a shaft was sticking in its side, from which the life-blood was flowing. It went on a little way, and then down it sank on the ground. We had no doubt that it was one of the creatures which had been speared at the hopo hunt when Stanley was present, and having escaped, had wandered thus far from its usual haunts. Scarcely had it disappeared, when we saw coming from a distance a large flight of crows, who with loud croakings descended to the ground. Presently a number of kites and buzzards approached from far and near, though an instant before not a bird was to be seen, and alighted on the same spot. We hurried on, wishing to get a sight of the spectacle; but before we got up, David pointed out, high above us in the air, a huge bird, which came wheeling round in a spiral line, seemingly out of the sky, towards the same spot.

"I know that fellow," he said; "he is an *oricus*. He builds his nest far up among the mountains, in the fissures of rocks. He equals in size the famed condor of America, and if we could kill one, we should find that across the wings when expanded he measures ten feet. No bird is bolder in flight. At daybreak he left his aerie, and mounting in the sky far beyond the reach of human vision, watched with telescopic eye the creatures wandering on the earth's surface. That poor zebra was seen by him probably long ago, and he knew well that he must shortly become his prey."

While David was speaking, numerous other oricus descended like the first. Their common name is the sociable vulture—*Vultur oricularis.* By the time we got up to the spot, the poor zebra was half torn to pieces by their powerful claws. The oricus having satisfied their hunger, and carried off what they required for their young, the buzzards approached, followed in a short time by the crows, who quickly denuded the bones of flesh.

On reaching home, we found that a stranger had arrived from the nearest village to the north of us, which Stanley had once visited. He came with a sad story. A young child had strayed out from the village the previous morning, and had been carried off by a lion, and the father and another man, going in search of the animal, had not since returned; but evident signs had been discovered that they also had been killed. A panic had seized the people, and they had sent to ask our assistance to destroy their fierce assailant with our guns. They knew well, from the way the lions attacked them, that they were accustomed to human flesh, which, when once a lion has tasted, it is said, he will always attempt again to obtain. The poor people declared that there would be no safety for them unless the lions were killed, for night after night they would come, and no one would be able to go beyond their enclosures without the risk of being seized. The difficulty was to find the lions, for they were as cunning as ferocious, and the blacks declared that, by eating men's flesh, they had obtained some of the sense of human beings.

"We will soon put that to the test," said Stanley, jumping up. "Tell him, Igubo, if he will go with you and I, and show us where we can fall in with the lion, we will soon give an account of him."

The stranger expressed his gratitude, and Igubo at once consented to accompany Stanley. I confess I felt somewhat unwilling that he should go, for he would thus completely put himself in the power of the strangers, of whose honesty we had had no proof. Igubo, however, fully believed them faithful, and would, I was sure, not desert him. I proposed that we should all go out in the day-time, and attempt to fall in with the lion man-eaters; but the stranger black said that would be useless, as they were sure to keep out of the way. He knew, however, he told us, of a spot which they were likely to visit in the early part of the night. It was a pool in a small stream which ran into the river, where numerous wild animals came to drink.

"But, dear Stanley, what is the use of you exposing yourself thus at night," said Kate. "The lions will surely visit the village, and could you not shoot them when they come? At the spot the stranger speaks of, you will be surrounded by ferocious creatures, and though you may kill one or two of them, the others may set upon you, and your life may be sacrificed."

Stanley laughed at the notion.

"In the first place, dear sister, the lions will not show themselves till some unfortunate person passes," he said. "Thus I might have to wait day after day without killing one. Now,

our friend here declares that every night they go down to the
water, so that I am sure to meet them. Let us manage it, and
do not be afraid. We shall return in safety, and probably have
been of service to these poor people, by getting rid of their
savage enemies."

" Oh ! let us accompany you," cried Leo and Natty. " We
will take care of Stanley," said Leo ; " so do not be afraid,
Kate."

" Thank you ; but the man-eaters might carry one of you
off," answered Stanley ; " so I must decline your company. I
would rather have my two black-skinned friends as companions,
for depend upon it they know more about the matter than any
one else."

" Massa," said Timbo, " I ever go out shooting wid you. I
no take care of you ?"

" Yes, indeed you have," answered Stanley ; " but I want
you now to stay at home and look after the camp. If there is
any risk, it is better that one should run it than both."

This answer satisfied Timbo, and Stanley having partaken
of the supper which Kate and Bella insisted on preparing for
him, set off with Igubo and the stranger. They carried the
two best rifles, with a supply of powder and bullets. I found
that Jack and Timbo had been busily employed in manufacturing a sort of infernal machine for the destruction of wild beasts.
They had selected a musket with a large bore, and they proposed using this as a sort of spring-gun. Jack told me that
while we had been away, a huge hyena had been seen in the
neighbourhood, and as they are cunning animals and not easily
overtaken, they thought it would be the best way of getting
rid of so dangerous a neighbour. There was still sufficient
light by the time they had finished preparing the gun to plant
it in the neighbourhood. The boys and I accompanied them
out. Timbo selected two trees, to which they lashed the gun

in an almost horizontal position, the muzzle only pointing slightly upwards. A piece of wood about six inches long was fastened to the gun stock so as to move easily backwards and forwards. A piece of string connected the lower part of this with the trigger. To the upper end a long piece of cord was fastened, which was carried through one of the empty ram-rod tubes, and then tied to a lump of flesh, fastened round the muzzle of the gun. As can thus easily be understood, an animal seizing the flesh pulls the lever which draws the trigger, and at the same moment that it has the meat in its mouth, the probabilities are that its brains will be blown out. However, that it should not take the meat sideways, or come behind it and thus escape, Timbo formed a fence round the spot, leaving only a narrow opening just in front of the muzzle of the gun.

"Now," said Timbo, "here are five bits of meat to tempt de hyena to come up to de trap. You go dere, you go dere, you go dere;" and we all, as he pointed out, went in different directions round the spot to some distance, and then dragged the tainted meat up towards the trap. "Now, we go home; and to-morrow morning we find hyena dead," he said.

It was indeed time, as darkness was coming on, and it was just possible that the hyena might prefer one of us to the bait which we had so kindly left for him. Scarcely, however, had we reached home, when a loud report was heard.

"If dat hyena, I bery glad we did come away," said Timbo; "but we not go now. Perhaps other hyenas dere. We kill anoder to-morrow night."

It was quite dark when we got home. Our anxiety for the return of Stanley prevented any of us from going to bed. Three hours had passed away since nightfall, and still he did not make his appearance. I saw that Kate was becoming very anxious; indeed I could not help feeling so myself. At last

I proposed to Timbo that we should go out and try and find him.

"Dat I will, Massa Andrew," he answered. "Dough he not let me go wid him, he no say dat I not to come afterwards."

With our rifles in our hands, and our long knives at our belts, we sallied forth.

"Thank you, Andrew," said Kate, as I was going out. "I cannot help fearing that some accident may have happened to Stanley, and you will do your utmost to find him. I am sure you will."

Timbo, who had several times accompanied his master to the village I have spoken of, was tolerably certain of the direction we should take. As we walked on, feeling our way in difficult places with the long poles we carried in our hands, our ears were assailed by the screeching of night-birds and the occasional roars and mutterings of wild beasts. A feeling of awe gradually crept over me, produced by the wild sounds and the peculiar scenery through which we were passing. On one side rose the hills, with dark rocks cropping out amidst the thick foliage; while, on the other, the river flowed by with a murmuring sound, reflecting the bright stars from the dark sky overhead. Far away to the right were sombre forests, with openings here and there, across which phantom forms were seen flitting to and fro, though so indistinct were they that we could not tell what animals they might be.

"I t'ink we get near where de captain come to shoot," said Timbo in a low voice. "We go slow now, and take care dat no lion or 'noceros see us."

We moved on, but could hear no sounds. Presently we saw, a little way below us, the stream of which we were in search.

"Can the captain have left it, and passed us on the way?" I whispered to Timbo. We were now close down to the stream. "What is that?" I asked, pointing to a huge mass on the opposite side. "Surely there lies the body of an elephant; and what are those creatures near us on the left?"

"Dey leopards," whispered Timbo. "De captain hab been here and killed dem, no doubt about dat."

Just as he was speaking, emerging from a clump of low wood, there appeared directly before us a magnificent lion. The creature stopped and lifted up his head, moving his tail slowly to and fro, as if about to spring forward. Now he crept on and on. Presently he uttered a loud roar. I stepped back, instinctively bringing my rifle to my shoulder; but at that moment there was the flash of a gun, and a loud report came, apparently out of the ground close in front of us, and the huge lion sprang high up into the air. Scarcely, however, had the report ceased echoing in our ears, than from another clump, a little way on our right, I caught sight of an enormous rhinoceros, who seemed at that moment to have discovered that he had an enemy close to him. I felt sure it was Stanley who had fired. I shouted out to him. He answered me, "All right!" not apparently perceiving the approach of a new assailant. On dashed the huge rhinoceros, dipping his snout, as he descended into the water, beneath the surface, his eyes alone remaining above it. He was making directly for where I supposed Stanley lay hid. There was no time for him to reload, and I felt sure that the monster would gore him or trample over his body. I had never prided myself on my shooting, but I felt now or never was the time to take steady aim, or the life of my cousin might be sacrificed, while Timbo and I, indeed, were placed in no little danger. Aiming at the creature's head, near its left eye, I fired. Instantly it rose up, uttering a loud bellow, but still came floundering on across the stream. "Up, Stanley, up!" I shouted out. "Timbo, do you fire, or the captain may be killed!" Timbo drew his trigger

Again the creature was hit, but still his progress was not stopped. Wading or swimming, it had just reached the bank, close to where Stanley lay. Again I shrieked out to him. He was attempting to reload without getting up, for which, indeed, he had not time. In another instant I expected to see the sharp horn of the rhinoceros plunged into his side, when it suddenly stopped and rolled over into the stream.

"A capital night's sport!" exclaimed Stanley, springing up, his nerves in no way shaken by the fearful danger he had gone through—for I fully believe that had he missed the lion, which was on the point of springing on him, he must have been killed; and had we not been near to defend him from the rhinoceros, nothing could have saved him. Just as Stanley had finished loading his gun, a loud roar echoed through the woods, and we saw, coming out from behind the back of the elephant, another large lion. We could almost distinguish the grin on his features as he stood shaking his head, but yet not daring to approach. The ferocious beast, which we concluded from his size was one of the man-eaters, advanced boldly towards us. He seemed about to spring, and might have reached us across the stream with a bound, when Stanley, raising his rifle, fired, and the lion rolled over, shot through the heart. Igubo and the other black, uttering shouts of triumph, came running up. They had been concealed in a pit at a little distance, where it appeared that they also had shot a lion and a leopard.

"Why you go so far off?" said Timbo, when he saw them. "Is dis de way to look after de captain? Captain, you kill Miss Kate and Miss Bella wid fright if you go away like dis." Timbo had evidently scarcely recovered his alarm at the risk his master had run.

"Well, well, Timbo," answered Stanley; "you see we have done our duty and performed our promise. Three man-eaters lie dead, and I hope we may bag the remainder before many days are over."

The blacks were very anxious to get us all to go to their village, that they might treat us with honour, and thank us for the services we had rendered, and for the ample supply of meat which our success had procured. Not being hard pressed ourselves, we begged them to accept the whole of it, with the exception of a small quantity of the rhinoceros meat, which they undertook to bring up the following day. I urged Stanley, however, to come back, to relieve Kate of her anxiety; and telling our new friends that we would come and see them another day, we returned homewards. Having reloaded our guns, we took our way along the banks of the river. I was a little in advance, when I put my foot upon what I thought was the log of a tree, when what was my horror to see stretched out before me the long head and scaly body of a huge crocodile! I stopped; for though the creature could not instantly turn round, he might first knock me over with his powerful tail, and then have time, before I could recover myself, to ware ship, as Jack would have called it, and seize me in his fearful jaws. The thought that he might do this flashed across me, but I kept my presence of mind, and raising my rifle, levelled it at his ear. I fired, and without a struggle the creature turned on one side, and lay perfectly still. Timbo was instantly up with me.

"Me kill him well, Massa Andrew!" he exclaimed. "You no do dat, him gib ugly bite."

As we had no wish to have any crocodiles' meat (although the natives have no objection to eat it), we hurried homewards.

"There they come!—there they come!" we heard Leo and Natty shouting out; and they brought torches down the hill to give us welcome. My kind cousin had not gone to bed, but insisted upon sitting up to prepare a meal for us all, as she declared (which was indeed the case) that we should be very hungry. Not till then did Stanley give us an account of his adventures.

"The first thing we did," he said, "was to dig some shallow pits, with boughs over them, in which we could conceal ourselves from the beasts which might approach the stream. We saw by the spoors that numerous animals were accustomed to come there. For some reason, however, none appeared at first, except hyenas and jackals, which came round staring and laughing at us in the most impudent manner. We threw stones at them, but this only tended to increase their mockery. At length I hurled a lump of wood at the head of one of them, which, hitting him on the nose, made him cry out, and the whole scampered off as fast as their legs could carry them. They were, I hoped, the forerunners of more noble brutes. I was not disappointed, for in a short time the ground shook with the heavy tramp of elephants hurrying down to the water. Nearer and nearer they came. At length I could see their dark phantom-like forms moving amid the trees. Next their shapes were distinguishable, and then an enormous elephant stood out in bold relief against the sky. Another and another followed, till the bank of the river was lined with them. They could easily have crossed the stream, had they been so disposed, when few people would have given much for my life or that of my companions. I felt a little nervous, I confess, but soon recovered my presence of mind. I raised my gun to take aim at their leader, who stood conspicuously forth from among his fellows. Of course, Kate, you will say I was very wrong to think of shooting him, but I could not help it. I allowed them to go on drinking, which they did, dipping their trunks into the water, and pouring it down their throats. I hesitated even now, however, about firing, lest I might warn the lions, whom I most particularly wished to destroy. Sud denly they all began to move off, and I was afraid that I should miss the chance of hitting one. I therefore gave a low whistle, which immediately attracted their attention. Once more turning round, they slightly raised their huge ears, and moved their trunks in eccentric circles through the air, as if they wished to ascertain the cause of the strange noise they had heard. I could resist no longer, but pulling my trigger, the loud thud of the bullet as it struck the animal's head showed me I had hit him fairly. He turned round, and staggered back a few paces. I was afraid that I might not have mortally wounded him. I fired my other barrel behind his ear, and without a struggle he sank down dead, the other elephants going off into the forest at a great rate, uttering notes of terror. I was about to rush forward across the stream to examine him, when my companions urged me to remain quiet; and in a short time I saw a leopard stealing over the ground. Then another came. I shot one with one barrel, and one with the other; but still the object of our hunt, if so it could be called, was not accomplished. Some time passed away, when I saw a creature moving towards me; and soon, as it came out of the darkness of the forest, I distinguished a fine lion. I let it get quite close before I fired. I drew my trigger. The brute turned round and bounded off, and I thought that it had escaped me, though the loud and peculiar roar it uttered made me hope that it was mortally wounded. Still Igubo urged me to remain quiet, and after some time another lion came. It seemed as if he was about to spring across the stream towards me. It was the one I shot just as Andrew arrived. The rest he has told you."

"Oh, brother, I wish you would not undertake such dangerous expeditions!" exclaimed Kate, when Stanley had finished.

"But surely, my dear sister, in this case I was fighting in a good cause," said Stanley, laughing. "If we have rid the country of these man-eaters, we shall have rendered an essential service to our neighbours, and the blacks, I hope, will show their gratitude."

We soon retired to rest, and slept more soundly than we had done for many nights, though we kept a guard as usual, as our fortification was not entirely completed. The next morning we set to work to finish it, and by noon had entirely surrounded it with an impenetrable hedge. It took us some time longer to fasten the prickly branches to the top of our fence. While we were at work, a party of blacks arrived from the village, bringing with them a large quantity of elephant and rhinoceros flesh. They came to thank our chief, they said, for the service he had done them, though they feared that there were still other lions in the neighbourhood. Stanley promised to do his best to look out for them, should any again appear.

The young koodoo was by this time completely tamed, and even the little zebra had lost all fear, and would come up when called by Kate or Bella to be fed, and allow itself to be stroked and petted by them; but when any blacks came near it, it would scamper off and kick out with its heels, or, if they pursued it, would turn round and try to bite them.

"I am sure it would let me ride it," said Bella, "if we could make a saddle to fit its back."

"I think I could do that for you, Miss Bella," said Jack; "but it might be a hard job to put it on."

"If you will make the saddle and bridle, I will try to put them on," repeated Bella.

We had no lack of skins, which I should have said Timbo and Jack employed themselves in dressing. Out of these, the former, who was very ingenious, in a short time contrived to make a very respectable-looking side-saddle. We had some iron wire, with which he formed a bit, as also a stirrup.

Bella was highly delighted when he produced it completed. She, meantime, had allowed no one but herself to feed the little creature, and every day when she did so she threw a piece of hide over its back. In a little time she placed a still larger hide on the animal, till it was thoroughly accustomed to the weight, and seemed in no way to mind it. To introduce the bit into its mouth was a more difficult task. However, it allowed her one day to slip it in, after it had been eating ; and she kept it there for some time, leading it by the bridle about the yard.

"Now bring me the saddle, Jack," she cried out. "I am sure it will let me put it on its back."

Jack brought it, and the zebra stood perfectly still while he tightened the girths. Next to Kate and Bella, Jack was evidently the zebra's favourite, and it never seemed to object to his playing with it.

"Now lift me up, Jack," said Bella; "and I am sure it will let you lead it about."

In a short time the little creature seemed perfectly contented with its new employment, and Bella was able to ride it round and round the yard, without its showing any wish to throw her off. The koodoo ran by her side, every now and then looking round into the zebra's face, as much as to ask how he liked it. She, however, did not try it too far ; and after riding about for half an hour or so, she jumped off its back, and relieved it of its saddle, patting its head and talking to it all the time. She then, leading it back to its pen, took off its bridle and gave it some more food. The following day she tried it in the same way ; and though at first it seemed rather disinclined to allow the bit to be put into its mouth, after she had coaxed it, and talked to it for some time, it allowed her to put it in ; and Jack again bringing out the saddle, it went through the duty of the previous day.

" I think now," said Bella, " if we have to make a journey, that I shall have a steed ready to carry me. I wish, Kate, we could find an animal for you."

" No fear about dat, Miss Bella," said Timbo. " If we no get horses we get oxen, and dey do better dan any other animal in dis country."

Timbo had been making inquiries, it appeared, about the natives further to the south, and had been told that at some distance there were herds of oxen, which the people were accustomed to ride. This gave us hopes that we might be able to procure some, and that we might proceed on our journey without waiting for Senhor Silva and Chickango. As yet no news had been received from them, though we were now in daily expectation of the arrival of a messenger whom they had promised if possible to send back to us, with an account of their progress. Our days were beginning to grow somewhat monotonous, from the fact that we had no great difficulty in supplying ourselves with food, and were unwilling to go out and kill creatures merely for the sake of amusement. Stanley made a second excursion to assist our friends in the northern village, and succeeded in killing two more lions, which the people declared were man-eaters.

CHAPTER XV.

LEO and Natty had been frequently begging me to accompany them to visit our friends to the south. We agreed that we should greatly shorten the land journey by proceeding along the lake, and landing at a spot on its borders nearest the village, which we thought we could then reach in a few hours' march. Stanley had no objection to our going, provided we did not remain away more than three or four days. Mango was to accompany us as interpreter. From the experience we had had of the natives, we hoped that the garrison, though thus decreased, was still sufficient for the protection of our fortress, especially as the lions and leopards had for some time kept at a distance, finding out, probably, that we possessed ample means for their destruction. It is extraordinary what instinct wild animals exhibit, and how soon they desert a neighbourhood where they are frequently attacked. It is said that even hippopotami and crocodiles become more wary after being hunted ; and though in the wilder districts they come out fearlessly to feed or to bask on the sandbanks, when hunters come to the neighbourhood they learn to conceal themselves in their watery retreats, and will only show their nostrils and eyes above the surface, keeping always in the most secluded parts.

The boys were greatly pleased at being allowed to take the proposed expedition. They made wallets to carry their food at their backs, and the articles they proposed to present to the natives, or to exchange for meat and other provisions should we not be able to supply ourselves. The village we were to visit, we learned from Igubo, was called Kabomba, and he seemed to consider it a very important place. To be sure, as Leo observed, he had never been in London, or even at Cape Town, so it was not surprising that he should look upon it with respect.

Our preparations were soon completed. Igubo gave his son charge to behave well, and to bring no discredit upon his white friends. Kate urged us all to take care of ourselves, and not to run into unnecessary danger. The whole party accompanied us down to the canoe. We had chosen the *Guzelle*, as the best of the two. As the wind was fair, we hoisted our sail and steered merrily down the river towards the lake. We had no difficulty, as we passed along, in supplying ourselves with food. Wild ducks of all sorts abounded. Among them were numbers of the Egyptian goose. We saw several of them ahead, and made chase. Being heavy of wing, we found they could not rise out of the water, and we caught four or five with our hands as we passed by. A little further on we neared a bank on which a large flock of ducks were seated. Leo and I fired at the same time, and on landing we picked up a dozen ducks and three geese which we had knocked over. Among them was a large black goose, which we saw in great numbers walking slowly about and picking up their food. The specimen we killed had a small black spur on its shoulder—as has the armed plover—and as strong as that on the heel of a cock; but the birds, it is said, never use them except in defence of their young. They are said always to choose ant-hills for their nests. The ants cannot hurt the eggs, and the material of which the hills are composed assists probably in hatching the eggs, as the sand does those of the ostrich.

I had hitherto held very little conversation with Mango. He had, however, picked up enough English to make himself understood, and during this trip I was able to ascertain some of his peculiar notions.

We kept for some time along the north shore of the lake. We were nearing a point when we saw a beautiful water-antelope, known under the name of *mochose*. Before I could stop him, Leo had lifted his rifle and fired. The poor animal was hit, and, as is always the case, instead of flying along the shore, leaped into the water and began to swim across the lake. We immediately made chase, for though we had ducks enough for food, venison was not to be despised. I saw Mango waving his hands and muttering in a peculiar manner. The mochose swam well, but we soon gained upon it; and I was anxious to put it out of its sufferings, for a red mark which appeared in its wake showed that it must have been badly wounded. Just as we neared it, a long snout projected above the water. It was that of a crocodile. The next instant the poor mochose and the hideous monster sank together. Mango uttered an expression of disappointment; and when I questioned him, he said that he had been praying to his fetich, who was himself a crocodile, that we might obtain the venison, but that the fetich would not hear him.

" That is a curious sort of religion," observed Leo ; " for to my certain knowledge he and his father and brother supped off the crocodile Igubo killed the other day, and still he worships the beast."

I have not before mentioned it, but we had tasted the flesh Leo spoke of. It had a strong musky odour, which did not tempt us to try it again ; though I do not know what we should have done had we been pressed by hunger. In a short time we came to a wide bay, across which we stood. The wind was fresh, and we flew rapidly over the water. The pure air raised our spirits, and we anticipated an interesting visit to our Kabomba friends. Mango pointed to a spot some way ahead, where he thought we might land; but at the same time said that if we continued further, we might possibly have a still shorter land journey to the village.

"It would be a pity to leave the canoe, as long as we can sail along so pleasantly," said Leo. "Do, Andrew, let us follow his suggestion."

As I saw no objection to it, we stood on down the lake. The breeze was increasing. I took two reefs in our sail, but still it was as much as the canoe could bear. Suddenly a strong blast came sweeping over the lake. I shouted to Natty, who was at the halliards. Almost before the words were out of my mouth, he had let them go. It was fortunate that he did so, or the canoe must inevitably have been upset. As it was, she heeled over so much that we took in a quantity of water. We set to work to bail it out; but the wind from that moment blew stronger and stronger, and in a few minutes the whole lake, which had hitherto been so calm, was covered with foaming seas. They increased every instant, and I saw that it would be dangerous to expose our light canoe broadside to them. Even as it was, they continued breaking over the sides, and it required active bailing to free her from water. Our only course, therefore, to escape being swamped, was to keep her directly before the gale. This carried us further and further down the lake, and drove us also off from the north shore. I told Natty and Leo to get out the paddles, while we set Mango to bail. We thus ran before the seas, and kept the canoe tolerably free from water. Night was approaching, and still there was no cessation of the gale. We could only see the land dimly on our right side, while we flew on, surrounded by the hissing and foaming waters. Much depended, I knew, on my steering well. The slightest carelessness might have allowed the canoe to broach to, when she must inevitably have been upset. Even had we clung to her, we should have lost our provisions, and we might have been picked up by some crocodile exploring the deeper water in search of prey; for I could not tell whether the monsters did not swim occasionally thus far from land. The boys plied their paddles energetically, as if they fancied our safety depended upon their exertions. Seeing this, I told them not to exhaust their strength, as it was only necessary to keep the paddles going sufficiently to assist me in steering the canoe. I tried to pierce the gloom ahead, but nothing could be seen but the troubled waters. It was different to any scene we had yet witnessed, for hitherto the lake had been calm as glass, unless when occasionally a ripple played over its surface.

"I say, Andrew, I wonder whether we are ever coming to an end of this?" exclaimed Leo. "If we go on at this rate, we shall be hundreds of miles away from Kate and the rest, and they will not know what has become of us."

"Not quite so far as that, I fancy," said Natty. "We must pray to be preserved, and hope for the best. I do not think we can do anything but that just now."

"Right, Natty," I said. "Do our best, and hope for the best. That is a right principle, and people who act thus are seldom led far wrong. Storms, in these latitudes, though they are very violent, do not last for any length of time; and I hope we may soon fall in with some island, under which we may take shelter."

"Suppose, though, we run against it. What shall we do then?" asked Leo.

"We must jump out and haul the boat up," answered

Natty. "The shore is not dangerous like that of the sea-coast, and we shall have no great difficulty in saving ourselves, even if we are driven on it."

"We need not talk of such a contingency," I remarked. "I hope we may keep clear of all dangers till the gale drops, or till daylight returns."

Though I said this, I could not help feeling very anxious, particularly at the thought of being driven so far from home, for I knew that Kate would become alarmed should we not return at the time we proposed. Still we kept on; but often as I bent my head forward, trying to make out any object ahead, nothing could I see but the curling waves as before. I had no idea that the lake was so long, and expected every minute to find that we were approaching the end of it. Still on and on we went. Hour after hour passed by, and I calculated that morning must be approaching. The gale still increased, and as the light canoe flew over the foaming seas I dreaded every instant that they would break on board. She behaved beautifully, however, and though occasionally the top of a wave tumbled over her, we took in no great amount of water. At length, as I cast my eye towards the east, a faint light appeared in the sky. I hailed it as the harbinger of morning. At the same time the wind began to fall, and in a few minutes had evidently greatly decreased. I began to hope that our dangers were coming to an end, and that we should only have the trouble of paddling back again without visiting our Kabomba friends.

"I see the shore!" cried Leo, "on my right hand."

"And I see it on the left!" exclaimed Natty.

Just then Mango, who had been sitting quiet at the bottom of the canoe, lifted up his head as if listening, and then pointed to the south evidently in a state of alarm. He uttered a few words, but what he meant to say I could not make out. There

was still so much sea that I was afraid of hauling the boat up
to attempt to reach the north shore. I therefore stood on as
before, and in a short time found that we were entering either
a narrow part of the lake or the commencement of a river flowing out of it, and I hoped every instant to reach some point
where we could safely land. We had stood on some little way
further, when I began to suspect, by the rapid way we passed
the land, that we must have a strong current with us as well as
the wind. Scarcely had I made this discovery when the loud
roar of waters reached my ears. It was the deep, solemn sound
which proceeds from a cataract. Now for the first time the
truth broke on me. We were in a rapid current, which was
hastily hurrying us on towards a waterfall. Not a moment
was to be lost. I told the boys to lower the sail and to endeavour to get the canoe's head round so as to pull in for the
shore; for as to making any way against the current and the
wind combined, that I knew was impossible. They did their
utmost, I helping them with my steering paddle, and Mango
working away with a spare one; but still so heavy were the
waves that they threatened every instant to capsize us, and I saw
that we were being carried down almost as rapidly as before. In
vain we paddled. We appeared to make no way. "Hope for the
best, hope for the best!" cried Natty, exerting himself to the
utmost. The perilous position in which we were placed pressed
heavily on my mind. The loud roar of the cataract sounded
louder and louder, and as daylight increased I made out in the
distance a cloud of spray rising in the air. Down it there
appeared every probability we should be carried, and what
hope was there then of our escaping with life? I looked
anxiously round on every side, and at length the increasing
light revealed a small island a little way further down the
stream. I trusted that by our exertions we might reach it.
We continued straining every nerve. Rapidly the canoe was

borne down sideways towards it. " A few strokes more and we shall be there," I cried out. " Work away, boys, work away." In spite of our exertions down glided the canoe, and the end of the island was passed. Still, we might reach some part of the side of the island. Had I been alone I might almost have leaped on shore. The moment was a fearfully anxious one. I could distinguish the southern end of the island. If we failed to reach that we must be lost. Trees overhung the banks. I gave a few more desperate strokes, and drove the canoe forward till her bows just touched the shore. " Leap out!" I cried. The canoe swung round. Natty seized the branch of a tree which hung down close to him, and swung himself up. I thought Leo and Mango had done the same, for I saw Leo clinging to a branch of a tree, and the black springing with the painter in his hand towards the shore. I therefore, seizing my gun and ammunition, leaped to the bank. What was my horror the next instant to see Leo fall back into the boat, the branch he had caught hold of breaking, and the black boy still holding on to the painter floating after the canoe. Leo seemed scarcely conscious of his own danger, but rushing to Mango, assisted to drag him in. My impulse was to spring into the water and try to regain the canoe, but just then Natty's voice reached me, crying, " Oh, help me, Andrew! help me!" and I saw that, though clinging to a branch, he could not manage, laden as he was, to climb along it so as to gain the shore in safety. I hurried to assist him, my heart sinking at the thought of what would become of Leo and Mango. I clambered along the tree, and at length got hold of Natty, but it required some caution to prevent us both falling off into the water. I got him, however, safe on shore, and then we hurried together to the south point, anxiously looking for the canoe. Leo and his companion had got out their paddles, and were working away in what appeared an utterly vain attempt to reach the north bank before the canoe would be hurried down the cataract. Natty wrung his hands in despair.

"Oh, how could it have happened?" he exclaimed. "I would have done anything rather than let Leo go. What is to be done? what is to be done?"

I had no consolation to offer him. Still the increasing light showed me that there were other islands intervening between the falls and the one we were on. It was barely possible, however, that the canoe would drift against one of them. We stood watching them with the deepest anxiety as the canoe was carried further and further down the current. Already she appeared to be in the rapids, from her quicker movement; and gliding faster and faster away, she soon was almost out of sight. It must be understood that there was a considerable distance between us and the cloud of vapour which I supposed to mark the situation of the fall. At length the canoe was hid from us altogether by a tree-covered island; but whether Leo and his companion had managed to reach it or not we were left in fearful doubt. It was some time before I could rouse myself. Poor Natty sat down on the ground with his head resting on his hands, completely overcome.

"But perhaps, after all, they may not have been lost!" he exclaimed, starting up, "and they may manage to tow the canoe along the bank of the river and come back to us. What do you think?"

"I dare not offer an opinion," I answered. "It is possible, just possible, and we must hope for the best."

Still we waited, looking in the direction we had last seen the two boys, anxiously hoping that they might reappear; but in vain. At length I began to feel somewhat faint, and Natty at last exclaimed, "Oh, I am so hungry!" It recalled us to the necessity of trying to find something on which we could support life. The island was so small, that had any birds been on it they would have flown away when we landed. I had, fortunately, a tinder-box in my pocket, so that we might light a fire if we could find anything to cook. At length Natty discovered a small fruit like a plum, growing on a tree covered with dark green leaves. He called me to it, and on examining it it struck me that it must be the *moyela*, which David had found near the banks of the river only a day or two before. This would at all events assist to satisfy the pangs of hunger, though it might not do to support us. I helped Natty up the tree, and he threw down to me as many as we thought we should require. We then sat down on the ground and discussed them, but the recollection of Leo made us too sad to talk.

"I am very thirsty," said Natty, "and must get a draught of water."

He went to the shore, and was stooping down to fill his hand full, when at that instant I saw a ripple in the water rapidly approaching. I had just time to spring up and pull him violently back, when a huge snout projected above the surface. The monster, startled by the fearful shriek Natty set up, and the loud cries I uttered, did not venture to approach, and slunk back again beneath the surface. I confess I was completely unnerved, and stood trembling all over, while Natty would have sunk to the ground had I not supported him. It was some minutes before I recovered.

"I must not again run the risk of being caught like that. I ought to have remembered the crocodiles," he said at last. "But I say, Andrew, don't you think it very likely that the creature may have its nest somewhere about the island? I will have a hunt."

Forthwith we began poking about in all directions with pieces of bamboo—a small grove of which grew on the island.

" Here is a hole," cried Natty at length, " and full of eggs, too. We will pay the crocodile off now for the fright he gave us."

I confess at first I could scarcely bring myself to think of eating crocodile's eggs. Natty had no such scruple. We filled our hats, and brought them to the beach, where, clearing away the grass to prevent an accident, we soon had a fire burning. As we had no pot to boil our eggs, we put them into the fire to roast, stirring them round and round with a stick. In spite of my repugnance, so excessive was my hunger that as soon as we thought the eggs were done, and Natty had pulled them out, I cracked one. The yolk alone had set, but that looked tolerably tempting; and on putting it to my mouth I could scarcely distinguish it, except by a peculiar flavour, from the yolk of a bird's egg. A couple, however, satisfied me.

" 'They will last the longer for not being too nice," observed Natty; " and we do not know how long we may have to stay here."

" We must think of means of getting away," I said; " for it is not likely that any canoes will pass by, and it is very certain that we must not attempt to swim on shore, though, were it only for the distance, I think I could do it, and carry you on my back."

" No, no, indeed!" exclaimed Natty. " We have had experience already of what would be our fate if we ventured into the water. But do you not think that the captain will come to look for us in the *Giraffe* when we do not return? He will never give us up without a search."

" But you forget," I said, " our friends do not expect us back for two or three days, so that they will not think of setting out till after that time, when they find we do not return."

" And what shall we do in the meantime?"

Although an idea had occurred to me by which we could

reach the shore, yet it was so perilous that I thought as long as we could find food on the island it might be prudent to stay there without attempting it. The day passed slowly away, and as evening approached I bethought me that we should wish to sleep.

"But what if a crocodile comes and picks us off?" said Natty. "That will not be pleasant."

"Too true," I said. "Then we must try and form a house in the trees."

There were not many on the island. We selected one with wide-spreading branches, into which we could without difficulty climb.

"But when we are there," said Natty, "how are we to sleep? As we cannot cling on like birds or monkeys, we should tumble off, for certain. I have it, though. Let us build a platform of bamboo; you have your hatchet, and we can soon form one large enough to hold us both."

The idea I thought excellent, and immediately set to work to cut down a good supply of bamboos. As I cut them I handed them up to Natty, who fastened the ends with flexible creepers, of which there was an abundance around us. Before it was dark we had formed a flooring about six feet long and as many broad. We now climbed up, and sat ourselves down to contemplate our performance.

"Suppose no canoe passes, how shall we ever be able to get from this," said Natty. "We are not going to live here for ever, I hope."

"I have thought of forming a reed raft, on which we can ferry ourselves across the narrowest part of the stream towards the north shore."

"But surely the current will carry us down?" he observed justly.

"I have thought of that: we must wait till a strong wind

blows up the river, and then I have hopes that it will keep
back the waters of the lake and probably greatly lessen the
current.   If so, and we can manufacture a mat sail, I think we
shall be able to reach the nearest bank.   It is dangerous, I
grant, but I see no other way."

" Nor do I, indeed," he said ; " but, by-the-by, I left our eggs
near the river, and probably the mother crocodile will come
to look for them and carry them off."

Without waiting for my reply he climbed down the tree,
and was soon back again with our provisions.   " I think I saw a
snout just coming out of the water," he remarked ; " and a
minute later the creature would have got hold of them, I fancy.
I did not stop to look a second time, however, for I was afraid
that it would have caught me had I delayed."

" I am very glad you have brought the eggs, and still more
that you escaped the monster.   It is evident that we must be
careful not to stand carelessly by the side of the bank, or go
into the water," I answered.   We were silent for some time.

" I think we should have prayers," I heard Natty observe.
" Will you say them, Andrew ? "

" Gladly," I replied ; and to the best of my power I offered
up a prayer for protection before we lay down to sleep.   I was
soon in the land of dreams, for I was thoroughly tired with the
exertion I had gone through during the previous day.   I was
awaked by feeling Natty touch my arm.

" Look down there, Andrew," he whispered.   " See ! it is
just as well we are safe up in the tree ! "

As I cast my eyes down on the ground below us, I saw three
huge crocodiles crawling slowly about ; and though they
generally take their food in the water, I had no doubt that
they would not have objected to seize us for their suppers had
they found us unprepared for resistance.   It was rather difficult
to go to sleep again with the knowledge that such creatures

were in our vicinity. However, after watching them for a time, I felt my eyes closing, and shortly forgot all about them and everything else present. When I awoke the sun was shining through the branches of the trees. The crocodiles had disappeared, the wind was light, the sky blue, and the smooth water shone in the beams of the rising luminary of day. Voices reached my ears. A faint hope rose in my heart that they might proceed from Leo and Mango. I quickly descended the tree, and made my way to the edge of the island on the side whence they appeared to come. There I saw, at some distance, a canoe with four blacks in her, engaged in combat with a hippopotamus. One of them, standing up, was about to plunge his spear into the animal's neck. Several more animals were standing on the nearest reed-covered bank, while the heads of others protruded from among the reeds in the distance. Here was a means of escape, if we could make the blacks hear, and they were inclined to assist us. I called to Natty, who, descending the tree, was soon by my side. We shouted with might and main, but the blacks were so eagerly engaged in attacking the hippopotamus that they did not hear us. The monster, as he received the wound in his neck, turned round and attempted to seize the canoe; but the blacks, plying the paddles quickly, got out of his way, holding him, however, by a rope attached to the spear. Spear after spear was darted into his neck; and in a short time the blacks, taking him in tow, dragged him on shore, where, in spite of his struggles, they hauled him up, and several other people hurrying down to the bank, soon despatched him with their clubs. No sooner had they done so, than they set up loud shouts, and began dancing away in frantic joy at their success. I thought this was a favourable opportunity for again trying to attract their attention. We shouted and shouted, but still they did not hear us.

" I think, Andrew, you must fire your gun.   They will heai that, at all events," said Natty.

I was about to do as he suggested; but then the question arose in my mind, whether we should be better off with the savages than we were by ourselves.   Still, should we lose this opportunity of getting to the mainland, another might not occur.   At length I fired.   The effect was curious.   The blacks ceased dancing, and looked about them with glances of astonishment.   Presently five of them leaped into the canoe, and having pulled out from the shore, so as to allow the current to carry them directly towards it, began cautiously paddling down to the island.   They, of course, knew its strength, and the necessity for care.   As they approached, Natty and I each took a branch in our hands and waved it, hoping that they would understand it as a signal of friendship.   As they drew near they stopped rowing, and gazed at us with looks of curiosity.   I again waved to them, and showing them my gun, I placed it by my side, that they might understand I had no intention of using it.   Except the usual small waist-cloth, the strangers had no clothing, though the man who sat in the stern guiding the canoe had a few ornaments about his head and on his neck, which showed that he was a chief.   They began jabbering away to us, but of course we could not understand a word they said.   I replied to them, therefore, by signs that we wished to be ferried over to the opposite shore.   Natty fortunately recollected just then that he had a few beads, a clasp-knife, and one or two articles which he had put into his wallet just as we were coming away.   He showed these to signify that we would pay them for the service they might render us.   They seemed to understand our signs, and beckoned to us to step into the canoe, carefully turning her round, so that they might instantly paddle off again up the stream.   We stepping in, they shoved off, exerting themselves to the utmost to stem the current. We made, however, but little way. As their backs were turned towards us, I could only judge of their disposition by watching the countenance of their chief. It was not particularly prepossessing, and the exertions he was making added not a little to its natural ugliness. He seemed to be regarding us with looks of intense curiosity, as if he had never before seen white people. After paddling along for a considerable time with the greatest exertion, they suddenly turned the canoe round and paddled across the current towards the shore. At length we got into a counter eddy, and now without difficulty they made way; the people who had been surrounding the hippopotamus running down along the bank to look at us. We soon reached a place where we could land, but for some minutes we were kept in the boat, while the tribe collected on the high banks above us, grinning down and gazing at us much as we should at a wild beast in its den in the Zoological Gardens. They were, I think, the ugliest savages we had yet met with.

"Well, I do declare I think poor Chico was a beauty to them," exclaimed Natty, as he looked up at them squatting in all sorts of attitudes on the bank. The women (I must not call them the fair sex) were even less attractive than their lords and masters. Two or three of them had huge necklaces hanging down over their breasts and rings round their arms, which in no way added to their beauty. Some of them carried children slung to their backs by straps of buffalo hides, and the little creatures, as they looked down upon us, grinned from ear to ear, though, when their mothers approached nearer than they liked, they set up the most terrific cries, such as I should have thought no human beings could have uttered. At last the canoe-men allowed us to land, when the female portion of the spectators hurriedly retreated, as if we were some wild creatures likely to do them harm. My first object was to in-

quire whether they could give us any information about Leo
and Mango; but they only shook their heads, and we could not
tell what that intended to signify. I explained to them, as
well as I could, that we had come in a canoe, which, with our
companions in it, had been drifted down towards the cataract.
Each time they only replied as before, with an ominous shake
of the head. Their countenances brightened a little as I dis-
tributed among them the articles which Natty had brought,
giving the chief a knife and a double allowance of beads.
Some who had been in the background, and had not received
any, now pressed forward, and looked very indignant when
we showed them that we had no more to give. I now made
signs to them that we wished to go down the banks to try and
ascertain what had become of the canoe; but they put themselves before us, and intimated that they did not wish us to
move to a distance. "But we must go, and we will go!"
cried Natty. "We must find out what has become of Leo, and
the whole tribe together shall not stop me!" It struck me,
however, that probably they wished to cut up the hippopotamus
and distribute it among the people; and that perhaps after
this operation they might be willing to accompany us. Without hesitation, therefore, we walked along the bank towards
the spot where the creature had been drawn ashore. I concluded that I was right, by seeing the chief and several men
instantly begin to attack the monster. In a short time they
had it skinned and cut up, each one taking a portion. The
chief took none himself, but several men, whom I supposed to
be slaves, were laden with larger portions than any of the rest,
which, I have no doubt, were his share. This done, I again
signified our wish to go down the banks to look for the canoe;
and at length, greatly to my satisfaction, the chief and six of
his companions began to move in that direction. Natty and I
hurried on as fast as we could walk, though, indeed, had we

not restrained our eagerness, we should soon have got ahead
of our companions. The distance to the falls was far greater
than I had supposed, for after we had gone some way we could
still see the cloud of mist rising above them. When we got
abreast of the islands to the south of the one we had landed on,
we examined them narrowly; but no sign of the canoe could
we discover. It was difficult, however, at all times to see across
the river, on account of the thick wood which in many places
fringed the banks and overhung the water. " Oh, they cannot
be lost! they cannot be lost ! " Natty exclaimed every now
and then. I could only reply, I hoped not; and still, as I saw
the rapid current rushing by, I dreaded to find my worst
apprehensions fulfilled.

At length we got near the edge of the cataract. A dark
ledge of rocks ran, it appeared, across the stream, some rising
high above the water, which flowed with terrific force between
them. There was, however, from the western shore on which
we stood a point which ran out for some distance, and within
this the water circled round, forming a back eddy. My only
hope was, from not having seen the canoe on any of the
islands, that she might have drifted into this eddy. We
searched the shore on every side, but nowhere was she to be
seen. That she could possibly have floated down the cataract
without turning over, I feared was impossible. We, however,
continued our way, not without difficulty, till we reached the
lower level, and looking back, saw the stream rushing over its
rocky bed, making a fall and leaping madly downwards to a
depth of fifty or sixty feet, where it bubbled and foamed in a
vast caldron, which sent up unceasing clouds of spray high
into the air. Then after a time it began to flow more calmly,
till it went gliding on as if fatigued by its hurried course
We now more narrowly examined the banks, not with any
hope of finding our companions, but in the possibility that

the canoe might have been drifted on shore, and that we might
thus ascertain to a certainty their fate.

We went on till we reached another stream which ran into
the main river. Here our companions placed themselves
before us, and signified that we must go no further. I could
only conjecture that they looked upon it as the border of their
territory, and were afraid that should we pass it we should
attempt to make our escape. I saw them looking out eagerly
over the country beyond the stream, as if to ascertain whether
any enemies were in the neighbourhood. They then signed
to us that we must accompany them back again. As we
returned we still continued examining the banks, in case
we might have passed any spot where the canoe could be
drawn up. We had not gone far when Natty, who had run
through a narrow pathway leading down to the water, exclaimed, "Here, here! Andrew. I think I see the canoe a
little way up the bank! Come and look!" I hastened to
him. There, under some bushes at the end of a little point
some few hundred yards from us, I saw an object which looked
very like a canoe. Still, it might be that of some of the
natives. We marked it well, and then hastened up again
along the bank, examining the bushes that we might discover
if there was any path through them. We searched about,
however, for some time before we could find a pathway. At
length one appeared, and Natty darting down it, made his way
towards the water as fast as he could run. It was like the
former one, formed, I concluded, by elephants or rhinoceroses to
reach their evening drinking-place. There was the canoe.
The paddles, however, and everything in her, had been taken
away. My heart beat with satisfaction and gratitude. Leo
and Mango had escaped destruction in the cataract; but what,
then, had become of them? We could discover no trace or
sign, nothing whatever to give us any clue to their fate.

CHAPTER XVI.

"WHAT can have become of them?" exclaimed Natty for the twentieth time as we stood examining the canoe. "But here come the blacks. Perhaps they will find out."

The chief and several of his followers assembled round the canoe, and began to talk eagerly to each other. They arrived at length at some conclusion, but what it was we could not divine. Then they examined the ground round, and seemed to discover certain marks, as one called the other to look at them. Then away they ran up the path, and began beating about in the surrounding wood. They came back shaking their heads, and when we by signs asked them what had happened, they pointed to the south.

"Depend upon it, Andrew," exclaimed Natty, "Leo and Mango have gone in that direction. Let us set off after them."

"I will try and make the blacks understand that we intend to do so," I answered.

I succeeded in explaining my wishes; but the blacks only shook their woolly pates, and made signs that if we did, we should be knocked on the head, or that daggers would be stuck into us.

"Oh, dear! Oh, dear!" cried Natty, "can that have been the fate of Leo and Mango?"

"I hope not," I said. "If anybody has carried them off, they would not have done so for the sake of killing them. What I suspect is, that the neighbouring tribe is at war with our present friends, and that they happened to be making a raid into the country, and falling in with Leo and Mango, carried them off captives. It was evident from the signs these people made to us that they did not wish us to cross the stream, which they probably considered the boundary line between their territory and that of the tribe with which they are at war. I may be mistaken, but we must try and return as soon as possible, and let Timbo and Igubo know what has occurred. If we can ascertain from them to what place Leo and Mango have been carried, we must lose no time in endeavouring to release them."

"Oh yes," said Natty, "I am nearly sure they must have been carried away prisoners, or they would have come up the river and endeavoured to release us, as I said they would do."

When, however, I explained to the chief that we wished to return to the part of the country from whence we had come, I found that he had no intention of letting us go. Still I hoped, of course, that we might find the means of escaping. At present, indeed, as we had no food, we were not unwilling to accompany the chief to his village, to which by signs he invited us. As we walked along I saw him eyeing my gun. I showed it him, holding it, however, pretty tightly, lest he should think fit to appropriate it. I saw by the way he looked at the weapon, that he was unacquainted with its use. First he examined the lock, and then put the muzzle to his eye and looked down the barrel. I hoped that before long, by means of my gun, I might be able to gain the respect of the people. I determined, therefore, not to fire until a favourable opportunity should occur. I in the meantime took great care that no one should wrest the weapon out of my hands. The people as we went along gathered round us, some coming up and touching our clothes, others putting their hands on our faces, evidently unable to understand the light colour of our skins. When the people began to press too closely round, the chief ordered them angrily to keep at a distance; and some still persevering, he swung his spear round and round, hitting them, without much ceremony, either on their shins or heads, when they quickly retreated to a more respectful distance.

I was very glad when at length the village appeared in sight. It was situated on the borders of a small gulf, I will call it, or the mouth of a stream near the lake, surrounded by a belt of elegant fan-palms and a number of gigantic wild fruit trees. Beyond it the lake was seen extending far as the eye could reach, though in some places the water was concealed from sight by vast masses of reeds and rushes of every shade and hue, several beautiful little islands being dotted over it, adorned with the richest vegetation, its many beauties heightened by the brilliant rays of a tropical sun, somewhat softened by the silvery mist which rose from the lake. The village was very similar in character to that which Stanley had visited,—so I concluded from his description. A plentiful repast was placed before us by the chief's wife and her attendants soon after we arrived. The principal dish consisted of hippopotamus flesh, but there were plantains and cassava porridge, with an abundance of wild fruits, the best of which was the moshoma, both fresh and dried. We had seen the tree growing outside the village. It grows to a great height. The trunk was beautifully straight, and the branches did not begin to spread out till very nearly at the summit. The fruit can thus only be gathered when it falls to the ground. It is then collected and exposed to the sun for some time. After it has been dried it is pounded in a mortar, when it is fit for use. In that state it will keep for some time. It is generally mixed with water, and made into a sort of jelly, which tastes and looks not unlike honey. It is especially useful for giving a flavour to the otherwise tasteless cassava porridge.

The chief seemed very well disposed towards us, and now, as the day was drawing to a close, he pointed to some mats in a corner of his hut, and signified that we might sleep there. Having been in exercise all the day, in spite of our anxiety we slept very soundly. The village was astir at an early hour. Though the appearance of the people was not attractive, they were more civilized than I had expected, and in the neighbourhood of the village we saw a wide extent of fairly cultivated ground. A bowl of cassava porridge, sweetened with moshoma, was placed before us for our morning meal, and Natty and I did ample justice to it. We now thought that we might let the chief know we were in a hurry to go away, but he shook his head, saying that that could not be. What his object was in keeping us we were unable to comprehend. It was very evident that he had made up his mind we should stay with him, for some time at least. The more we urged him to allow us to take our departure, the more determined he seemed to keep us. At last I thought it wise to give up the point for the present. We were allowed, however, to walk about, but were always accompanied by either the chief himself, or four or five of his attendants armed with spears, or bows and arrows. Some of the latter were blunt-headed, and others were barbed, and, I suspected, poisoned. We found a party of them setting out for the forest at a short distance, and wishing to see what they were about, we accompanied them. We found them engaged in making what we afterwards discovered to be bee-hives. They first took off the bark of some small trees, fifteen or eighteen inches in diameter and about five feet in width. They managed this by making two incisions right round the trunk five feet apart. A longitudinal slit from one to the other enabled them to detach the bark from the trunk. The bark immediately assumed the shape it had before, and where the slit was made it was sewn together again with the fibre of a tree called the *motuia*. It thus formed a wooden cylinder. A top and bottom were next fixed in it, formed of grass rope, the lower one having a hole in the centre for the ingress and egress of the bees. When the hives, as I shall call them, were completed, they carried them off, and placed them high up in lofty trees in different parts of the forest. The bees which were flying about seeking habitations soon discovered them, and even while we were there some were already taken possession of. A piece of cloth was tied round the trunk of each tree, which, when I came to inquire about the matter, I learned was looked upon as a charm, and was sufficient to prevent any thieves robbing the hives. In the woods a number of beautiful yellow birds were flying about, some of which I afterwards saw in cages in the house of the chief and several of his people. The cages were very neatly made, and had traps on the tops to entice their still free companions. The chief called the birds *cabazo*, and I found that they were a species of canary. They fed them on a plant called the *lotsa*, of which they cultivate a considerable quantity for food, and wild canaries come and help themselves, much as sparrows do to the seeds which the gardeners have sown at home. I saw also several tame pigeons. Early in the morning numerous other little songsters were chirruping merrily away, some singing quite as loudly as our English thrushes; and one, the king-hunter, *Halcyon senegalensis*, makes a clear whirring sound like that of a whistle with a pea in it. As we were returning, suddenly we saw streaming out of a hole innumerable flying creatures. They formed a dense column. Out they came; there seemed to be no end of them. No sooner did our companions catch sight of them than they made chase. As the insects began to scatter, they appeared like snow-flakes floating about in the air. They were, I suspected, white ants. After flying a considerable distance they alighted on the ground, when, as we watched them, they bent up their tails, unhooked their wings, and began immediately digging away with wonderful rapidity into the earth. They had good need of haste, for birds were seen assembling from all quarters, numerous hawks being among them, who began snapping them up with the greatest avidity. The natives, too, immediately set to work to collect them, giving them a pinch and putting them into baskets which they carried at their sides. They were quite as eager to obtain them as the birds were. On picking some of them up I found that they were fully half an inch long, as thick as a crow-quill, and very fat. One I caught had its wings on, and fancying from the ease with which its fellows got rid of these appendages themselves that I could help it, I made the attempt, but the wings appeared to me as if hooked into the body, and I tore away a piece of the flesh at the same time. As long as an ant was to be found, the natives continued picking them up; and I suspect, out of the whole brood but a small number could have reached places of safety beneath the earth.

Our companions hurried home with their prizes, when they immediately lighted fires, and roasted the ants, much as they might have done chestnuts. All hands gathered round and ate them eagerly, evidently considering them among the greatest of delicacies. When they saw us watching them, they offered us some. "No, no," said Natty; "I do not know what I may do, but I have not come to that yet." The chief, who had the larger share brought to him, sat on the ground, rolling his eyes round as he dropped insect after insect into his mouth, evidently enjoying the repast, and seemed to look with an eye of pity on us when we declined partaking of it.

Soon after this we observed a number of the men dressing themselves up in a curious manner. Some had covered their heads with caps made of the skins of water-antelopes, with the horns still attached, part of the skin hanging down over their shoulders so as to conceal the upper part of their bodies. Others had manufactured the heads and beaks and long necks of white cranes into coverings for their heads. Carrying their bows and arrows in their hands, and quivers and darts at their backs, they set forth to the bank of the lake. We watched them crawling along amid the reeds, their heads alone being visible, and looking very like the animals they intended to represent. I could see in the distance on a sedgy bank several dark objects, which I guessed were crocodiles. The hunters approached them cautiously, now stopping, just as an antelope or crane would do to feed, now advancing again, now stopping, till they had got within bow-shot of the creatures. Then, quickly raising their weapons, they let fly at the same moment. The result at that distance I could not ascertain, but it appeared to me that, although I saw some movement among the objects, yet two or more remained on the bank. The hunters rushed on, now careless of exhibiting themselves, and in a short time returned with some of the flesh of the creatures they had killed. They immediately set out again, and as I watched to ascertain the direction they took, I saw in the far distance several buffaloes going down to drink at the lake. They were not back till dark; but, from the quantity of buffalo flesh they brought with them, I had no doubt they had killed one or two of the animals. Their plump cheeks and bodies showed that they had an abundance of food; and they were liberal in bestowing on us as much as we could desire.

Our friends remained for a couple of days, enjoying, after the

African fashion, the abundance of food they had collected Whenever we signified our wish to depart, the chief, as before, strenuously opposed it. In vain I protested against being detained, and made signs that we were determined to go, whether he wished it or not. This made him very angry, and from his manner when he left us, we feared that, should we really make the attempt, he would use force to prevent us. We therefore, as other people have done, had to yield to circumstances, and to make the best of our position. At last we agreed that we would appear to be contented with our lot, so as, if possible, to throw our captors off their guard. They were the most active and persevering hunters of any people we had yet met with. The morning after the last piece of buffalo flesh had been eaten (it had been rather too high for our stomachs), we found that they were preparing to set off on a hunting expedition, and we were not sorry to find that they expected us to accompany them. I carried my gun : I should have said I never let it out of my hand by day, and always placed it under my mat by night, that no one might take it from me. The chief, I fancy, looked upon it as my fetich, and certainly regarded it with considerable awe. Whether or not he had discovered that it had made the noise he had heard, I could not as yet ascertain. Among the hunters was a young man, whom we found to be the chief's son. He was one of the best-looking of the tribe, though that is not saying much for him. He was, however, good-natured, and seemed inclined to make friends of us. We therefore kept by his side. About thirty hunters set out, headed by the young chief. They were armed with long spears and bundles of javelins, on which they appeared to depend for killing their prey, trusting to their activity and the knowledge of the animals they might attack to get out of their way. We passed through the wood we had before visited, and continued across an open prairie till we

arrived at a forest of considerable size, extending on either
hand as far as the eye could reach. The band at once entered
it, spreading themselves out so as to beat a large part of the
wood, but yet continuing within call, if not always within sight
of each other. Natty and I followed the young chief. After
proceeding some way one of the men came up, and presently
we saw that they were all closing in towards a point a little
way ahead. As we advanced I saw, just over the bushes, the
back of a large white rhinoceros. The monster had come there
probably to enjoy the shade of the wood. It seemed to be
alone. The men all approached cautiously, concealing them
selves under the brushwood till they were close upon the
creature, then, starting up, they hurled their darts at it. The
rhinoceros started forward, pursued by the hunters, the young
chief taking the lead. Suddenly the creature seemed to
stagger forward. Its front feet had sunk into a hole or artificial pit, I could not ascertain which. As it did so, instead
of struggling, it remained perfectly quiet. At this juncture
the young chief, with his spear in his hand, leaped on the
animal's back, intending apparently to plunge the spear into
its head behind the ear. At that moment it suddenly reared
itself up, and before our friend could leap off again began tearing away at a rapid rate through the forest. He clung to his
seat in a wonderful way. His spear, however, before he could
strike it into the animal's neck, was hurled by a bough from
his hand. The hunters pursued, shrieking loudly through
fear of the life of their young chief. I too dreaded lest he
should be thrown off, when the animal would too probably
turn round upon him, and, before assistance could arrive, might
transfix him with its terrible horn. I was also afraid to fire,
lest I might wound the young man. His companions followed,
shrieking and shouting as fast as they could. Natty and I followed after, but could not make way through the thick and

tangled underwood so rapidly as the blacks. We were therefore left behind. Presently the rhinoceros turned, and came tear
ing towards us, forcing its way through the underwood. Still
the black kept his seat, when the rhinoceros, swerving on one
side, passed under the bough of a tree, and in the same manner
that he had lost his spear he himself was hurled to the ground.
He attempted to rise, but his ankle had apparently been
sprained, and before he had gone many paces down he fell.
The enraged creature seemed aware that it had got rid of its
rider. It stopped, and eyeing him with a savage glance,
rushed towards him with its horn pointed at his body. Now,
I felt, was the time for me to fire, or the young man would
certainly be killed. I had, providentially, a rest for my gun,
and pulling the trigger, my bullet hit the rhinoceros directly
behind the ear. The impetus it had gained sent it on
several paces. A loud shriek rent the air; but just before
it reached the young chief over it fell, and lay perfectly still.
We ran forward to help up our young friend. He glanced up
in my countenance with a look which showed that he was
grateful for the service I had rendered him. He then took
my hand and pressed it to his lips. In a few minutes the
rest of the hunters came up, when he addressed them, and,
I concluded, was telling them what I had done. I certainly
never fired a shot with so much satisfaction. The men came
round Natty and I, their whole demeanour completely changed,
evidently looking upon us as heroes worthy of renown, while
some begged to examine the wonderful weapon which had done
the deed.

As soon as the hunters had cut up the rhinoceros, we returned
in triumph to the village. The chief showed that he appre-
ciated the service I had rendered him in saving the life of his
son by warmly embracing us—a ceremony, by-the-by, with
which we would gladly have dispensed. We were now, instead

of being looked upon as prisoners at large, treated with every consideration; and when I signified that the only reward we required was to be allowed to return to our homes, I understood him to beg that we would remain one day longer, when he would accompany us as far as he could venture to go.

I suspected that his tribe were at war with their neighbours, as scouts were constantly coming and going, and that this was the reason why he could not accompany us in our search for Leo and Mango. We would gladly at once have set off to look for them; but when we showed a wish to go to the south, he made us understand that they were already carried a long way off, and that, coming from his village, we should be looked upon as enemies, and probably murdered. This we thought so likely, that we agreed it would be prudent to return home to obtain the assistance of our friends.

There was a grand feast at night on the flesh of the rhinoceros, and dancing and singing were kept up till a late hour—an amusement we would willingly have avoided.

Natty and I talked over the possibility of returning in the canoe, but there were no paddles; and we could scarcely have propelled her, even had we made some. We begged the chief to take care of her till our return, and this he promised, as far as we could understand, faithfully to do.

Next morning we again expressed our anxiety to set off, but the chief showed no inclination to let us go; and each time that we pressed him, he signified that we must remain a little longer. We were the less unwilling to do this, in the hope that we might, in the meantime, gain some news of Leo and Mango, and we once more urged the chief to try and discover where they were. He let us understand that he wanted first to have another hunt, and that I must bring my gun to assist him. I, of course, expressed my readiness to comply with his wishes, but resolved not to expend much of our powder, as we should require it on our return home. We were allowed to wander about the village wherever we liked, but we observed that all the time we were carefully watched. The women and children always started up with looks of astonishment when we came near them, the young ones running away, frightened at our white skins, just as European children would be alarmed at the sudden appearance of a black man among them. On the outskirts of the village, near the river, we came upon a group of people employed in burning large quantities of a coarse-looking rush and stalks of a plant which I had seen growing in a marsh near at hand. I had, the day before, by chance tasted the water in the marsh, and found it slightly brackish. On examining the proceedings of the people, I found that they were employed in manufacturing salt. Before them were a number of funnel-shaped baskets formed of grass rope. These were filled with the ashes, and water being poured into them, percolated through the basket-work into calabashes placed below to receive it. They were then put out in the sun, and the water evaporating, left a small amount of salt in each. Although there was not a sufficient quantity for salting fish or meat, the supply was ample for ordinary use, and we were glad to purchase some with a few beads which we had remaining in our pockets. Amply supplied as we are in England with that necessary article, we can scarcely appreciate its value in a country where it is not to be obtained without great difficulty. Natty and I agreed to husband our little stock carefully, as for the last few days we had felt the want of it when eating rhinoceros flesh. We had observed several animals coming down to this salt marsh to chew the coarse grass or to lick up the salt collected on the reeds.

As we were walking along we heard the chief calling to us, and found that he was prepared to set out on his proposed expedition. We saw as we proceeded many large animals in the distance, but they had evidently learned caution from the attacks made on them by the natives, and would not approach the village. As we appeared they took to flight, keeping always a long way out of range of our companions' arrows. Once I got near a rhinoceros, but was unwilling to fire without feeling tolerably sure of hitting the animal, as I had determined not to throw away a shot if I could help it. At length we got into a region where we could obtain cover among low bushes, and occasionally clumps of trees. The natives took advantage of this, and hiding themselves under bushes, clumps of tall reeds or grass, proceeded for some distance. Natty and I followed their example. At last I saw, a little way from a grove of trees, a herd of cameleopards quietly feeding. The blacks lay like logs of wood on the ground, every now and then creeping slowly on when the heads of the animals were turned away from them. Still they were too far off for me to make sure of a shot. I saw, a little way on, a solitary bush. I thought if I could reach it I might be able to bring down one of the nearest giraffes. The natives watched me eagerly as, trailing my gun after me, I cautiously approached the bush. I was very anxious to kill an animal, in order still further to establish our credit, hoping thereby also more speedily to obtain permission to depart. I could not help constantly thinking of the alarm our prolonged absence would cause our friends.

As I crept on I saw the giraffes turning their heads, raised high in air, now in one direction, now in the other, as if they suspected danger. I should have said that they were near a small grove of trees, from the branches of which some of the herd were plucking the leaves. This grove had partly concealed our party, or we should not have approached so easily. I had never prided myself on being a sportsman; but I had steady nerves, and of late had given good practice to my eye,

and thoroughly knew the range of my rifle. The bush was
gained. A large bull cameleopard stood the nearest, every now
and then turning his head to pluck a bunch of leaves from a
branch which no other animal could have reached, but still
apparently on the watch for danger. I raised myself on my
knee, and lifting my rifle, took a steady aim at his breast. At
the report the whole herd moved off, swinging their legs over
the plain at a rapid rate. I thought that I must have missed,
and yet my bullet seemed to strike the creature at whom I
had aimed. Away he went with the rest. Before, however, he
had proceeded fifty yards down he suddenly fell, and lay prostrate on the earth. The blacks, with loud shrieks and shouts,
rose from their hiding-places and darted forward, and in a few
minutes the wounded giraffe was surrounded by a band of
dancing, shrieking, shouting blacks, delighted at the thought
of the meal he was about to afford them. Natty and I stood
at a little distance, when suddenly we saw the giraffe raise his
neck high above the heads of the shrieking band. Presently
out went his legs, and the chief and his followers were seen
scattered here and there on every side, some prostrate on the
ground, others scampering off to avoid the fury of the kicks of
the dying animal. I thought some of them must have been
killed. It was his last effort, however, and again sinking
down, he lay perfectly quiet. The blacks picked themselves
up, showing that at all events no mortal injury had been done,
and again assembled round the body of the animal, though
keeping at a more cautious distance till they had ascertained
that he was really dead. On finding this to be the case,
they sprang on the body, and began hacking away at it with
their knives, till, in a short time, it presented nothing but a
mass of mutilated flesh. The chief seemed highly delighted
at our success, and I took the opportunity of again urging him
to allow us to go, trying to make him understand that I would

return, if he wished it, with companions who were still better able to kill game for him than I was.

As a large portion of the day had been expended, without attempting to seek for more game the chief led us back to the village.

"What do you think he will do?" asked Natty as we walked along. "If he will not let us go willingly, I propose that we take French leave, as Leo would say, and I do not think he will attempt to stop us by force."

At a little distance from the village there stood, under a grove of trees, a hideous idol, at the top of a stout post. It was elaborately carved, representing rather the face of an ape than that of a man, and covered with red, yellow, and black paint. The hunters placed some of the meat of the giraffe before it, on a block of stone ; but only a small quantity, and that of the least valuable parts. I guessed by this that they had no great respect for their idol. "Poor people," said Natty, "perhaps they guess that they can cheat even it, and that it will not be able to distinguish between the best and worst parts." Natty and I were also tempted to stop. He made signs to the chief, touching his own ears, and then shaking his head and pointing to the ears of the idol, to signify that it could not hear. Then he pointed to its mouth, and in the same way tried to explain that it could not eat the meat placed before it. Then he touched its head, to show that it could not understand. We fancied that the chief comprehended his meaning, for he laughed, and cast a contemptuous look at the ugly block. Although he did this, however, in our presence, it is possible that he still had some superstitious fear of the idol, or of the evil spirit it might have been intended to represent.

"The poor Africans have no knowledge of the powerful, kind, and merciful God," observed Natty. "The beings to whom they pay respect they believe to be malign spirits, who will do them harm if they do not attempt to propitiate them by gifts and observances."

I may observe here that we never paid the slightest respect to the negro idols, and never were treated worse in consequence; indeed, I believe that they would have despised us if we had done so, for though they may fancy that their idols have something to do with them, they believe that they have no power over the white men.

There was great rejoicing in the village on the arrival of the flesh of the giraffe, the greater portion of which was consumed long before the night was over. While seated with the chief, I again asked him to let us go, and he seemed to intimate that he would do so the following morning. While we were at supper, Natty proposed that we should hide as much food as would last us for the following day. "A good idea," I observed. The pockets of our shooting-jackets were capacious. Whenever the chief was looking another way, we contrived to slip in large pieces of meat and cassava cake, besides pieces of plantain. They made somewhat of a mess in our pockets, but we could not be particular. As the chief consumed double as much as we hid away he was not surprised at the rapid disappearance of the food, and had not observed our manœuvre.

Natty and I lay down to rest, hoping that before another sunset we might be far on our way homewards.

CHAPTER XVII.

AWOKE just as day broke, and roused up Natty.

"Where are we?" he exclaimed. "Oh, I was dreaming, and so happy!"

"We have realities before us," I remarked. "Are you prepared for starting?"

"Yes, yes," he whispered; "by all means. Probably the people, after their debauch, will sleep soundly, and we may get some way before we are overtaken."

We put on our jackets, which we had placed at our sides, having slept covered up with mats provided for us. We then cautiously pushed open the reed-formed door, and stood looking out up and down the street of the village. The stars were still twinkling overhead, though gradually growing dimmer as the gray light of morning advanced. I carefully marked the course we were to take, and observing all the doors closed, we now sallied forth, and crept cautiously along towards the end of the street which opened out in the direction of our home. Every moment we expected to be pursued. If we were, we agreed to put a bold face on the matter, and to claim the right of departing. Fortunately the inhabitants, from having sat up the greater part of the night eating, were sound asleep, and we hoped that our flight would not be discovered till we were a considerable distance on the road. We stole on, treading as lightly as possible in the centre of the street. We could hear loud snores proceeding from some of the huts. The sound gave us confidence. It also showed us how easily a native village might be surprised by enemies. The careless, thoughtless people seemed to have forgotten that they were at war with their neighbours. We reached the end of the street. There was a gateway, but the gate was closed. On examining it we found that it might be easily opened; but I feared that while we were doing so the proper guards might pounce out on us. They too had left their posts, and we were reassured by hearing loud snores coming forth from a hut close at hand. I did not like to leave the gate open. Natty whispered to me that he thought he could climb over it. There was no great difficulty in doing that; the only fear was that on dropping on the other side we might be heard. However, there was no time to be lost. I helped Natty up, and he scrambled down without making any noise on the opposite side. I followed, and reached the top. I might without danger have dropped down, but, for the reason I have mentioned, I thought it better to lower myself gradually. My foot, however, slipped when half-way, and the wood-work creaked loudly, while the noise I made in falling would, I feared, arouse the sleeping guard. We stopped for a minute. Still the snoring sounds came loud as before. There was no necessity for further delay. We therefore, walking as noiselessly as we could, hurried on towards the north-west. We followed a well-beaten path, which I had before noted as leading in the direction we wished to go. As soon as we had got far enough from the village to make it unlikely that our footsteps would be heard, we began running, I leading, and Natty following close at my heels. I had been a good runner, but was out of practice. Natty, however, was very active, and easily kept up with me. We ran on for an hour or more without stopping, till we were bathed in perspiration, and I felt that I could not go much further without rest.

"Do stop," said Natty. "Even if the blacks discovered our escape directly after we left the village, it must be some time before they can overtake us."

"You are right," I answered; "a little rest will enable us to go faster afterwards."

We sat down under a wide-spreading tree. The shade was pleasant; for the sun, which had already shot up high into the heavens, sent down his rays with great force. The air was full of life. Insects were buzzing about, and gaily-decked parrots flew from bough to bough, while the monkeys came out of their leafy covers and looked down upon us with astonishment. We took the opportunity of eating some of the food we had brought in our pockets. It was not very nice, but it satisfied our hunger. I was soon ready to proceed. Natty, however, urged me to rest a little longer, thinking that I should be over-fatigued by such unusual exercise.

"Come along," I said; "the further we are off, the less likely our friends at the village will be to insist on our returning with them."

We went on as before. Sometimes a snake glided across our path, but quickly got out of our way, more frightened at us than we were at seeing it. Now we heard a rustling in the underwood, and a panther or hyena dashed away amid the foliage without thinking of attacking us. We had gone on at a slower pace than at first, when, by the appearance of the sun in the sky, I saw that it must be noon. I now once more called a halt, for I felt tired myself, and was afraid that Natty must be equally so. We had for some time been crossing the open prairie, steering, by the sun and the distant line of mountain, as I hoped in the right direction. Before us lay a thick wood. Natty proposed that we should take shelter within it, as, even should the natives closely pursue us, we might there have the prospect of remaining concealed while they passed by.

"I am afraid that you have scarcely yet recovered your strength sufficiently to march on all day without rest," said Natty; "and as I could not find my way without you, I hope, for my sake as well as yours, that you will stop for an hour or so here."

This argument prevailed with me; for I was so anxious to reach home, and felt so strong, that I should have gone on till night would prevent us from proceeding further. We accordingly entered the wood, and after making our way a short distance into it, came to a small open spot, free of trees or thick underwood.

"I propose that we build a hut of boughs here," said Natty. "We can take our turns to go to sleep in it, and it will help to guard us against the attacks of lions or other beasts."

With our axes we cut down a sufficient number of boughs to form a shelter, and having planted them close together, a hut was formed which would, I hoped, afford us ample protection. As the sun struck down into the woodland glade with great force, we took our seats within the hut for the sake of the shade, and discussed a further portion of our provisions. I saw that Natty was very sleepy, though he was trying his utmost to keep awake. I therefore told him to lie down, and that I would watch. Finding a branch of a tree torn off, perhaps by lightning, I chopped a piece of sufficient length to serve as a pillow, and having examined it carefully to see that no scorpion or other stinging insect lurked within, I placed it under his head, and sat down at the entrance of our leafy bower to keep watch, with my gun by my side, ready for action at a moment's notice. I felt somewhat drowsy, but made every effort to arouse myself, feeling the importance of keeping awake. Presently I heard a slight rustling, as if some animal was moving among the bushes near me; but without shifting my position I could see nothing. Then I heard a sound as if creatures were nibbling grass or leaves. This made me sure that no savage beast caused the sounds, and I sat quiet, expecting soon to discover what creatures they were. Presently two beautiful fawns came in sight. They did not perceive me, but went on quietly grazing, unconscious of the presence of one of their many enemies. At length they came full before me. " Shall I fire?" I asked myself. " One of them would afford us ample food till we could reach home." I was afraid, however, that should I move to raise my rifle, and get into a better position for taking aim, they would instantly be off, and bound into the thicket before I could even fire. While I was considering (though sportsmen may laugh at me, I own I was unwilling to kill one of the beautiful creatures), another, of a very different character, appeared on the scene. Suddenly I caught sight of a pair of glaring eyes amid the thick gloom of the thicket on my left. I saw a large tiger-looking animal of a fawn colour, the back variegated with round black spots. I guessed at once that the creature was a *cheetah*, or hunting leopard, and thought it was lying in wait for the deer till they should approach within distance of its spring. I had no idea, however, that it could make so prodigious a bound as it at that moment did; for scarcely had I seen it, when it sprang out of its ambush, and alighted on the unfortunate buck, which it struck down with one tremendous blow. Seizing my rifle, and throwing myself on my knee, I took a steady aim, and the ball entered the cheetah's head. It sprang up, dragging its prey with it, but instantly sank down, and rolled over dead. Natty sprang to his feet at the report. The cheetah had settled the question whether the buck should die, for the poor creature was so mangled, though not killed outright, that we saw it would be a mercy to put it out of its sufferings. This we immediately did. Its more fortunate companion had escaped into the open ground.

We lost no time in cutting up our prize, for the meat we had brought with us was already scarcely fit to eat, and we both confessed to feeling very hungry again.

"We may as well light a fire and cook it," I said. "We must take care, however, not to set the wood in a blaze."

There was ample fuel about, and choosing a spot where the grass was green, and did not readily burn, we piled up the sticks we collected. I had a tinder-box and matches in my wallet, and thus we soon had a good fire burning. In a short time we had some pieces of venison roasting in wood-land fashion on forked sticks before the fire.

Having selected the best parts of the venison, and wrapped them up in leaves to carry with us, we recommenced our meal on the portion we had cooked. The salt we had purchased a few days before was now particularly acceptable, and we both had meat, we hoped, sufficient to sustain us for many hours. Now greatly refreshed, we prepared to proceed on our journey. We first put out the fire, however, that there might be no risk of setting the forest in a blaze.

"We must leave the cheetah and the rest of the deer to the birds and beasts of prey which are sure to visit it before long," I observed. "And now, Natty, let us be off."

Scarcely had I uttered the words when he touched my arm. "Stay," he said; "I am sure I heard voices in the distance."

We listened. There could be no doubt of it. The sounds drew nearer. The tones were very similar to those we had heard during our stay at the village.

"They must be our late friends come to look after us," I observed. "If we are discovered, we will put a good face on the matter."

"Would it not be better to go and meet them at once, and present them with the game we have killed?" said Natty.

I agreed with him; and peering out from our shelter, I recognized the chief and his son, with a band of followers. Loading ourselves, therefore, with as much venison as we could carry, we sallied boldly forth, and were soon face to face with our late hosts. Their look of astonishment when they saw us and the meat we carried was very great. By the expression of their countenances, however, we saw that they were not offended. As far as we could understand, the chief only reproached us for going away without bidding him farewell. I felt myself somewhat embarrassed, for I could not help seeing that he had intended to let us go as he promised. We showed that we had enough meat for ourselves, and presented him with the larger portion. Having done this, we led him and his companions to the spot where the cheetah and the remainder of the deer lay. His companions quickly cut up the cheetah as we had done the deer, and divided the flesh among them. We then pointed in the direction we wished to go, and the chief taking my hand, and his son Natty's, we proceeded onwards in the most friendly way. At length my conductor came to a full stop, and, looking me in the face, seemed again to be reproaching me for having left his village by stealth. I tried, as before, to explain that we were in a hurry to reach our friends; and as he had detained us longer than we wished, we were afraid he might still keep us prisoners. Whether I was right in my conjectures as to his meaning, I am not quite certain. He, at all events, showed that he was friendly disposed towards us.

He now signified that he could go no further, and, pointing ahead, intimated that we should find enemies in our path if we went direct. He then, pointing to the left, advised us, as I understood him, to make a circuit so as to avoid the danger.

Having satisfied himself that we clearly understood his advice, he and his son warmly shook our hands, their followers imitating their example. Such a shaking of hands I had never before gone through. I observed as I turned away that there was an expression of sorrow in their countenances, which arose partly, perhaps, from parting with us, and partly from the dangers which they apprehended we should have to encounter.

We now took our way to the south-west, skirting the edge of the forest, which appeared to extend towards the lake. We had not gone far, when, turning round, I saw the young chief stopping and gazing at us. When he found that he was observed, leaving his party, he darted after us, and once more took our hands, pressing them warmly, intimating that if his father would give him leave he would accompany us. His father's voice, however, called him back, and, with a look of regret, he again left us.

"At all events," observed Natty, "we must acknowledge that gratitude can exist in the breasts of these Africans."

"It does certainly in that of our friend," I said. "They generally have a very different character bestowed on them."

As long as the blacks remained in sight, we could see the young chief every now and then turning and casting lingering glances towards us.

We now pushed on, hoping to find some secure place where we might pass the night. We were fortunate in finding a tree with wide-spreading branches, radiating so closely from a common centre that they formed a wide and secure platform on which we could rest without fear of falling off. We climbed up as soon as we had supped, and passed the night in perfect quiet.

I need not mention the incidents of the two following days. We were cheered as we trudged on with the expectation of soon rejoining our friends.

It struck me, on the third day of our journey, as we walked on, that Natty was less inclined to speak than usual ; and looking at his face, I saw that he was deadly pale. He did not complain, however. I asked him if anything was the matter. He said no ; he only felt a little fatigued, and thought that he should be revived by a night's rest. I proposed that we should stop at once ; but to this he would not consent, declaring that he was well able to get on as long as daylight lasted.

The country though which we passed was similar to what I have already described. We proceeded in as direct a line as we could steer, keeping the distant hills on our right, instead of going towards them.

I proposed the following day to begin circling round more directly for them, as I hoped that we had now gone far enough south to avoid the village against which the chief and his son had warned us. I should not have hesitated, however, to have gone amongst the people, had I not feared that we might be detained by them as we had been by their neighbours.

The forest as we advanced grew thinner, and we found the trees at length standing so widely apart that we could see the plain beyond them. As the wood might afford us more shelter than the open plain, and the sun was already sinking towards the blue hills in the distance, we agreed to halt. As I saw that Natty was not able to exert himself, I bade him sit down while I cut branches to form a hut, and collected wood for a fire. As I could not tell what wild animals might be roaming about around us, I determined to make our hut sufficiently strong to resist an attack.

I selected a tree of a considerable diameter, which served as a back to the hut. I stuck the uprights in the soft ground among the roots. There were plenty of vines, with which I bound the cross pieces to the trunk and to the up-

rights. The intervening spaces I filled up with light perpendicular poles. While I was gathering a further supply, I
found that Natty had interwoven them with branches and
vines, thus forming tolerably substantial walls. Some of the
boughs thrown over the top served for a roof, which, however,
would not have kept out a tropical shower; but there was
no fear, we thought, of rain. Darkness was now coming
rapidly on, but I had not yet a sufficient supply of wood to
keep up our fire all the night; and I told Natty to make it
up and light it while I went to collect more broken branches,
of which there were numbers lying about, torn off probably
by a hurricane. While I was engaged, I saw the fire blaze
up, and hoped that Natty would have some venison roasted by
the time I had finished my work. Having brought a couple
of loads and placed them down by the side of the hut, I went
away for a third. I had got as many branches as I could
carry, and returned with them towards our encampment, expecting to hear Natty hail me as I drew near; but as I
approached the fire I could not distinguish him. I called, but
no answer came. My heart sank within me. I was afraid
that some accident had happened. Again and again I called.
Throwing down the branches, I hurried on towards the hut,
when what was my grief to see him extended on the ground
at the entrance, and some little way from the fire! I knelt
down by his side and put my hand to his heart. It beat,
though feebly. I examined him, but could find no wound.
He had swooned, apparently from exhaustion. Our water-
bottles were full, as we had replenished them at the last stream
we passed, knowing that we might afterwards have to go many
miles without finding more. His whole dress was so loose
that there was no necessity to undo any part of it; but I
sprinkled his face with water, and then poured a few drops
down his throat. Still he lay without moving.

"Natty, my dear Natty, what has happened to you? Speak to me! Speak!" I could not help exclaiming.

I had no stimulant, no medicine of any sort. I must trust, I knew, alone to Nature, or rather, I should say, to the kind Being who directs its laws. To Him I looked up and prayed that my young friend might recover. Forgetting everything but Natty, I continued kneeling, holding his head on my arm. At length, by the light of the fire, as night came on, I saw his eyes opening.

"Push on, Andrew," he whispered; "we may still keep ahead of them! I will run as fast as you do!"

I saw that his mind was wandering. Then he heaved a deep sigh.

"Are you better, Natty?" I asked.

"Oh yes. Where am I?" he asked, staring about him.

I told him I thought he had fainted, and begged that he would take some food and then lie down. He had already torn off some of the leaves from the boughs, and had made a sufficient bed for us; but, of course, we intended that only one should sleep at a time.

At length, to my great joy, he was able to sit up by himself on the ground. Finding this, I went to the fire to get the venison, which had been left roasting before it. As may be supposed, it was somewhat burned, but I was able to cut as many small slices from it as he could eat. After tasting a piece, he said, "Do you take it, Andrew. I do not think I want it." I pressed him, however; and in a little time he was able to make a tolerable meal. I then placed him inside the hut, telling him that I would sit up and keep watch till it was his turn, of course intending to let him sleep on the whole of the night, if he could do so.

I then made up the fire, finished the piece of burned venison, and sat myself down in front of the hut. I looked in several times, and was thankful at length to find that Natty was asleep. I felt a strong inclination to sleep also, and had the greatest difficulty in keeping myself awake. Whenever I felt myself nodding, I got up and walked about; but I was tired, and certainly required rest. At last I did what many a sentinel has done under similar circumstances. Though believing I was quite awake, I fell fast asleep. Even in my dreams I thought I was getting up, walking about, and then sitting down again, and then going to look in at Natty. Then I thought I made up the fire. I was somewhat surprised that it did not blaze as readily as before. By this time I was fast asleep. At length I thought I went in to look at Natty again, when what was my horror not to find him.

I awoke, to find myself leaning against the entrance at the end of our hut. The fire was very low, a few glowing embers alone remaining. The night was dark. As I looked round me, trying to open my eyes wide, what was my dismay to see numerous pairs of shining orbs gazing at me through the gloom! That they were the eyes of wild beasts I was convinced, though of what description I could not tell. The usual night sounds of an African forest alone reached my ears. The eyes seemed to be drawing nearer and nearer; and now suddenly a chorus of loud sharp barks and snarls burst forth, and by the faint light cast by the fire I could see a number of animals approaching the spot. I now guessed that they were wild dogs, a species of hyena, which hunt in packs like wolves; or perhaps true hyenas, and would prove, I dreaded, formidable assailants. Through the gloom I saw just then another body, which I guessed was a second pack arriving, thus causing the angry remonstrances of the first. A pile of firewood lay near me. I threw some of the sticks on my fire, hoping, if it blazed up, they would not attempt to pass it. My gun I had ready by my side; but as I could only kill one at a time, I was afraid, should I begin the assault, I should find it a hard matter to drive them off. I did not like to wake Natty; indeed, in his weak state he would have been of little assistance. The effect of throwing the sticks on the fire was, at first, to dull it, and I was afraid I had put it out altogether. This made the creatures draw still nearer. I rose to my feet and stood at the door of the hut, resolving, should they come, to defend my young companion to the last. If they seized me, I knew that my fate and his would be sealed. The brutes kept rushing backwards and forwards within a few yards of the fire, growling and yelping furiously. I was surprised that the noise did not awake Natty. His sleep, doubtless, was produced by utter exhaustion. I was afraid, however, that if I fired, Natty would be startled. I therefore called out to him, " Do not be alarmed when you hear the report of my rifle ! Natty ! Natty ! awake !" I called out several times. I began to fear that he was senseless, or even that worse had happened. " Natty ! my dear Natty ! what is the matter ?" I again shouted out.

The effect of my voice was what I had not expected, for my savage assailants on hearing it began to retreat to a more respectful distance. I thought that I might venture to enter the hut to see what was the matter with Natty. The brutes, however, directly I was silent, again came on. I was relieved too by hearing Natty ask, " What is it all about, Andrew ? Have you found Leo and Mango ? I have been dreaming about them so much." Greatly relieved, I replied that some wild animals were in the neighbourhood, and that I was going to fire at them; and once more I turned my face towards our enemies.

The brutes were again drawing nearer. Advancing a pace or two to the fire, I gave it a kick with my foot. This made the flames leap up. By their light I saw that a fresh actor had come upon the stage and attracted the attention of the savage brutes. A huge serpent had crawled out from among the bushes. It sprang upon one of the dogs, which immediately, writhing in agony, sank on the ground. Instead of taking to flight, however, they rushed at the creature, one of them seizing it by the back, but not before one or two others were bitten. The rest then set on it, and tearing it to pieces, quickly devoured the greater portion, leaving the head, on account I concluded of the venom it contained. Not satisfied with their victory over the snake, they once more advanced towards me with hideous growls and yelps. Seeing that it would be dangerous to allow them to approach nearer, I took aim at a large animal, which appeared to be the leader of the pack. I knocked him over, and he lay struggling on the ground yelping loudly. His companions came round him, and gave me time to reload. I did not wish to expend my ammunition uselessly, so, stooping down, I seized a burning stick, giving another poke to the fire as I did so, and then waved the brand round and round, shouting loudly in a gruff voice, and ordering the dogs to be off. Though they did not understand what I said, the tone of my voice had the effect I desired; and, greatly to my relief, barking and yelping, they scampered away, I shouting after them. The animal I had shot kicked his last as they disappeared in the gloom of the night, and I hoped that I was rid of them.

Having thrown some more sticks on the fire, I went back to Natty. I felt his hand; it appeared very feverish, and I was still more alarmed by hearing the incoherent expressions he uttered. Weary as I was, I could not venture again to go to sleep. I sat down, therefore, by the side of my poor young companion, moistening his fevered lips every now and then with water, and bathing his forehead. Still it was with the greatest difficulty I could keep my eyes open. Sometimes I got up and walked about in front of the hut, and threw a few more sticks on the fire. I myself, it must be remembered, had scarcely recovered from my illness. Having again made up the fire, increasing it to nearly double the size, I once more sat down by Natty's side. I talked aloud, and kept pinching myself, in the hope that by so doing I might keep awake. But exhausted nature at length had its way—my head dropped on my bosom, and I was asleep, so soundly indeed, that I doubt if the loudest noise would have aroused me.

In spite of my intentions, I must have had some hours' sleep. I was awaked by a bright light striking my eyes, and opening them, they were dazzled by the almost horizontal rays of the rising sun coming across the plain. My ears were assailed also by a loud barking and yelping, and I saw close to me the pack of savage dogs which had paid me a visit the night before, setting furiously on the body of their companion whom I had shot. The light of the sun had awaked me in time, or they might have made an attack on the hut before I was ready for their reception. I let them devour their companion, which they speedily did, leaving not a particle of skin or bone behind them; one running off with one piece, and one with another. The remainder, disappointed of their share of the prey, then turned their savage eyes towards me. Once more I shouted loudly, and taking off my jacket, waved it at them. Again, to my satisfaction, off the creatures scampered; and I hoped that I had seen the last of them. They had not touched the bodies of their companions bitten by the serpent, which had already become putrid. As I dragged the carcases to a distance, I felt thankful that the dogs had visited us, as, had they not come when they did, the snake might have found its way to the hut, and bitten Natty or me. I could not tell its species,. but thought that it was probably the same which had made its appearance on the island when we were escaping from the Pangwes.

CHAPTER XVIII.

N re-entering the hut I found that Natty was still
sleeping; but his slumbers were greatly troubled,
and he had evidently much fever on him. Oh, how
I wished that David had been with us; for, with all
my anxiety, I did not know how to treat him.
One thing was certain, he was utterly unable to travel.
I was unwilling even to go out of sight of the hut, lest some
wild beast might in the meantime come near it. I must do
so, however, before long, I saw; for our slender stock of water
was already almost exhausted, and cold water, I felt sure, was
absolutely necessary for him. In what direction I was most
likely to find it I could not tell. The last stream we crossed
was some distance back, and I might have to go a long way
across the plain before coming to another; indeed, in no direction did the appearance of the country indicate a stream or
fountain. This thought caused me the greatest anxiety. I
would have endured any amount of thirst, I thought, rather
than not give Natty what he required. I remembered that
the orphan boy was committed to my charge by his father,
and as a father would treat his son, so was I bound to treat
him.

After sitting by his side for some time and eating a slender breakfast, I took my gun and walked about the hut, now going in one direction, now in the other, in the hope of finding indications of water. Perhaps, I thought, I may kill a parrot or pigeon, or some other bird, which may be more palatable to him than stronger meat. I went further and further, but still could find no signs of water. While I was at the furthest point the dread seized me, that although the hut was in sight some creature might have stolen in, and I hurried back, dreading to find my fears realized. Not till I had entered the hut and knelt down by his side was I satisfied that he was safe. He was still sleeping, and I hoped he might thereby recover his strength. After sitting for some time by his side, I again got up and cut a number of boughs. These I stuck in round the entrance, so that no creature could possibly get in. I now ventured to go rather further from the hut, but could not bring myself to lose sight of the tree under which it was situated. I continued looking about for birds; for though I saw some at a distance, I could not get near enough to be certain of a shot; and as I said before, I could not venture to throw any of my ammunition away. I was beginning to feel very thirsty, and had recourse to chewing leaves, hoping that it would relieve me. It had, however, but little effect. At last, greatly out of spirits, I returned to the hut. Natty awoke as I pulled aside the boughs. He scarcely seemed to know me, however. I gave him a little water, and I thought, after taking it, he looked rather better, so I gave him more. I had been sitting by his side for some time, when I heard him whisper—" You had better go on, Andrew; I will follow by-and-by, but do not stop for me."

" That will never do," I answered, thankful to hear him speak. " You will get well shortly, and then we will go on together."

" I will try to go with you now," he said, trying to rise;

but he sank back immediately, unable to lift himself from the
ground.   He uttered a sigh on discovering his weakness.

I passed the remainder of the day as I had the commencement.   As I saw evening approaching I collected a large supply of broken branches to serve as firewood, and then made up
a semicircular heap, which I intended to keep blazing all the
night.   I was sorry that I had not slept during the day, that
I might the more easily keep awake while on my watch.   I
took some supper, though, in consequence of the thirst from
which I was suffering, I felt little disposed to eat; but still I
was unwilling to exhaust our water by drinking more than a
few drops.   I knew that the next day I must inevitably go in
search of some.   My young companion's life might depend
upon my finding it.   To avoid the risk of being surprised
should I fall asleep, after I had lighted the fire and seen that
it blazed up thoroughly, I took my seat inside the hut, and
secured the boughs as before.   In spite of my resolution to
keep awake, I had not been seated on the ground more than
an hour or so before I felt sleep stealing over me.   At length
I tried to arouse myself.   I was completely overpowered,
though I still retained a consciousness of where I was, and of
the necessity of being on my guard.   Suddenly I awoke, feeling an undefined dread.   I could hear Natty breathing, but all
was dark inside the hut.   On looking out I discovered that I
must have been asleep for some time, for the fire was entirely
extinguished.   I sprang up, leaving my gun on the ground.
My first impulse was to re-light the fire.   I hurriedly felt
about for the sticks, which I had placed on one side, and
carried them to the spot where the fire had been burning.   I
placed them as before in a semicircle.   Finding that I could
not strike the light in the open air, I retired into the hut to
do so.   Whilst thus employed I fancied I heard some creature
moving over the ground.   I got the match lighted, and then

set fire to the bundle of twigs which I had collected. With these in my hand, I went to the pile of wood. I tried to light it. At last I set it on fire in one place. I was then moving round to another, when I saw at about twenty paces off a dark object creeping slowly towards me. On it came; and while blowing away at the wood to cause it to ignite, I began to distinguish the outlines of a huge lion. In a few seconds the savage monster might be upon me. Already he was near enough, I thought, to make his spring with fatal effect. I knew that my chief safety would depend on the fire blazing up quickly. Taking the torch, therefore, and mustering all the nerve I possessed, I tried to light the pile at another spot between the two which were already beginning to burn, though feebly. Now I bent down and blew, now looked up towards the lion. To my horror, I saw him crouching down, and slowly creeping towards me. I knew he was doing so preparatory to making his tremendous spring. Just then a breeze fanned my cheek. It came stronger and stronger, and up blazed the fire. The lion stopped. Giving a stir to the fire as I passed along it, I rushed back to my hut and seized my gun. As the fire blazed up the monster gave a tremendous roar of rage and disappointment, but still held his ground. The sound awoke Natty, who asked, in a trembling voice, what was the matter. " Remain quiet," I answered. " We have an unwelcome visitor, but I hope to drive him away." Again the lion roared and lashed his tail, but he could not bring himself to dash through the fire, though he must have seen me moving about on the other side of it. I stood up with my gun, which I had loaded with a bullet, hoping to hit him should he make a spring. Still he did not move; and remembering the effects of my shouts the night before, I suddenly rushed towards the fire, kicking it about, so as to make the flames rise up more briskly than before, and at the same time shouting out at the top of my voice. The lion roared in return. The louder he roared, the more wildly I shouted and shrieked; and then, seizing a number of burning sticks, I sprang over the fire towards him. The effect was satisfactory, for, turning round, away he bounded into the darkness, whilst I shouted out, "Victory! victory!" I had heard that if lions are thus met by a bold front, they often prove cowardly; and I hoped, therefore, that my visitor would not return. I now made up the fire, and went back to Natty. I found him trembling with alarm, but in other respects far more like himself than he had been all the day. This raised my hopes of his recovery. I gave him a little water and a few mouthfuls of cassava; and I was glad to find that in a short time he again dropped off to sleep. As may be supposed, I had no inclination, after my encounter with the lion, again to close my eyes. Should Natty be better in the morning, I resolved to start off at an early hour in search of water. I was therefore thankful when the cheering light of day again returned. I gave Natty some more food, and almost the last drops of water we possessed. I had a small drinking-cup; into this I poured the remainder, and told him to husband it carefully.

"I must go out, Natty, and try and find some more," I said. "I will imprison you as securely as I can, and you must try to wait patiently till I return. I will not be absent a moment longer than I can help."

Natty looked anxiously up at me. "Is it absolutely necessary?" he asked.

"Yes, indeed," I said; "but I hope that before long I shall find what we want, and in a day or two you will be able to accompany me home."

"I will try to get well; but it is not my fault, Andrew. I would walk if I could," he said, in a faint tone.

I was not content with merely closing the entrance, but

getting some strong vines, I intertwined them round the walls, and then got some large boughs, and placed them over the whole building. I trusted that thus no animal could possibly enter. I knew that sufficient air would be obtained through the roof. All that I could do was to pray, for his sake and my own, that I might return in safety to him.

" Good-bye, Natty," I said, when I had finished the work. " Keep up your spirits, my boy. I hope soon to be back; but if I do not come as quickly as you expect, do not be alarmed. I may have to go some way for water."

My wisest course perhaps was to have gone back to the last stream we had passed; but then I could not have returned the same night to our hut, and what would poor Natty have done all that time without me? I therefore determined to push on in an opposite direction, hoping that I might meet with a fountain or rivulet. On and on I went. The sun, as he rose in the sky, grew hotter and hotter. I had not a drop of water to cool my dry tongue. I had never before really known the feeling of want of water. I had been very thirsty; but now the whole inside of my mouth and throat seemed to consist of a dry horny substance, or as if I had swallowed some of the contents of a dust-bin. Still on and on I went. I hoped by continuing in a direct course that I should obtain water more speedily.

A considerable portion of the day had passed away. The sun had attained its greatest heat, when I thought I saw in the distance a line of trees, which I felt sure indicated the presence of water. I pushed on more eagerly, but as I advanced they changed their outline, and suddenly disappeared. All I could see before me was a low line of grass and bushes, which had evidently been magnified by a mirage into the proportion of lofty trees. I went on, but continued to be deceived time after time in the same way. In every direction the

mirage danced on the plain. I found that in reality the range
of my vision was restricted to a very moderate distance. Suddenly a herd of animals appeared, lifted completely up in the
air. They were deer of some species. I hoped by killing one
that I might somewhat quench my raging thirst with its blood,
but before I had got up to where I had seen them they had
scampered off. At length I saw what I felt sure was a pool
of water. Eagerly I hurried towards it. It was a long way
off, I thought; but I was willing to go any distance for the
wished-for fluid, hoping that my sufferings would find relief,
and that I might return before nightfall to my young companion. I was confirmed in my opinion by catching sight of
several gnus going in the same direction. "They are going
there to drink," I thought; and I felt ready even to encounter
lions or any other savage beasts for the sake of the water. The
gnus did not perceive me, as they were to windward. There
was, however, so little wind that I had to wet my finger and
hold it up to discover the point from which it came. I hoped
that I should be able to get close up to the animals. Now
they stopped and fed, now they moved on again slowly.
Presently I saw them stop, when they began switching their
tails, and sniffing the air, and scraping the earth impatiently
with their hoofs. As I was concealed by the ground, which
here was sufficiently uneven for the purpose, I did not think
that they could have discovered me. Presently I was startled
by the fierce growl of some animal at no great distance. I
stopped; and looking round, I saw to my horror a huge lion
and lioness at a short way off, just above me. It was evident
that they had been following the gnus, who had only at that
instant begun to suspect their presence. The lion must at the
same time have discovered me, and uttered the roar which I
had heard, while his companion was still creeping on after the
gnus. I stopped and knelt down, holding my rifle ready to

fire should the lion approach me. Still there was the lioness, and being sure that the report of my gun would attract her even should I kill the lion, I determined not to fire till it was absolutely necessary. The growl which the lion uttered at seeing me must have been heard by the gnus, which now set off at a rapid pace to escape from their pursuers. The lioness darted forward in pursuit, and the lion, uttering a few more savage roars at me, turned round and followed her. I was free from their company for the moment, but the knowledge that they were in the neighbourhood added greatly to my anxiety. I could not help fearing, too, that they or others might find their way to poor Natty's hut during my absence. I had for the moment forgotten my thirst, but now again the sufferings I had been enduring returned, and I turned my eyes once more towards the spot where I had seen the pond. Both the gnus and the lions had disappeared. I went on, thinking that I must soon reach the water. After hurrying on till I felt ready to drop, I found myself standing on an extent of hard-baked earth, while the glittering pool I had hoped to reach had disappeared. I looked round. Similar pools appeared in various places on the very ground I had come over. I knew therefore that they were but deceptions caused by the mirage. What had become of the lions I could not tell. I only hoped that the gnus had led them a long chase, and that they were far away from me.

Wearied out, I sat down under the shade of a rock, which just sheltered my head and shoulders. My spirits were sinking. I began to fear that I had death alone to expect as a termination to my sufferings. And poor Natty, he would die too; for weak from fever, and unable to help himself, he must inevitably be starved. " I will go back and die with him," I exclaimed. " While a particle of strength remains, I will push on. I cannot let him suffer alone !" While these thoughts were passing through my mind, I saw some birds flying through the air, uttering as they went a soft melodious cry, which sounded somewhat like "Pretty dear! pretty dear!" I watched them anxiously. They were too far off for me to hit them, but I judged from their flight that they were a species of partridge which I had before seen. They came from the south-east, directing their course towards the north-west. Presently I observed, as I watched them anxiously, that they neared the ground, and then seemed to me settling down at no great distance off.

I remembered having heard that springs have been discovered by travellers in the desert by watching the flight of birds, and I hoped that these were on their way to some fountain. I arose, and hurried on as fast as my weakness would allow in the direction they had taken. Still I could not help dreading that I might be again disappointed. I caught sight at length of some rocks, on the other side of which they had disappeared. The rocks rose high above the dry, hard ground. As yet there was no indication of water. My heart sunk within me, but I persevered. I had not strength to climb the rocks, which rose high up before me, but I circled round them. I got to the other side, when my eyes were gladdened by the sight of green herbage and luxuriant shrubs, which I knew delight in water. Hurrying on, I saw beneath the rocks a calm, clear crystal pool. Oh, how delighted I felt! But on getting to the edge I found that the water was too far below me to be easily reached. I scrambled along the rocks, till at length I discovered a spot which appeared not more than a foot or two above the water. I reached it at length, and throwing myself on the ground, bent over till I could dip my hands in the pure liquid. I eagerly lapped it up. I felt that I could never drink enough. By degrees my parched tongue and mouth began to feel cool, and I rose like another person. After resting a few minutes, I filled my water-bottle. Evening was approaching, but I could not bear the thought of leaving Natty all the night without attempting to return to him. Once more I drank my fill. While drinking, I saw several other flights of birds arrive at the water. A covey of them pitched thickly on a rock near me. They would afford valuable nourishment to my young friend. I withdrew the bullet from my rifle and loaded it with small shot. It was an ungrateful act I was about to perpetrate, I confess; I thought so even at the time. The birds, too, seemed fearless of me. I raised my gun and fired. Greatly to my delight, I saw three lying on the rock, and two others fluttering near. I hurried forward to secure them. Scarcely had I done so when, looking round, I saw a lion and lioness—probably the same which had pursued the gnus—approaching the pool. Strange to say, I felt but little fear of them. Still I thought it unwise to stand in their path should they be on their way to drink, as I had no doubt they were. I accordingly scrambled along the rock to a high point, whence I could look down upon them as they passed. On seeing me they stopped, and seemed to be consulting together whether they should attack me. " I will be ready for you, old fellows," I said aloud, as I reloaded my rifle and carefully rammed down the bullet. " If you do not interfere with me, I will let you enjoy your draught unmolested ; but if you attack me, look out for the consequences—Ha ! ha ! ha ! " My own voice struck my ear as strangely loud and wild. The effect was to make the lions decide on letting me alone ; and while they went on towards the water, I scrambled down from the rock, and began to make the best of my way towards where I had left Natty.

I hurried on, though I scarcely expected to reach the hut before dark. Still I hoped that even at night I might find my way. I will not say that I was very sanguine about it, as

the mirage had deceived me, and often made objects appear
very different to what they really were. The sun in a short
time sunk behind me. Still, as long as I could move over
the ground, I determined to persevere. I was keeping,
I believed, in a direct line. At length the stars came out,
and the moon rose and shed her pale light over the scene.
I knew that lions and other wild beasts will seldom attack a
person while the moon is shining. This encouraged me to
proceed. The stars, whose brilliancy even the moon could
scarcely dim, assisted me in steering my course. I own, however, that now and then I cast an anxious look over my
shoulder, lest the lion and the lioness might be following me.
Where the ground was open I hoped that I might be able to
discover them, should they approach; but in some places it
was rocky, scattered over with thick bushes, within which
beasts of prey might lurk. I was somewhat heavily laden,
with my water-bottle and birds. While suffering from thirst
I had no inclination to eat, but now I began to feel the pangs
of hunger, and my knees trembled from the exertion I had
been for so long making. I therefore sat down with my back
against a tree to rest, and to eat the few mouthfuls I had in
my pocket. I scarcely knew till then how tired I was. Anxious
as I was to get on, I yet could not help indulging in a short
rest. " I shall be able to move the faster after it," I thought
to myself. Whilst thus sitting and meditating, what was my
dismay to see the two lions stalk slowly up to me, while behind them appeared a vast troop of the savage dogs I had encountered on the previous night ! I felt spell-bound—unable
to fly, or even to move. The lions whisked their tails and
ground their teeth as they uttered low savage growls, while
the dogs kept barking and yelping behind them. Nearer and
nearer they drew. In vain I tried to lift my rifle and have
one shot for my life. No ; I could not even do that. There I

sat. In another moment their sharp fangs would be planted in my throat. Suddenly I gave a start. The whole panorama of savage eyes and the two central monsters disappeared, and to my infinite relief I found that I had been asleep, and that the whole was a phantom of my brain. I really think I must have slept some time, for after I had recovered from the alarm into which my dream had thrown me, I felt sufficiently strong to resume my journey. As long as the moon shone, it was far pleasanter travelling at night than in the day. Again I went on, but still I could not help acknowledging to myself that I might very likely after all not be on the right road. Still I should not gain it by hesitation, and I tried to make up my mind to be prepared for a disappointment, should I be mistaken. I was doing my best; I could do no more. At length I saw in the distance a line of dark trees, which I hoped was the wood on the borders of which our hut was situated. As I marked the outline, I stepped on with more elastic tread, thinking of the delight my reappearance would give my poor young companion. As I was thus walking on, I felt my foot sink into the earth, and before I could recover myself I fell flat on my face. I quickly sprang up, for the thought seized me that I might have stepped into the hole of some snake, and that in another instant he might be issuing out to attack me. I ran on for some paces, when I stopped and looked back, but nothing appeared. Not till then did I discover that I had sprained my ankle. It might be a slight matter under ordinary circumstances, but, in my case, if it stopped my walking it might be serious. It pained me considerably, still I found that I could walk. I went on, but soon began to limp. There was no elasticity in my step now. My great consolation was that I was near Natty, for I was sure the wood I saw was that I had left in the morning. The pain had damped my spirits, and I now began to fear that perhaps after all Natty had grown worse, or that some wild beasts had found out our hut, and managed to penetrate into the interior. I was wrong to allow these thoughts to enter my mind, I know, but under my circumstances it was but natural. At length I caught sight, under a tree, of what in the moonlight looked like a mound. It was our hut; but just then I observed several objects moving about round it, and as I drew near a loud barking and yelping saluted my ears. I rushed forward. "Those brutes of dogs have found out Natty!" I exclaimed. Even then I thought that I might be too late to save him. Shouting out in a stern, strong voice, which I had found successful before, I ordered them to depart, waving my gun with furious gestures before me. The dogs saw me, and began to retreat; but some of them, I thought, seemed to come out of the very hut itself. "Natty! Natty!" I cried out, "are you safe? Tell me! oh, tell me!"

I got no answer, but the barking and yelping might have drowned Natty's voice. I dashed frantically forward. I could not fire without the risk of sending the ball through the hut. I doubted, indeed, whether the sound of my rifle would have much effect on them. The yelping, barking pack retired as I advanced. "Natty, Natty, speak to me!" I again cried out.

My heart bounded with joy when a faint voice proceeded from within. "O Andrew! have you really come? I was afraid you must have been killed."

"I am all safe," I answered; "but I must drive these brutes to a distance before I come to you."

There was a good supply of sticks. I hastily drew them together, and lighting a match, quickly had a brisk fire burning. The light and my shouts finally drove off the pack, and I now ventured to open the entrance to our hut. Natty was sitting up. He pointed to his mouth. I hastily poured out some of the water, and gave him an ample draught; and then

I sank down on the ground, overcome with fatigue and the pain which my sprained ankle gave me. I recovered sufficiently, however, to exchange a few sentences with him, when he told me of the anxiety he had been suffering, and of the dread he had had that the dogs would force their way into the hut. I then briefly narrated my adventures. He seemed, I thought, somewhat better. Having secured the entrance, I lay down by his side, and, in spite of the pain I was suffering from, was soon asleep.

CHAPTER XIX

MY ADVENTURES WITH NATTY CONTINUED.

WHOLE day had passed away. Although I husbanded our water with the greatest care, I could not expect it to last beyond a second day. Still my ankle gave me great pain, and I felt utterly unable to walk. Natty, too, was far too weak to proceed on our journey. The fever, however, had subsided, and he required less water than at first. Still, it was almost as necessary for him as food, and I did not like to stint him. Though suffering from thirst myself, aggravated by pain, I refrained from taking more than a few drops at a time. I did everything I could think of to restore strength to my limb.

"I am afraid there is only one thing, Andrew, will do it; and that is perfect rest," observed Natty at last.

I did not like to alarm him by telling him of my anxiety about water; but as I sat on the ground with my poor sick friend by my side, darker forebodings than had ever yet assailed me oppressed my mind. It might be many days before Natty would be able to move, and if I could not go to the fountain to procure water, we must both die of thirst.

Two more days passed away, and when I lay down to sleep, scarcely a pint of water remained. I had remained perfectly quiet all day, hoping that the long rest would cure the sprain. I had made the hut so secure, I did not think it necessary to light a fire outside. On again rising, I put my foot to the ground. Oh, how thankful I felt when I found that it gave me but little pain, and that I could walk without difficulty! I told Natty that I would go back at once for water, leaving him our scanty stock, and the remainder of our birds after I had satisfied my hunger. The flesh, however, though roasted and dried, was scarcely eatable.

"Will you not let me go with you, Andrew?" he said. "I think I could walk as far, if I rested now and then."

He made the attempt, but sank back again on the ground. I persuaded him to have patience, and to remain quiet; and closing the hut even more carefully than before, with the thickest sticks I could find, I set off on my expedition. Though at first I walked with pain, I got on better than I expected. The air was cool, for the sun was not yet above the horizon, and I hoped to get to the fountain in time to kill some birds collected there for their morning draught. The way, I trusted, would appear shorter than at night, and I believed that I well knew the direction I should take. My feet were, however, very weary, and the rocks were not yet in sight. I was weak from want of food, and soon became as thirsty as on the previous occasion. I was anxious, too, for I could not be quite certain that I was on the right way. How I longed for a beaten track which would lead me without fail to the fountain! It would have made all the difference to me. I could have endured double the fatigue had I been sure that I should arrive at the spot at last. At length I caught sight of a flight of birds winging their way over my head in the direction I was going. This gave me more confidence, and I now pushed on with greater energy. At length I saw the rocks before me, and flights of birds rising in the air, and flying off in different directions. I was afraid that I should be too late to shoot any; though I might obtain water, food would be wanting. Just as I reached the rocks, I saw a covey apparently about to take wing. I fired, and four lay on the rocks fluttering about. I rushed forward to seize them, when, to my horror, I saw my old enemies the lion and lioness just taking their departure from the water! I had already got some way up the rock. It was better to lose the birds than my life; so I stopped, faced my foes, and began loading my rifle. The brutes looked at me with astonishment, as much as to ask how I dared come into their territory again. I replied by ramming down the bullet. "If you will go your way, I will let you alone," I shouted out; "but if not, beware of this leaden pill!" The lion seemed to understand me, and looked at the lioness; and then, perhaps considering discretion the better part of valour, began leisurely to walk away from the fountain. I shouted after them, to show them that I was not alarmed; and, greatly to my satisfaction, they at length disappeared in the distance. I secured the birds, which were unable to fly, and then eagerly hurried down to the water. I drank my fill, and sitting down, bathed my burning feet. The water seemed to give strength to my ankle. Having filled my bottle and rested a while, I felt so much better that I determined to take a swim, hoping thus entirely to recruit my strength. Never have I so much enjoyed a bath. On getting out, however, I felt so hungry that I was compelled to light a fire and cook one of the birds. I could not have proceeded on my journey without it, though anxious to get back as soon as possible to Natty. Thus thoroughly recruited, I again set off, looking about, as I went along, in the hope of finding some other animal to shoot for food. Though I saw many at a distance, I could not get sufficiently near one to have a fair shot.

It was late in the day before I got back, and when I shouted to Natty, as I drew near the hut, he answered me in a stronger voice than before. I soon had the bottle of water to his lips, a fire alight, and a partridge cooking. Enough of the day remained to allow me to search about for wild fruits and roots which might assist our meal. I could now leave him without fear; invariably, however, closing the hut when I went out. I was successful in finding some fruits such as I have before described, and returned well satisfied to the hut. Natty declared that he felt able to sit outside by the fire to take his supper. He crawled without my assistance to the entrance. After he had taken his seat, as I happened to look inside, I saw the leaves on which he had been lying moving slowly. Presently the hideous, black, swollen-looking head of a snake emerged from under the leaves, its bright eyes glaring at us. In another instant I believed that it would spring at Natty or me. Without speaking, greatly to his alarm, I threw him on one side, and then, seizing a heavy stick which lay at hand, I rushed at the creature and struck it a blow with all my force on the head. It had the effect of knocking it over; and before it could recover itself, I dealt it another blow on the tail. Poor Natty, not seeing what I was doing, thought I had gone mad, I believe. I repeated my blows, till I felt sure that the creature was dead. I now dragged it out by the tail, prepared, should it give signs of life, to renew my attack. As I brought it into the light, I saw that it was a black variety of the puff adder, which is among the most poisonous serpents of Africa. It is said that if a person is bitten by it, death ensues within an hour. To make sure, I threw the body into the fire. Not till then did Natty sufficiently recover the effects of his fall and alarm to see what had occurred, and to be aware of the fearful danger in which we had both been placed; for had the creature come out while we were sitting together in the hut, unable to

defend ourselves in so narrow a space, nothing could have pro-
vented one of us being bitten.

We sat for some time before we could begin our meal, and
we did not fail to return thanks for our merciful deliverance
from danger. We naturally talked about what we should have
done had either of us been bitten. It was a subject which I
had discussed with David on several occasions, for we had had a
great fear of the bites of serpents when we first arrived in the
country. However, we had hitherto met so few, that we had
lost all alarm about them.

"If you had been bitten, I should have tried to cut away
the flesh immediately round the wound, and sucked the blood,"
Natty said to me; and from the look of affection he gave me,
I was sure that he would without hesitation have made the
attempt.

"I should have first tied a ligature above the wounded part,
so as to prevent the venom spreading," I observed. "Had we
been with David, we might have found remedies in his medicine-chest. It is said that *eau de luce* is often effectual. Five
drops are administered to the patient in a glass of water every
ten minutes till the poison is counteracted. It is also applied
externally. I have heard that Dutch farmers attempt to counteract the effects of serpents' bites by making an incision in
the breast of a living fowl, and applying it to the bitten part.
If the poison is very deadly, the bird becomes drowsy, droops
its head, and dies. It is then replaced by a second, and so
on till the bird no longer shows signs of suffering, when the
patient is considered out of danger. A frog is sometimes applied in the same way; and turtle blood, prepared by drying,
when applied to the wound produced by a venomous serpent
or a poisoned arrow, is supposed to be efficacious. The
wounded person takes a couple of pinches of the dried blood
internally, and also applies some of it to the wound. It is

said also that the Brahmins in India manufacture a stone which has the virtue of counteracting the poison of serpents. They alone possess the secret, which they will not divulge. The stone is applied to the wound, to which it sticks closely without any bandage, and drinks in the poison till it can receive no more. It is then placed in milk, that it may purge itself of the poison, and is again applied to the wound, till it has drawn out the whole of the poison."

"Yes," observed Natty, "I remember hearing of those stories; but David said they were merely pieces of the bone of some animal, made into an oval shape, and burned round the edges. If they have any power in drawing out poison, it is in consequence of being porous; and he said he believed any substance made up of capillary tubes, such as common sponge, would be equally efficacious. After all, I believe that my remedy is the only one on which dependence can be placed, except, perhaps, the immediate application of *eau de luce*, and of course, when a person is bitten by a snake, in rare instances only is he able to obtain any."

As may be supposed, we hunted about the hut thoroughly before lying down, in case any other snakes might have crawled in; and I stopped up every crevice by which I thought it possible the one I had killed could have entered.

Natty was so much better by the time our last supply of water was nearly finished, that I no longer refused to let him accompany me to the fountain, intending to proceed from thence towards our ultimate destination. Clouds had gathered in the sky, and the air was cooler than it had been for some time, as we set out. I insisted on his frequently stopping, and wished him to allow me to carry him at intervals; but to this he would not consent. We each of us had a long stick in our hands to support our steps, and I assisted him on with my arm. Our progress was, however, but slow; for in spite of his efforts, I saw that he was still very weak. Thus it was not until the sun was already sinking before us in the west that we got within sight of the fountain. We had exhausted our water, and I was anxious to get a further supply before the night closed in. Again I begged Natty to let me take him on my back, for I thought it would rest him, and enable us to get on faster. At last he consented, and though he was but a light weight for his age, reduced as he was by sickness, yet I found, after proceeding a couple of hundred yards or so, that I was myself beginning to get fatigued. Perhaps he discovered this, by finding the slower pace at which I was going, and he insisted on again getting down, declaring that he was much rested by the ride. Giving him my arm, therefore, we again pushed on. The dark rocks which surrounded the fountain now rose up clearly before us. I looked round carefully, but could see no trace of the lions. We reached the spot, and soon I had the satisfaction of seeing Natty swallowing an ample draught of water. I then took some myself, and filled our bottle. I felt a longing to take another swim, but afraid that the lions might come upon us while I was in the water, I refrained. I was fortunate in killing five more birds, out of a covey which rose just as we sat down by the water's brink.

Having rested for some time by the side of the pond, we continued our journey. We saw herds of animals in the distance—gemsboks, steinboks, gnus, and cameleopards—but they were too far off to enable me to get a shot at any of them. We stopped frequently, for Natty was unable to proceed without doing so. Thus the day had come nearly to a close before we had made much progress. I was looking out anxiously for some spot where we might camp for the night, when I saw on our right what appeared to be the fallen trunk of some giant of a former forest, for no other trees were near it.

" I dare say we may there find shelter," I observed, pointing it out to Natty.

" But see!" he said, " there are some animals moving about round it."

As we got nearer, I saw several heads rising among the roots and fallen branches. They appeared to me to be hyenas, or hyena dogs, similar to the pack which had visited us. They, however, with their ears pricked forward, were so eager in watching some object on the opposite side of them that they did not perceive us. We were thus able to move on without being discovered. Presently we perceived what had occupied their attention ; for the leaders of a herd of buffaloes appeared in sight, going along a shallow valley on the other side of the fallen tree. Even at that distance we could hear the hollow sound of their feet as they dashed over the ground. On they went with their heads lowered, and tails in the air, faster and faster, a regular stampede. What had caused their flight we could not ascertain. Whether it was alarm at some danger behind them, or whether they were driven by an impulse which sometimes makes the bovine race dash headlong over the ground without any apparent cause, we could not tell.

" One thing I am very thankful for," observed Natty,—" that we are not in their way, or we should have but a poor chance of escaping them. Perhaps the dogs expect one of them to fall, and are looking out for a feast."

" At all events, we must take care not to allow ourselves to be attacked instead of them," I observed. " I am far from certain indeed that they are dogs. They appear to me larger, and rather more like hyenas. I suspect that they are spotted hyenas, which are among the fiercest of the race ; and though I believe they seldom attack a man on his guard, I do not know what they might do if they found us asleep. They are said to have an especial liking for human flesh, and I know that in

some parts where they are numerous, they frequently carry off
the children from villages. I have heard it said that they
will even steal noiselessly into a hut at night, and drag a
sleeping child from under its mother's kaross or rug, so that
the first intimation she has of what has occurred is from the
cry of her infant as it is borne away in the jaws of the monster.
They will sometimes break into villages, leaping over high pal-
ings; and so great is their strength, that they will carry off
any animal they find loose. In one respect, however, they
are of use, as they act as scavengers, and clear the neighbourhood of villages of the carrion which they find scattered
about. This makes it necessary to protect graves, by raising
over them piles of thorns, or of the prickly pear, as they will
otherwise scrape away the earth to reach the newly-interred
corpse."

"Horrid creatures!" said Natty, shuddering. "I do not
think I could go to sleep if I thought that any were likely to
pay us a visit."

"I do not know that they would be more formidable than
the dogs we have already encountered," I remarked. "Indeed,
I believe these dogs are their cousins, if not their brethren;
for though complete dogs, as to the character of their skulls
and teeth, they have, like the hyenas, only four toes on the
front feet. However, I hope we may be able to take precau-
tions which will guard us from any annoyance those brutes out
there are likely to offer, should they be hyenas or simply
hyena dogs, such as the visitors to our late camp. There is
a wood, I see, on the left; we must try and push through it,
and build our house on the other side."

On went the herd of buffaloes, and were soon lost to sight
across the plain. As we went on, I looked back every now
and then to see if the hyenas were following us; but though
I fancied that their heads were turned in our direction, they

perhaps could not make out what we were, and at all events remained in their fortress. I should have preferred, however, being further off from them at night. While preparing our camp, the sky gave indications, I feared, of a coming storm. I therefore made the roof of our hut thicker than usual, in the hope of keeping out the water should the rain come down. In spite of my fears, neither did the storm break, nor did we receive a visit during the night from our canine neighbours. Natty was greatly fatigued by his long journey; and from the way he talked in his sleep, I was afraid that the fever had again returned on him. This made me resolve, should he not be better in the morning, to remain there another day. My worst apprehensions were fulfilled. But still it was satisfactory to be near the water, so that I might obtain as much as we required.

We remained two whole days. Though we several times heard the roars of the lions, I did not see them. Each day I made a trip to the pool, and took a refreshing bath, which greatly restored my strength. Natty declared that he was now ready to proceed. Having obtained in the evening some more birds from my preserve, as I called it, we went on in the morning in the same direction as before. Natty, however, was still very weak, and I saw that the next day we should make but little progress. We were now again in a completely open plain, the only trees being far away in the horizon, though the mountains rose up in the north-west, towards which we were proceeding. The signs of a storm again appeared, and I was afraid that it would break upon us where no shelter could be obtained. Push on therefore we must, as long as Natty could continue moving. I gave him a lift every now and then, very much against his will; indeed, it was only by persuading him that we could thus get on faster, that he would allow me to carry him. Soon the wind began to blow in fitful gusts, and heavy drops of rain fell. I constantly looked behind me, dreading every instant that the deluge would burst upon us. " It will kill poor Natty, I fear," I could not help saying to myself. Presently the rain began to descend more heavily, and clouds collected, and flashes of lightning darting from them went zigzagging over the ground. Just then I caught sight in the distance of what looked like a low clump of trees. I directed our course towards it, taking Natty up and running along as fast as I could move. Although I well knew that it is dangerous to take shelter in a thunderstorm under a tree, I hoped to be able to obtain wood and leaves to build a hut by which Natty, at all events, might be partly sheltered. I saw, as I got nearer, that the grove consisted chiefly of one enormous tree, from the branches of which descended numerous slight stalks, apparently supporting them as they spread out on every side over the ground. I now recog-nized a magnificent specimen of the baobab-tree, of immense girth, and with numerous branches and almost countless off-shoots. On one side was a Guinea-palm, its graceful fan-like branches rising from a centre stalk—a mere liliputian plant it looked in comparison to its lofty neighbour. On the other side was an acacia, the size of an ordinary oak, though a little way off I took it for a diminutive shrub. A very few other trees only were scattered about. Getting still nearer, I ob-served a hollow in the trunk of the baobab-tree—a wooden cavern, capable of containing a dozen or more persons. Re-membering to have heard that the baobab does not attract lightning, I made my way towards it, resolving to take shelter within. I hurried to the mouth, and looking in, was thankful to find that it contained no inhabitants. Here, at all events, we might rest secure from the storm.

Putting Natty down, I examined the interior to see that no snakes lurked in the crevices of the wood. I could discover none : so I cleared out a spot where Natty could rest more at ease ; and as the wood and leaves under the tree were still dry, I collected a sufficient supply of both—one to form our couch, and the other for our fire. The rain had begun to pour down in torrents outside, but within the trunk we were completely sheltered. As there was ample room to light a fire inside, I soon had one, and some of our birds roasting before it. Natty agreed that we were better lodged than we had been since we left home. There we sat watching the storm, which howled and raged outside. The rain came down literally in a deluge.

The tree in which we had taken shelter was evidently of great age. I have since heard that some people suppose that the patriarchs of these trees may have been alive before the Flood. The natives cut off and pound the bark, from which they thus obtain the fibres for making a strong and fine cord. Although the bark of many of the trees near their villages is completely torn off in a way that would destroy any other tree, the baobab does not suffer, but throws out a new bark as often as the old one is cut off. Trees are either exogenous — that is to say, grow by means of successive layers on the outside ; or they endogenous—which means that they are increased by layers in the inside. Thus, in the latter, when the hollow is full the growth is stopped, and the tree dies. The first class suffers most severely by any injury affecting the bark ; the second, by an injury in the inside. Now the baobab, from possessing all these qualities, may have the bark torn off, and may be completely hollow, and yet continue to flourish. The cause of this is, that each of the lamina possesses a vitality of its own, the sap rising through every part of it. I had seen some trees, from which the natives had so often stripped the bark that the lower part was two or three inches in diameter less than the higher portion which they could not reach. The wood was of a particularly spongy and soft nature; and I was able to cut off enough with my knife to assist in keeping our fire burning.

The storm still continued raging without, the wind howling among the branches above our heads, although we sat secure as in a mansion of granite. I was not free, however, from anxiety; for it occurred to me that I might be mistaken as to the tree we were in not attracting the lightning, and that the account I had heard about it might be incorrect. I did not, however, express my misgivings to Natty. He, poor lad, looked very pale and ill, and I regretted having allowed him to walk so far; indeed, I felt it would have been better to have remained at our former abode a couple of days more, or even longer, although it might have made one or more journeys to the fountain necessary. I determined, therefore, to secure the entrance, and make the inside of the tree as comfortable as I could for him, and to remain there till he was better able to proceed.

The rain continued to come down in torrents; the thunder roared, and the lightning flashed vividly. I was afraid that the fine weather was breaking up, and that the rainy season was about to begin. This would make travelling more diffi-cult than before, and give Natty less chance of recovery. I made up my mind, however, to be resigned to whatever might occur, and to do my best. Courageous as Natty generally was, he at length became alarmed at the loud roaring of the thunder, and the fearful crashing sound which ever and anon reached our ears as the electric fluid, darting from the clouds, came zigzagging through the air, and snake-like darted over the ground, sometimes, it seemed, within a few yards of the tree. I did my best to reassure him, and was thankful that it was daylight, for the storm would have appeared even more terrific at night.

Although there were no large inhabitants in our woody cavern, I discovered several insects. The ground inside it was covered with earth, and almost level. I observed a large reddish spider running in and out with wonderful rapidity among the uneven parts of the wood. Now it darted out on a small insect, and quickly devoured it ; immediately setting forth again in search of another, which it pounced upon in the same energetic way. I had seldom seen so large and hideous-looking a spider, and felt a horror lest it should come near us. It moved so quickly that I in vain attempted to reach it. Presently I saw it run along the ground, when it entered a small hole which I had not before observed. Though I had exactly marked the spot, I in vain searched for it. After a time I saw the earth lifting, and out came the spider again. I sprang down to the spot, and there I found a small circular substance, of a pure, silky white, like paper, about the size of a shilling. On touching it, I discovered that it was a regular trap-door with a hinge, and, on turning it down, that the outside was coated with earth, so exactly like that in which the hole was made, that when shut it was impossible to discover it. I observed inside a substance which I took to be eggs; and I had little doubt, therefore, that this was the nest of the spider I had seen. I pointed it out to Natty, who was, however, too weak to feel inclined to rise and examine it ; and when I again looked, I could nowhere discover the hole, and the spider had disappeared. I could not help having an uncomfortable feeling that the creature might come out again and attack us. But I may as well say here that it did not do so ; and on making inquiries since, I found that though people are often frightened at its appearance, it has never been known to do any harm. There is another spider which builds a regular nest with a lid, and attaches it to a wall or the branch of a tree. Whether it is of the same species as the one I have described or not, I am uncertain.   There are spiders in Africa which are said to inflict poisonous wounds.   One is a very large, black, hairy creature, fully an inch and a quarter long, and three-quarters of an inch broad.   It has a process at the end of its front claws like that of a scorpion's tail, out of which poison, when it is pressed, is seen to ooze.   I have also observed another spider, which can leap a distance of several inches on its prey.   When alarmed by my approach, I have seen one spring nearly a foot away.

The thunder, which had for some time been roaring louder and louder, at length gradually began to grow less frequent and more faint, and by degrees rolled further and further away, though its mutterings were still heard in the distance. The rain ceased, and the bright rays of the western sun penetrated beneath the wide-spreading branches of our baobab-tree.   The change raised my spirits, and the air already felt cooler and more refreshing.

CHAPTER XX.

THE bright rays of the sun, which streamed into the hollow tree, had a good effect upon Natty ; and feeling that I could leave him, I proposed cutting some stakes with which to secure ourselves during the night from the attacks which wandering beasts of prey might be inclined to make on us. Taking my hatchet, I accordingly went out and set to work. I easily cut a sufficient number of stakes for the purpose from the branches of the neighbouring trees. I should have been better off with a good supply of nails ; but as they were wanting, I had to do without them. Pointing the stakes, I drove them into the ground just inside the mouth of the hollow, placing other pieces crossways, and jamming them as I best could into the sides of the entrance. I left only a small hole, through which I could just creep in and out. I made the grating so high that I hoped no panther or lion could leap over it. I had gone to the outer edge of the grove to get some firewood, and was returning by a path through which I had not yet passed, it being already dusk, when suddenly I found my face covered with what I can only describe as a long veil; while just at my nose I saw a horrid monster, of a bright yellow colour, with long legs and claws, struggling violently, and in its fright I thought it would scratch out my eyes. I rushed forward, throwing down my load, and dashing into our cavern, entreated Natty to relieve me from my fearful tormentor. Even he, ill as he was, could scarcely help laughing at my alarmed countenance. The spider—for such the creature was—was as much frightened as I was, and crawled away in a great hurry before we could kill him, the instant Natty had assisted me in tearing off part of its web. It took some time to clear my face of the remainder, and several minutes passed before I could entirely recover my equanimity. I had seen such webs before, but had never run tilt against them. This was suspended between two of the stalks of the baobab-tree, in a perpendicular position, by lines the thickness of coarse thread. The fibres of which it was composed radiated from a central point, where the creature was lying in wait for its prey, when it found the tip of my nose instead of an unwary moth or butterfly. The web was about a yard in diameter, so that it completely enveloped my face and head. Though very disagreeable to me, the occurrence, I really believe, did Natty good. It was pleasant to hear even a faint shout of laughter from him.

The spider I have mentioned is a solitary individual : but I have seen others which live in society ; and industrious creatures they are, too, for their webs frequently cover the entire trunk of a tree, so as literally to conceal it from view. I have seen a bush in the same way completely covered up, as if a table-cloth had been thrown over it.

I was thankful we had so secure a house, for I saw that Natty could not possibly proceed for some time. I therefore made up my mind to remain where we were till he was better, even though it might involve the delay of a whole week. My chief anxiety arose from the small amount of ammunition I now possessed. Should that fail me, I could not tell how I might obtain food. Water I had in abundance; that was one comfort. The immediate neighbourhood of the baobab-tree afforded neither roots nor fruits; so even as it was I must visit the fountain, or go to a yet further distance, to obtain food. Notwithstanding the interruption I have described, I had time to collect some leaves for Natty's bed, and a supply of fire-wood, in case I might find it necessary to light a fire.

Several times during the night the distant roars of lions and other wild beasts reached my ears; but as none were near, I went to sleep without any unusual feeling of anxiety. In the morning, however, I found the marks of a lion's feet in the soil, made soft by the rain, just outside the tree. Probably he had come up to our sleeping-place; but, finding the entrance barred against him, had not attempted to make his way in. I was thankful that I had guarded it securely.

I am obliged to make a long story short. Three days passed by, during which there was a storm and a fall of rain. I went to the fountain for water, and shot more birds, and made expeditions in the neighbourhood of the grove; but Natty continued so weak that I did not like to leave him for any length of time by himself. I was one day attracted by a mound a little way off, which I suspected to be an ant-hill. On approaching it, I found that such was the case; but it was ornamented in such a way as I had never seen one of those curious nests adorned before. It was covered with enormous mushrooms. They were perfectly white, their tops nearly eighteen inches in diameter. They looked very tempting; and on examining them, I found that they were genuine mushrooms. I ate a piece, which was very palatable, and I accordingly slung several over my back to carry home: they would, I hoped, prove useful to eat with our roasted partridges. Not far off was another ant-hill, and on this were growing a number of other mushrooms. Some were of a brilliant red,

and others of a dull light blue. I examined them; but from
their consistency and general appearance, I was afraid of eating them lest they might prove poisonous, for such I knew is
the character ordinarily of coloured fungi. I carried a couple
home, however, to show to Natty; but he agreed with me that
it would be unwise to eat them.

Another day, when further from home than usual, I saw
before me a lagoon, in which water-plants were already rising
up. I was convinced, however, that it had only been filled by
the late rains. From its appearance, it was probably not more
than a few inches deep in any part. As I passed by I
observed some odd-looking black lumps on the top of some
tall stalks of grass, which rose above the level of the surrounding edges. I was tempted by curiosity to examine one of
them. It was about the size of my thumb; and as I held it
it broke, when what was my surprise to see emerge from it a
whole army of ants, which began to attack me furiously! I
brushed them quickly off, though their bite was not particularly severe. On examining others of the black lumps, I
found them inhabited in the same way; and I now came to
the conclusion that the ants which had their usual abodes in
the dry season underground on the spot, taught by experience
that at a certain season it would be covered by water, built
these aerial abodes in order to secure for themselves a refuge
as soon as the waters should flood the ground around them.
Many of these houses were as large as I have described, but
others were considerably smaller, though all built of the same
material and in the same firm manner. Taking up one by the
stalk, I carried it home to show to Natty. He declared that
he thought some of our black friends would swallow them,
if baked, as a delicious mouthful. I carried it out again, and
stuck the stalk in the ground, when I saw the inhabitants
crawling down, evidently under the belief that the waters had

subsided, and that they might now descend into their subterranean habitation.

I need scarcely say that I looked out anxiously all the day in the hope that Stanley or some of our other friends might pass in that direction on a hunting expedition. Natty asked how it was they had not come to look for us. I accounted for it from their naturally supposing that if we had not lost our lives, we were detained somewhere on the lake, and that they would therefore search for us there.

Natty grew no worse, but still he did not appear to gain strength. Often he urged me to set off without him; but to this I would not consent. The journey might occupy me two or even three days, and it would take as long a time to return to him. "No," I replied; "until you are well enough to move, I will stay by you." I thought that if I could but procure some variety of food, he might improve faster; but I had now only five or six charges of powder left, and I was anxious to preserve these for any emergency. One of my fears was, that from so frequently shooting the birds in the neighbourhood of the pool, they might grow wary of me. However, they did not appear to be more alarmed when I came near them than at first. Sometimes I went in the evening, sometimes in the morning, and never failed to bring down three or four birds. I think that I must have frightened away the lions, for I never saw them again, though I heard their roars in the distance. I suspect that they waited to visit the pool till they saw me take my departure.

I was one day about half a mile from the baobab-tree, when I saw, perched on a bush near me, a little bird about the size of a chaffinch, of a light gray colour. It seemed in no way afraid of me, but continued chattering and twittering in a state of great excitement. Then it got up and flew backwards and forwards before me, apparently endeavouring to attract my

attention. As I approached it flew on a little in front. I
followed it. On seeing this, it went on and on in a wavy
course, a few yards before me, alighting every now and then
on a bush, and looking back to see if I was still following, all
the time keeping up an incessant twitter. Though I had no
idea at the time of its object, I continued following it. At
length I saw a short distance ahead the huge trunk of a fallen
tree. The bird appeared still more excited; and when I happened to turn aside, apparently to take an opposite direction, it
came flying back, and twittering louder than before, trying, I
was sure, to make me turn in the direction of the tree. I
accordingly did so, when, satisfied, the bird went on as before.
It now hovered for a moment over a part of the trunk at which
it pointed with its bill, and it then turned and pitched on
the top of a decayed branch which rose in the air out of the
trunk, and fluttered its wings and twittered still more violently
than ever. There it sat while I examined the trunk. I was
not long in discovering a hollow surrounded by wax, and the
idea at once occurred to me that this was a bees' nest, and that
the bird was the honey-bird of which I had heard. On a
further examination I was convinced that I was right. I therefore collected a number of dried leaves and twigs, in order to
light a fire, and with the smoke to drive the bees from their
habitation. I also manufactured some torches, which might
assist me in the operation, and would, I hoped, enable me
to defend myself should the bees take to flight and attack
me. As soon as I had got everything ready, I lighted a fire
under the nest, and taking a torch, waved it about in front of
it. No bees came out, and I began to fancy that the nest must
be empty. After a time, however, on looking in, I found that
the effect of the smoke had been to stupify the bees. I therefore, without fear, began to cut out the nest. It consisted
of cells of wax full of honey. The difficulty was to carry it.

However, as the wax was tolerably hard, I tied it up in a large handkerchief I fortunately had in my pocket, in which I hoped at all events to be able to carry home a good quantity of honey for poor Natty, trusting that it would be beneficial to his health. While employed in putting it up, I observed the honey-bird fluttering about in a state of great agitation close to me. " Oh, I almost forgot you," I said, turning to the bird. " You deserve some honey ; " and accordingly, taking some from the nest, I placed it on the trunk of the fallen tree. Instantly the bird dashed down, and began eating it with evident delight. As soon as he had finished the portion I had bestowed on him, he rose and began fluttering about as before in front of me. I whistled to him, to try and induce him to come with me, but I have since heard that whistling encourages the bird, and makes him more eager to go off in search of another nest " As you will not come with me, I must go and see what you want now," I said to the bird, following the way he led. In vain I whistled. On he went in a wavy course, as before, directly in front of me. I rather doubted, however, should he lead me to another honeycomb, whether I could carry it. Still, I did not like to miss the opportunity of obtaining what might prove so valuable. I therefore went on in the direction the honey-bird led. I could not help thinking of tales I had read in my boyhood of kind fairies or good spirits leading travellers who had lost their way to some enchanted castle, where a comfortable couch and an ample banquet was prepared for them. Perhaps the honey-bird may have been the origin of such tales. Sometimes, indeed, an evil fairy has appeared, and beguiled thoughtless travellers to their destruction. After the conduct of my honey-bird I had no doubt about his good intentions. I had gone on for twenty minutes or more, when the bird pitched on the bough of another decayed tree still standing upright. Seeing me approach, it began fluttering about, and pointing its beak towards a hole some way above my head. " I should have thought you might have known I could not reach that," I said, looking up at him. " However, I will do my best to accomplish the feat." The quickest way, I thought, would be to build a platform on which to stand whilst cut- ting out the honey. I accordingly chopped down some stout poles and drove them into the earth, securing cross-pieces with vines to the trunk. I thus formed an erection similar to a builder's scaffolding, and now climbing to the top, I made another small platform directly under the entrance to the nest. I then proceeded as before, by burning leaves and twigs, and having thoroughly smoked the unfortunate bees, took posses- sion of their habitation and store of food. With this further supply I descended, and having given the honey-bird a share, put the remainder into the handkerchief. I had to make it more capacious, by fastening a number of vines round it, so as to form a sort of basket. " Well, Master Honey-bird, if you will lead me to another nest, I think I could manage to carry it in this fashion," I said to my little conductor, who seemed to understand me, and off he flew as merrily as before. This time he did not appear quite so steady in his course. Sud- denly he made his way towards a small wood which I saw in the distance. I followed him, and every now and then he stopped and looked back to see if I was coming. It was a tiring walk, for the sun struck down with unusual heat after the rain, and I began to think that I should have acted more wisely had I returned at once with my sweet stores. Still, I did not wish to disappoint the honey-bird, as I was in hopes he would on another day be on the look-out for me, and help me to get a further quantity when we might need it.

At last the wood was reached, when, making his way into it, I saw him pitch on a bough as before; but the trees were small, and I could see none round likely to contain a cavity in which bees would have formed a nest. Still, I thought I would examine the spot, supposing that perhaps some decayed trunk of a fallen tree might lie beneath. I was advancing rapidly, when, to my horror, I saw before me a pair of glaring eyes, and there stood within the thicket an enormous lion with a huge mane. The king of beasts had just aroused himself apparently from his noonday rest, and was stretching himself, wondering who the bold intruder could be who had ventured into his domains. I gazed at the lion, and the lion gazed at me. I know I did not like the appearance of the monstrous brute. My rifle was loaded with ball, but still I dreaded lest, should I fire and not kill him outright, he might yet attack me. I therefore, keeping my face towards him, slowly retired, hoping earnestly that he would go to sleep again, and allow me to retreat unmolested. Still, from his attitude, I had some doubts whether or not he was going to spring at me. I dared not take my eye off him, for I knew that my best prospect of escaping was to continue facing him boldly. I suspect that he had gone into the wood to indulge in a nap, after having taken a full meal off some unfortunate gnu or antelope. I was very thankful when I at length managed to get to the edge of the wood without stumbling. I continued to retreat backwards, however, after this, fearing lest the lion might pounce out upon me. Every moment I expected to see his enormous head and shaggy mane appear amid the bushes. It would have been a very grand sight, but a very disagreeable one. As I retreated through the wood the treacherous honey-bird flew out also, twittering as before, just as if he had not played me a scurvy trick. "What, do you not like the last honeycomb I showed you?" he seemed to say. I began to think that he was an evil spirit instead of a kind fairy; but yet, perhaps, after all, he was as much astonished at finding a lion instead of a honey-comb as I was. At all events, he appeared regardless of the danger into which he had led me, and not aware that I might have shot him dead in a moment. I could not at the time account for the trick he had played me; but I have since heard that such is not at all an uncommon occurrence, and that honey-birds frequently take the natives who are in pursuit of honey in the same way up to some savage monster.

Having got to a considerable distance from the wood, I ventured to turn round and walk forwards, at the same time very frequently casting anxious glances over my shoulder to ascertain whether the lion was coming in pursuit of me. In vain the honey-bird tried to draw me off on one side. I declined after this accompanying my little friend any further.

I had taken the bearings of the baobab-tree grove, so that I could easily find it. When at length I reached it Natty was in a state of great agitation at my long absence, but was delighted with the delicious honey I had brought him.

"Perhaps the honey-birds want to have the wild beasts killed, and are not aware that when people are only in search of honey they are not prepared to encounter a lion or a rhinoceros," he remarked, when I described my adventure.

He might have been incredulous about my account, but I showed him the honey-bird, which had perched on a branch near us; and, as soon as I took out the honey, down it came and ate some of it with the greatest confidence. I then felt convinced, from his unsuspicious behaviour, that he had had no intention of leading me into danger.

We immediately ate some of the honey spread on the mushrooms. I wished that I could find some means of stewing those curious productions of nature, for they would be, I was sure, a valuable addition to our fare. Poor Natty still continued very weak. I did my best to forage for him, but, in spite of my exertions, the only food I could procure was not satisfactory for a sick person. As to leaving him, the more I thought of it the more dangerous for him did it appear. Even were there nothing to apprehend from the attacks of wild beasts, he was too weak to obtain even water for himself, and we had no means of preserving the food I obtained for any length of time. I should not have cared so much for myself, but I felt all the time how alarmed our friends would be on our account, besides which I felt very anxious to go in search of Leo and his companion. We had reason to be thankful that we were in so sheltered a spot, as for several days in succession violent storms burst over us, heavy downfalls of rain flooding the lower ground in our neighbourhood.

My honey-bird led me in the interval to more bees' nests, and I got an ample supply of mushrooms; but they, as may be supposed, were not sufficient to support life. The birds, getting an abundance of water elsewhere, no longer visited the pool, and I became greatly afraid of starving.

One day I had gone to the ant-hill in search of mushrooms, when I saw a troop of gnus coming across the plain. As they advanced towards me I remained stationary, hiding myself from them by the hill. I got my rifle ready to fire, earnestly hoping that my aim would be steady. On came the herd, frisking and prancing, till they got within thirty yards of where I lay concealed. They scented danger, I fancied, for they began to look about, and seemed ready to dart off in an opposite direction. I selected the nearest, and fired. I could scarcely say how delighted I was when over rolled the creature. He got up, however, and even then would, I was afraid, escape me. I dashed forward, and drawing my axe, struck him on one of the hind legs. Down he fell, and in another instant I had deprived him of life. I now understood the feelings of a famished hunter. Without a moment's delay I began to cut up the animal, and loaded myself with as much of the best parts of the meat as I could carry. The remainder I left for the birds and beasts of prey, and hurried back with my prize to Natty. I selected as much as I thought we could consume while it remained eatable. The rest I cut into thin strips, and hung them up to the boughs outside our cavern. Natty meantime made up a fire, with which we roasted a good portion. I felt no longer surprised at the way I had seen the blacks feed, so ravenous did the smell of the roasted meat make me.

"Don't you think that if we were to smoke some flesh it would keep longer?" observed Natty.

I followed his suggestion, and from the way it dried I was in hopes that the experiment would be successful. I was about to return for the remainder of the meat, to dry it in this way, when the rain came down.

Notwithstanding the more substantial food Natty had now got, he was still too weak to walk any distance. The flesh of the gnu, with the honey and mushrooms, enabled us to subsist in tolerable plenty for a week. The portions I had smoked and dried, at the end of that time became almost uneatable, and I saw that I must succeed in killing another animal, or that we should starve. That night I was awaked from sleep by hearing a low cry of distress. The dreadful thought seized me that a hyena had come into our cavern and carried off Natty. I anxiously put out my hands, and to my relief found that he was on his bed, breathing quietly. Then I thought that he must have cried out in his sleep. But again that low wail of distress reached my ears. It is some human being, I thought to myself, attacked by wild beasts, or fallen into a lagoon; indeed, it sounded exactly like the cry of a person in danger of drowning. Perhaps it may be one of our friends come in search of us. Again it came through the night air. I could bear it no longer, for I was certain that a fellow-creature was in danger. I awoke Natty. "Do not be alarmed," I said; "I hear some one calling for help. I must go out and see what I can do, but I will be back presently. Remain quiet till my return!" Seizing my rifle, and feeling the lock to ascertain that it was all right, I hurried out in the direction from whence the sounds came. Again that plaintive cry reached my ear. I thought I heard the very words,— "Come, come! Help, help!" I dashed forward, for I knew the ground thoroughly. It could not be a person drowning, for there was no lagoon in that direction. As I advanced the wails became lower and lower, and sobs alone reached me. I was afraid that I was too late to render help. Presently, bending down, to be more certain of the direction I should take, I saw against the dark sky the outline of a lion. His claws were on his prey, and his tail was moving round. "He has killed the man, I fear," I thought. Still, regardless of the danger I was running, and urged by an impulse I could not resist, I rushed forward, ready to fire should the lion advance towards me. I shouted at the top of my voice. I went on till I was within a dozen yards of the brute, and then once more raised a loud and determined shout. As I did so he turned his head, and then uttering a loud growl, slowly stalked away, and disappeared behind some bushes at a little distance. I hurried to the spot he had quitted, but instead of a human being, I saw before me an animal stretched lifeless on the ground. On feeling the head, I discovered that it had no horns, and then, taking one of the hoofs in my hand, I found that it was either a zebra or quagga. . To leave it there would be to ensure its being carried off by its destroyer. I therefore set to work as well as I could in the dark, and cut off the flesh, looking up cautiously every minute, as may be supposed, to ascertain whether the lion was coming back to reclaim his prey. The necessity of obtaining food only could have induced me to run so terrible a risk, for I could scarcely suppose that the monarch of the woods would allow me thus before his face to carry off his prize. He did not appear, however. I supposed that, never having before encountered a human being, he was more alarmed by my appearance than I had been by his. Perhaps he took me for a gorilla, which the lion is said to hold in wholesome fear.

I now hastened back to Natty. The lion must have returned and carried off the portions I left him, for the next morning not a particle of the zebra could I discover. Still, it was not pleasant to know that he was in our neighbourhood. I treated the flesh of the zebra as I had done that of the gnu, although it was not quite so palatable.

The following day we were seated at our dinner, when, looking out, I saw a troop of zebras trotting by, stopping occasionally to feed, and then again moving on. I remarked especially a young zebra following them at a short distance. They passed close to the thicket in which I had seen the lion disappear. "If the old fellow is there," I observed to Natty, " I should not be surprised were he to rush out and seize one of them." Scarcely had I spoken when the whole herd began frisking about, and scampering here and there. Just then I heard a loud roar, and, as I had been surmising might possibly occur, out dashed a gray old lion towards the little zebra. I had instinctively seized my rifle. " You shall not kill that pretty little beast if I can help it," I exclaimed. But the lion seemed determined that he would do so in spite of me. In another instant he was up to the zebra, and had struck him with one of his paws, which threw it staggering some paces from me. He was evidently, I saw, an old fellow, unable to leap as a young lion does. I ran forward, and before he had again come up with the little zebra, I had levelled my rifle and fired. The ball hit him in the head, and over he rolled. Greatly to my astonishment, the little zebra, instead of attempting to escape, rose to his feet, and, looking at me for a moment, came trotting towards me. "I am sure I know you," I exclaimed. "You are Bella's little pet." The poor little creature was very much hurt, but not, I hoped, maimed altogether. From the way he came up to me, I had not the slightest doubt that my conjecture was right; for when I held out my hand, he put his nose into it, and seemed to recognize me as a friend. He looked very thin, but as I examined him I was sure that he was an old acquaintance. The lion, meantime, giving a few struggles, fell over perfectly dead. Putting my handkerchief round the zebra's neck, I led him up to our tree. Great was Natty's delight at seeing him.

"O Andrew," he exclaimed, "now there is a way for us to rejoin our friends. Though you cannot carry me so far, Zebra, I am sure, can; and as soon as he is well, we will set out."

As there was ample room for the little animal inside our cavern, I brought him in, and closed the entrance. Having washed his side, I bound it up with a handkerchief, when the bleeding stopped. The rain had brought up an abundance of grass. I went out and cut some, which he readily ate out of my hand. Having done this, I went back to examine the lion. I found the mane thickly streaked with gray; and on examining his huge mouth, I discovered that the teeth were completely worn away, while his claws were broken and blunted. This accounted for the escape of the little zebra I had heard that when lions in their old age can no longer kill the prey to which they have been accustomed, they lie in wait for the young of animals, or take to robbing the poultry-yards of the natives, attacking their goats, and sometimes, indeed, try to carry off women and children. It was the consciousness, probably, of his weakness which made the old

fellow so easily render up his prey to me on a former
occasion. In spite of his age and probable toughness, I was
tempted to see if I could get any steaks out of him, to form a
supply of food should our stock of meat not be sufficient to last
us till we could get home. I cut off a few pounds; but the
smell of the flesh at last made me desist, thinking that neither
Natty nor I would be able to eat it, either smoked or dried.
I had thrown it down, indeed, but still I thought it might be
wiser to secure some; so I took up what I had cut off, and
returned with it to the tree. Without telling Natty, I lighted
a fire, and cutting it into strips, hung it up to the branches,
so that it might be thickly enveloped in smoke. By giving
the little zebra plenty of grass, in three or four days he had
entirely recovered from his injury. Natty also said that he
felt better, and was sure he could undertake the journey
homeward.

CHAPTER XXI.

N a bright morning, as soon as we had breakfasted, I mounted Natty on the zebra's back, and leading him with my handkerchief, set off in the direction of our home. I had manufactured some baskets, in which I stowed the honeycombs and the remaining portion of our meat, with several large white mushrooms. I hoped we might find provisions on our way; at the same time, as I had only three or four charges of powder left, I did not think it wise to abandon what we possessed. The little zebra bore Natty very willingly, but, unaccustomed to the burden on its back, could only proceed at a slower pace than I could have walked. However, I was very thankful to have this means of conveyance for my young friend. The sun came down with great heat, and I began to fear he would suffer from it. Accordingly I steered a course towards a clump of trees, where he might rest under the shade. I placed him on the ground, and told him to hold the zebra, which, I was afraid, might, following the wild instincts of its nature, scamper off. I then cut a stick and several boughs with large leaves, with which I manufactured a parasol to shelter him as we walked along. He was very grateful for the shade, and begged that I would make it sufficiently large to shelter my head also.

This I accordingly did.   I should have said that I had doubled up my jacket and placed it on the zebra's back for a saddle. I made also, out of some vines, a pair of stirrups, which enabled Natty to ride more at ease.

Having taken some dinner, we again pushed on.   I was greatly disappointed when, as the evening began to close in, I found that we were still at a considerable distance from the hill which we were anxious to reach.   Just as I had finished our hut, it occurred to me that should we leave the zebra tethered outside, it might very likely attract either lions or hyenas, or other wild beasts of prey.   I accordingly cut down a large number of stakes, with which I formed an enclosure by the side of the hut.   I covered it also with a tolerably strong roof, lest any animal might leap over the walls.   The little creature had, I suspect, learned so severe a lesson during his wanderings with his kindred, that he seemed fully to understand the necessity for these arrangements.   At all events, when I led him in he was perfectly quiet and contented, especially when I gave him as much grass as he could require. I also made up a large fire outside our hut, and although I did not attempt to keep awake all the night, I was able to rouse myself from time to time to throw on enough wood to keep it alive.   Although I heard the sounds of animals in the distance, the fire prevented them from making an attack on us.

The next morning we again started.   Natty looked somewhat better; but when, in order to relieve the little zebra, he got off and attempted to walk, he was unable to proceed many paces, and made no objection when I again put him on the animal's back.

Our pet was tamed entirely by gentleness and kindness, or it would have remained as wild and savage as its fellows. I believe there are no animals which cannot be made subject to man, provided they are treated in the right way. I have often wished that our horses and asses in England were treated more gently. I am sure they would be more faithful and useful animals than they often prove when subjected to a contrary system.

As we proceeded, we began to recognize more clearly the outline of the hills on which we had so long lived. Still, however, we were at a considerable distance, and I soon saw that, at the slow rate we were proceeding, another day must elapse before we could reach them. The arrangements of the previous night were repeated with similar success. We now hoped to reach our destination early in the afternoon. Once more the lake appeared in sight, the stream running into it, the woods on the other side, and the well-known hill, though we were much too far off to distinguish our village. The little zebra seemed to know it also, for he hastened his pace. We were anticipating the delight our reappearance would give our friends, though then the thought came across us of the disappointment they would feel at not seeing Leo. " But perhaps," said Natty, " they have gone in search of him, and discovered him and brought him back, and we shall find him all well; oh, how joyful that will be!" As we reached the hill I could not resist the temptation of firing off my rifle, to attract the attention of our friends, and give them notice of our coming. No one, however, appeared; still I was sure they must have heard the report. We wound our way up the hill, when we came to a point where I expected to see the huts; but no trace of them could I discover. The grass was green from the recent rains; the trees waved on the hill-side as before; but the huts, the habitations of our friends, where were they? I shouted out, but no answer came. My heart sank within me. I could no longer restrain my anxiety, and telling Natty to follow slowly, I rushed up the hill. There, on the spot where the huts had stood, were heaps of charred timber. I felt faint and sick! What had become of our friends! I scarcely dared to search about, lest I might find some dreadful traces of their death. Oh no, no! It is impossible! The dear, energetic, gentle Kate—such could not have been her fate! And sweet little Bella too! Still, I could not resist the temptation to search about. There were no traces of human beings. I saw, too, by the way the grass had sprung up, that some time must have passed since the fire took place. I roused myself as I saw Natty approaching. I was afraid of what the effect might be on him, and hurried down the hill to prepare him for the scene; indeed, I thought it might be better to turn the zebra's head, and let him proceed down the mountain again. Still, I did not like to leave the spot without a further examination.

"I should like to look at it," said Natty, when I told him. "I cannot believe that they are lost; and perhaps by an examination we may discover something to guide us in our future proceedings."

The little zebra did not object to come up the hill, but when he reached the black spot where the house had stood, he stopped, gazing at it, and I thought trembled.

"It seems to me," said Natty, after remaining silent for a minute or two, "that the zebra must have made his escape when the huts were on fire, and the other animals were set free. "Oh! I do—I do hope that our friends escaped! I will not believe that they did not!"

I would not let Natty quit the zebra, but allowed him to sit down on a stone, holding the rein, while I examined the ruins in the neighbourhood. Though I searched carefully in every direction, not a trace of any sort could I discover. Everything they had must have been destroyed or carried off by them I trusted that the latter was the case.

"It is of no use, Natty," I said at last. "Here they are not, and we must go in search of them."

"What do you think, Andrew?" said Natty. "Perhaps they have gone to Kabomba, where the people know the captain and Timbo, and would, I am sure, receive them kindly."

"I trust you are right, Natty," I said; "and we will set off there immediately."

Without loss of time we descended the hill. I had spent so much time, however, in examining the ruins, that we could get but a little distance before it grew dark. I made our camp as usual, and had only finished a hut sufficient to hold Natty before darkness overtook us. I made up a good fire, also, and hoped by tethering the zebra close to the hut, that no wild beast would injure him during the night. There was little fear of my fire going out, for my anxiety concerning our friends kept me awake. Over and over again I thought of all sorts of accidents which might have happened. We had but little food remaining, and all but my last charge of powder was expended. Still, my anxiety about our friends prevented me thinking of our own condition.

We travelled on all next day, and I began to fear that we must have passed the village. Just, however, as the sun was about to set, his rays lighted up the tops of some huts in the distance. We made towards them, though still doubtful whether they were those of Kabomba or not. Perhaps the inhabitants had themselves attacked and destroyed our friends. I had often heard of the treachery of the natives, and these might be as bad as others.

"Still, we must hazard everything for the sake of ascertaining the truth," I said to Natty.

"Oh yes, yes," he answered. "I do not fear them; and after all, Andrew, they can but kill us; and if they have killed our friends, were it not wrong, I should almost wish that they would kill us."

As we got nearer to the side of the village I had no longer any doubt that it was the one Stanley had visited. That we might not take the inhabitants by surprise, as I drew near I shouted out, and presently several people appeared at the chief entrance. As soon as they saw us they came running forward. Among them was an old man, whom, by his appearance, I took to be the chief. He had no weapon, and as he drew near, his countenance, which wore a friendly expression, reassured me. I therefore hastened on, leading the zebra, to meet him. He took my hands in his, and looking into my face, seemed to be inquiring whence we came. Then he seized Natty's hands and stroked his face, and exhibited every sign of regard. He cast, however, an astonished gaze at the zebra, and was evidently greatly surprised at seeing the docility of the animal.

"At all events, you see, they are friends," said Natty. "I do hope they can give us some account of the rest."

We were quickly conducted inside the village. The chief led us to his house. He then seemed to inquire what we would do with the little zebra, and pointed to a small enclosure on one side. I begged that it might be placed within it, and signified that I should be glad if it could be supplied with grass. Immediately several people set out with knives, I concluded for the purpose of cutting the grass. My disappointment was great, however, at not seeing any of our friends, and by all the signs I could think of I inquired of the chief what had become of them. I could get no satisfactory reply to my questions, and I could not help supposing that the chief had some reason for not informing me. We were taken at once into his house, and in a short time food was placed before us. How delicious the plantains and cassava tasted, and some well-dressed venison. As soon as our hunger was satisfied

I again began to inquire by signs about our friends. A stranger coming in might have supposed that I was performing some pantomimic play for his especial amusement. He, however, seemed greatly puzzled, and I concluded of course that I had not the right talent for my purpose. At length a sign of intelligence came over his countenance, and he now in return made a variety of gestures, which I must own were considerably more clear than mine. He first pointed to the north, and held up his fingers, counting the number of people of whom our party consisted. He then got up and ran across the room, and next opened his arms, and seemed to be receiving some phantom guests. He then lay down on the ground and pretended to be asleep, and got up seven times; by which I understood that they had come and remained at the village that number of days. He next pointed southward, and seemed to be mourning, as if regretting that they had taken their departure. I now told Natty I was sure our friends had come to the village, and after stopping a few days had proceeded to the south. The chief seemed to understand that Natty was ill, and he and his wives did their best to arrange a comfortable bed for him with mats placed over dried grass strewed on the ground. I hoped that after a day's rest he would be able again to set forward, as I wished to lose no time in following our friends. I spoke of my intention to Natty.

" Can you think of doing so without first trying to find Leo ? " he said. " Perhaps our new friends here will assist us."

" If you were better able to undergo the fatigue I would," I said ; " but I wish first to place you in safety."

" Oh, do not think of that," he answered, " leave me here. The people seem so friendly, that I am sure they will take care of me ; and though I wish very much indeed to go with

you, I am sure I should only be an impediment to your progress."

I immediately set to work to try and make the chief understand that two of our party were in captivity somewhere in
the east or south-east, and that I wished to go in search of
them. I was nearly sure that he understood me, and with
some hopes of setting off next day I lay down to get a sounder
sleep than I might possibly enjoy for many days to come.

The next morning, when I again entered on the subject, he
appeared to be unwilling to accede to my wishes. I was indeed not sorry to rest another day and night, hoping that in the
meantime something might occur to assist my project. I remembered the account Stanley had given of the idol like a
crocodile which he had seen. Curiosity prompted me to search
for it as I walked about the village. The chief divined my object, and, taking my arm, led me into a hut, where on the
ground lay a number of fragments of plaster, wicker-work, and
hair. On these he stamped, and then turned away with a
contemptuous glance, touching his ears and eyes, and then
shaking his head, as much as to say that the idol could neither
hear nor see. From several other signs he made, I came to the
conclusion that Timbo had carried out his project, and at all
events succeeded in showing the blacks the falsity of their
wretched faith. I had hopes, too, that he had also planted
the germs of a purer one in their minds. It was on that
and other accounts very vexatious being so utterly unable
to exchange ideas with them. One thing was certain,—they
were disposed to treat Natty and I with the greatest kindness. At last, by perseverance, I made the chief understand
what I wanted, and he signified his readiness to assist me. I
showed him also that I wished him to take care of Natty while
I was away. At this he seemed highly pleased, and brought
his son—a boy of about Natty's age—to show that he would

be his companion, and that he would take as good care of him
as he would of his own children. To show his still greater
readiness to assist me, he brought a number of articles which
had evidently been left by our friends, I could not make out
whether as gifts or not. He signified that we might ransom
Leo with them if he was detained as a prisoner. These, and
sufficient provisions to last me for several days, I placed on the
back of the zebra. The load, though not very heavy, was as
much as I thought it could carry.

I was doubtful whether I should venture to go alone, or obtain some attendants. If they proved faithful they would be
of great use, otherwise I would rather have trusted to my own
energy and watchfulness. The matter was settled by the chief
bringing up three young men, whom he signified were to accompany me. They were armed with shields, bows and arrows, and
spears; but these might alarm their countrymen, and I knew I
must depend for success only on pacific measures. It cost me
a good deal to part with Natty. He looked so sorrowful when
I bid him good-bye.

" But you will bring back Leo; I know you will," he said
" I cannot help thinking he is not very far off."

Just as I was parting the chief brought me a prize, which,
in my circumstances, was of the greatest value. It was a
powder-horn full of fine powder. I could not help fancying
it must have been left behind by accident. It was certainly,
however, not the one which Stanley had been in the habit
of using.

I think I have before said that the zebra would not allow
any of the blacks to come near him. I was therefore obliged
to lead him myself, they following at a little distance behind.
He then went on readily enough; but the moment they came
near his heels, he flung out in a way which made them always
keep at a respectful distance.

I must give a very brief account of my journey. It required a good deal of calculation to direct my course. I had
first to consider the position of the village where Natty and
I had remained so long near the lake. It was some distance to
the south-east of this that I might hope to find Leo, and yet
at no very great distance, otherwise my former hosts would
not have refused to go beyond the stream, at which it will be
remembered we turned back. The journey might, I thought,
occupy me three or four days, if I could manage to steer a
direct course for it. The weather was now again fine, so we
camped out at night, lighting the usual watch-fires; and I lay
down on the ground with the zebra tethered near me. We
saw two or three villages in the distance; but I understood
from my companions that they were sure no white men were
there, or they would have heard of it. At length, at the end
of a four days' journey, a village appeared directly before us,
situated on some rising ground. It was in the direction
where, by my calculations, I thought it possible the one would
be found to which Leo had been carried. A number of goats
were feeding on the side of the hill, and below my eyes were
gladdened by the sight of some horned cattle, which, by their
movements, were evidently tame.

My companions now made signs to me that I might go
on alone, as they did not feel disposed to trust themselves
within the village until they had ascertained the disposition
of the inhabitants. Leading the zebra, I therefore walked on
till I came in sight of a gate at the end of the principal street, if
I may so call it, it being always remembered that the houses
were only reed huts, and the gates were composed of rough
poles. As I neared it several people issued forth with javelins
in their hands, and, vociferating loudly, rushed towards me.
My gun was slung at my back, so I held up my hands to show
that I had no intention of attacking them. On this they

somewhat slackened their pace, though they still held their weapons in a threatening manner. I knew that my best chance of safety was to advance boldly without showing any sign of fear. This had the desired effect, and they now came on in a more friendly manner. They showed signs of astonishment at seeing the zebra in my company, and, I observed, paid me more respect from believing that I had the power of taming an animal so generally untamable. We were still at some little distance from the gates, when another person came out. Seeing me, he rushed forward, and breaking through the people who surrounded me, threw himself at my feet. Greatly to my delight I recognized young Mango. Tears dropped from his eyes as he took my hands.

"O massa, so glad! so glad!" he exclaimed, showing that he had not forgotten his small knowledge of English.

"And Leo?" I asked, taking him by the hand; "where is he?"

"Gone! gone!" he answered.

My heart sank as I heard this.

"What! dead?" I exclaimed, the thought of the grief his death would cause his sisters and Natty, indeed all of us, coming into my mind.

I was greatly relieved when Mango answered,—

"No, massa, not dead; but gone away," and he pointed south.

"What! did any one come to take him away, or did he go all alone?"

"Yes, massa, all alone," said Mango. "He run away. Dey catchy me, and bring back."

This was indeed disappointing. Still, I hoped that he might reach some place of safety, or that possibly I might find him. On making further inquiries of Mango, I ascertained that he had started only two days before. Then I thought,

(272)   24

perhaps he has gone towards Kabomba; I may actually have passed him on the road.

The inhabitants now conducted me into the village, accompanied by Mango, and I was led before the chief. He was an enormously fat man, and was seated on a pile of matting in a sort of verandah in front of his abode, and supported by a number of women, whom I took to be his wives. Determined not to be treated as a prisoner, I went up at once and shook him by the hand, and told Mango to explain that I had come from a distance to look for a young countryman, and that my people would be very angry if any injury had happened to him. The chief was evidently not addicted to making long speeches, indeed it was with difficulty he brought out his words. Mango interpreted what he said. He declared that he had no intention of injuring the white boy; that his people had found him and his companion some time back, and that he had since fed him and taken good care of him, and that of his own accord he had run away.

"Yes," added Mango, "what he say true; but when we want go away, he no let us, so Massa Leo run. He got rifle and powder, too, and dis make old rogue here wish keepy."

I concluded from this that Leo's case had been very similar to ours, and as my anxiety about him had somewhat decreased, I began to fear that the fat chief would detain me in his place. I therefore assumed a still more authoritative air, and declared that though my people were very much obliged to the chief for taking care of our friends, they would be very angry at his having detained them longer than they wished.

"Tell him I insist upon their letting you go immediately, and if they do so, I am prepared to make them a present; but that if not, I shall fight my way out of the place on the back of my wonderful steed there"—pointing to the zebra—"and very likely return and burn their village to the ground."

"Bery good," said Mango; and he began to interpret my address, adding, I suspect, not a few threats and boastings of his own.

The effect, at all events, was to make the old chief and his attendants treat me with great civility. His wives hurried off to prepare a banquet, and I was allowed to proceed through the village with Mango as my guide. I led the zebra all the time, for the little animal showed a great disinclination to leave me, or to go nearer the blacks than he could help; indeed, when any of them drew near, as was his usual custom, he struck out with his heels right and left at them, or, if they appeared in front, he ran forward and tried to bite them. He, however, appeared to recognize Mango, and though he would not allow him to touch his head, yet he showed no hostility when he came near.

By the time the banquet—which consisted of a variety of dishes of the meat of several wild animals—was over, it was almost dark. I had no doubt my attendants would camp out in the neighbourhood of the village, and I therefore told the chief that I would take my departure, accompanied by Mango, and camp with them, to be ready to start on the following morning. I found, however, that he had no intention of letting me go so easily, and insisted that I must pass the night in his village. Seeing how matters stood, I said that I had no objection to do this, but that I must have a house to myself, where my zebra would obtain accommodation, as I could not be parted from the animal; and that I wanted Mango also to attend on me. There is an old saying, "There is nothing like asking for a thing one wants," and I found the advantage of so doing; for my request, after the chief had consulted his wives, was granted. This arrangement being made, I told Mango to inform the chief that I required a supply of green grass for my animal. This also was brought me before night. I asked Mango whether he thought the chief intended to detain us. He did not think so; but expressed himself ready to try and get out of the village during the night, if I thought it advisable. I discovered, on further questioning him, that he and Leo had heard of the appearance of some white people at the distance of three or four days' journey off, towards the south-west, and though the account was not very exact, from that moment Leo had determined to make his escape. He arranged that if they could not get off together he should go first, and leave marks to show his route. Mango was to follow, or should he be prevented, Leo promised that he would return with his friends to his rescue.

"But, massa," added Mango, "long way walky. Dey got cows, big horns, for ridey. Me steal one for massa."

Perhaps I am making Mango speak even more clearly than he really did; but he made me understand his meaning by the help of words and signs.

"No," I replied. "I shall be very glad to buy one of their animals, though they must suppose it is for you to ride, and not for me, as they now believe that I could not possibly require any other steed beside my zebra."

While I remained in the hut, I sent Mango to the chief with an offer to buy an ox, provided he would bring several to the village early in the morning for me to choose from. Mango shortly returned to say that the chief agreed to my proposal; indeed, the old man was probably, as most Africans are, perfectly ready to do a stroke of business, particularly as Mango had told him that I was willing to pay a good price for the animal.

CHAPTER XXII

NEED not enter into the particulars of my purchase. The transaction was soon completed. I had brought articles sufficient, I hoped, to ransom both Leo and Mango. I told the chief that, although I did not consider myself bound to pay him anything for releasing Mango, yet I would make him a present in consideration of the kind treatment which he and my young countryman had experienced. All parties seemed well pleased, especially when I offered a further sum for some provisions—cassava, plantains, antelope flesh, and dried elephant meat—which I intended for my attendants, whom I hoped to meet in the valley below.

In case the fickle negroes should change their mind, I hurried off as soon as I possibly could without exciting their undue suspicions, and was glad to find that no one followed us. We took our way down the hill to a spot where I left my three attendants, but they were nowhere to be seen. There was their camp-fire, but it had long gone out; and I supposed that, having been alarmed, they had taken to flight. I hoped to come up with them further on. Still, no traces could I see of the deserters. As I had made up my mind to search for Leo before returning to Kabomba, I gave up the pursuit, and turned on one side for the purpose of intersecting the course I concluded, from Mango's account, that he had taken.

Leo had promised to make crosses on trees, and where no trees existed to cut the same mark on the grass, or to arrange stones in a like form, or to stick little crosses into the ground, to show his course. " I always thought that Leo had his wits about him, and this proves it!" I exclaimed, though Mango probably did not understand me. We accordingly examined the ground on either side as we went along. I could still see in the far distance the outlines of the village, and, judging by the sun, I calculated that it was about north-east of us, while I hoped by travelling south-west to come up with my young friend.

We had been searching for some time, and at Mango's suggestion I had mounted the ox. I have not before described the animal. It was clean-limbed, almost white, with long pointed horns projecting horizontally from its head; a thoroughly tame and tractable animal. It went on at a steady pace, sufficient to keep Mango and the zebra at a trot. We were searching carefully as we went for Leo's promised indication of his route, when Mango suddenly started off, and running a few paces, lifted up a small cross, formed of two pieces of wood, fastened together by the material of which the natives make their mats. Mango's delight was excessive. "See! see!" he exclaimed. "We now find—we now find Massa Leo!" and running on ahead, he lifted up a second cross made in the same way. The arms of both of them were pointing in the direction which we supposed Leo had taken. This fact also showed his forethought, for if a single cross only had been left, we should have had to search about perhaps for a long time before ascertaining his route. We now went on with more confidence. From the start he had had, I feared it would be some time before we could come up with him. Still, as he had his rifle and provisions to carry, I knew that he could not proceed as fast as we were doing. We travelled on till nightfall, when we tethered the animals with some rope which Mango had brought, lighted our fires, and made a slight shelter from the wind. As the weather was clear, there was no necessity for building a substantial hut. Having unloaded the zebra, I placed the packages under my head as a pillow, keeping my rifle as usual by my side, and told Mango that we would watch alternately during the night. I gave him the first watch, with directions to call me after a couple of hours, intending to allow him a longer rest than I took myself. I was awoke by a loud roar sounding in my ears. It was the well-known voice of a lion. I started up. So did Mango, for he had been asleep. A few glowing embers of the fire alone remained. I had seized my rifle instinctively, and with it in my hand looked around on every side. The ox stood near, though trembling violently; but the little zebra was nowhere to be seen. I caught sight, however, of the massive form of a lion bounding over the ground. The zebra, I hoped, had escaped, though the lion might be pursuing it, and I resolved to try and save the life of our little pet. I fired, and believed that I had hit the savage brute, for it stopped and growled more furiously than before. Meantime Mango was employed in throwing sticks on the fire, blowing with might and main to make them blaze up. The lion drew nearer. Again I fired, but missed. There might be scarcely time to reload before the lion would be upon me. I hurriedly began to do so. I never more eagerly rammed down a charge. Still the lion came on. Mango piled on more sticks, and blew and blew harder than ever, as if his existence depended on it. So, perhaps, it did, for had the lion made a spring, and had I again missed him, Mango's life must have been sacrificed. Just then the fire blazed up. Fortunately the sticks were very dry. A

few bounds would have brought the savage brute up to us.   I
shouted, and so did Mango, with might and main.   I refrained,
however, from firing till the lion had approached nearer, for
should I not kill him outright, he might, in spite of the fire,
rush towards us.   On he came roaring, but slowly, afraid of
the flames.   Once more he stopped.   He dared not face them.
Greatly to my relief, he then turned round and moved off, roaring furiously.   Fearing that he might still pursue the zebra,
which I hoped had escaped, and might, after making a circuit,
come back to us, I raised my rifle and fired again.   I fancied
I could hear the thud of the bullet as it struck the lion behind
the shoulder.   Fearful were the roars he uttered; but defeated,
he stalked off, evidently having had enough of the fight.
Mango, who had been thoroughly alarmed, seemed very peni-
tent for having gone to sleep.   There was no necessity to
point out to him the danger we had been in in consequence.
He tried to say he would never do so again.   At last I
persuaded him to lie down and rest, while I sat up.   I kept
looking round, in the hope of seeing the zebra trot up to
us, but when the morning came our little pet had not returned.

I had begun to cook our breakfast even before daylight,
that we might lose no time in starting, so as to take advantage
of the cool air of the early day.   We had not gone far when
we came to a small cross made of stones on the ground.   It
revived my spirits, for it was the sign that Leo had passed
that way.   Then again the fear came across me that the
lion which had scented us out might have attacked him.
During the day we passed several other crosses, some cut, as
he had promised, in the trees; but the greater number were
composed, as were the first we had seen, of sticks.   It took a
shorter time to erect them than to cut the marks on the trees
or the grass, or even to make crosses of stones on the ground.

Frequently during the day I turned back, in the hope of seeing the zebra following us, but I was disappointed.

The next night passed away, and then another, and Mango kept wide awake during his watch. Leo must have pushed on well, for still the crosses appeared. We came on all the spots where he had slept—his lean-to or hut, with the ashes of his fire before it; and generally midway between them a black patch alone, where he had stopped to cook his mid-day meal. We found the feathers of several birds which he had shot. It was evident, indeed, that he had exercised all the sagacity of an experienced hunter—remarkable in one so young. I was very thankful that I had an animal to ride, for the heat and the constant exertion I was undergoing tried me greatly.

On the third day we still found Leo's crosses, but several were out of the straight line. The country had become open, similar in character to that which I had passed over with Natty. Hitherto we had found springs affording sufficient water for ourselves and the ox. Now, however, we had to go a long way without meeting with any, though we carried enough in our bottle for ourselves, and a small quantity for the patient ox. Travelling on, I saw something lying on the ground a short distance off. I pointed it out to Mango, who ran towards it, and returned with a knapsack. "Yes," he said in a sorrowful tone, " dis Massa Leo's." I recognized it indeed as the one Leo had with him. Fatigue alone could have made him throw it aside ; and perhaps, hoping soon to reach the Europeans of whom he had heard, he would no longer encumber himself with it. Securing it to the ox's back, we went on still more eagerly, looking carefully about on every side. I expected every moment to overtake Leo. We went on for another mile or more, when to my dismay we found his rifle on the ground. That he certainly would not have thrown away unless greatly overcome by fatigue. Still, perhaps, he might have had no powder, and found it a useless encumbrance. I, however, dreaded that, weak as he must have been before he would quit his knapsack and rifle, he might have fallen an easy victim to some beast of prey. Though we looked anxiously about, we could see nothing of him. Presently Mango, who had gone ahead of me, began running very fast. I pushed on to overtake him, when I saw, lying on the ground, a human form, by the side of which Mango had thrown himself. Could it be Leo ? I urged the ox into a gallop, and did not stop till I reached the spot. My worst apprehensions were fulfilled. There lay Leo extended on the grass.

"Is he dead?" I exclaimed, in a faltering voice.

"Hope not, massa," answered Mango, looking up; "he 'till breathes."

The words somewhat relieved my fears, and throwing myself from the ox, I knelt down by his side. My first care was to pour some water down his throat, then to bathe his temples; to treat him, indeed, as I had Natty under similar circumstances. I cannot express my thankfulness when I saw him at length open his eyes. He gazed at me with a look of surprise, but he was still too weak to speak. He pointed to his lips, and I gave him more water. It was necessary to get him at once into the shade, for, exposed to the hot sun, it was scarcely possible that he could regain his strength. Mango accordingly lifted him up on the ox's back, and I supporting him in my arms, he urged the animal on towards a wood we saw in the distance. Leo was still too weak to speak, but he recognized me, and a grateful look lighted up his eyes as he gazed at my face. As I thought he might understand me, I briefly narrated some of my adventures in search of him, of course not telling him my anxiety about his sisters and brothers. How thankful I felt that I had come in time to save him, for it was evident that he would not have survived many hours lying out on the exposed plain. I was now doubtful whether we should proceed on in the same course we had been steering, or turn away to the west in search of Kabomba, where, I felt sure, he would be well taken care of. I should have to go there at all events for Natty, even if we could gain certain tidings that our friends were further south. Presently Leo's lips moved, and I heard him whispering, "On! on as before! You will find them, I am sure!" This decided me. Still, I resolved to rest at the nearest wood we could reach. I was thankful when at length we arrived at one—a little oasis in the desert. What was still more satisfactory, within it appeared a small pool, a bright stream rushing out of the bank on its side. We had tethered the ox. While Mango sat by Leo's side bathing his temples and wetting his lips, I was busily employed in collecting wood for our hut. Suddenly the sound of animals rushing across the plain reached my ears. I looked up, and saw a troop of giraffes galloping at full speed, and, closely following them, two horsemen. On they dashed! Shouting at the top of my voice, I called again and again. I rushed to the ox, in the vain hope of overtaking them. Even at that distance I fancied I recognized Stanley, though his companion's figure I did not know. Just as I was about to mount, there came tearing after them, as if in pursuit, a large herd of buffaloes, among which appeared several huge rhinoceroses. It seemed as if they were in pursuit of the horsemen. Another herd of buffaloes came out of the wood opposite, and stopping, gazed a few moments before joining the chase. The whole passed by like creatures in a dream. I saw at once that it would be impossible to catch up the horsemen ; besides which, I should have run a great chance of being gored to death by the rhinoceroses or buffaloes. On they went, tearing across the plain. Poor Leo lifted up his head.

Just then Mango called to me. " He say he sure dey're friends," said Mango. " We go after dem."

" Not just yet," I answered ; " but it is a great satisfaction to have seen them, for it shows that they must be encamped not far off, though in which direction it is hard to say."

Had I been alone, I should certainly have followed ; but it would have killed Leo to move. I therefore remained en-camped, hoping that he would soon be sufficiently recovered to proceed. In a short time not an animal was to be seen. How-ever, the incident greatly raised my spirits, especially as Leo was evidently getting better. Mango and I therefore went on building a hut, and collecting wood for a fire. We mean-time propped up Leo with the baggage and some piles of wood. While thus employed, I saw a couple of parrots on a bough near, and fortunately killed them; and by the time our fire was burned up, Mango had plucked them, and they were soon roasting before it.

Night came on ; but Leo was very restless, and declared that he could not sleep. I did everything I could to soothe him, but in vain. At length the moon rose and lighted up the whole landscape. " Me t'ink good time go on," said Mango. I thought so too ; indeed, I had become very anxious about Leo. The camp, I hoped, was at no great distance, and I thought it would be better to obtain assistance for him, rather than take a long rest and have to travel during the heat of the day. Accordingly, rousing our patient ox, which had lain down near the fire after cropping the abundant grass, I mounted and lifted Leo up, holding him in my arms. Mango carried my rifle, and led the animal, that I might be more at liberty to support my young friend. On we went over the plain. We had gone some distance, when I felt Leo resting more heavily on my arm. I asked him what was the matter. He did not answer. I feared that he had fainted. Telling

Mango to stop, we bathed his temples, and I poured a few drops of water down his throat. I had no other remedy. It slightly revived him, for he opened his eyes and spoke a few words; but his condition made me more than ever anxious to discover the camp, if such was indeed to be found. I had already gone through a great deal of anxiety, but nothing to equal what I suffered at present. It seemed so sad to think that Leo might die when succour was so near at hand. Eager, however, as I was to proceed, Leo's condition prevented me from allowing the ox to go out of a steady walk. Still, even thus, without any jolting, he got quickly over the ground. On and on we went, looking about in every direction for the light of a fire which might indicate the situation of the wished-for camp. I say wished-for, for I was not certain that our friends were actually in the neighbourhood. Perhaps the horsemen I had seen had come from a considerable distance, and were in light hunting order, with merely saddle-bags to hold their provisions and ammunition. If so, they could render us, even if we should fall in with them, but little assistance. These thoughts passed through my mind as we proceeded, while I formed a variety of plans, to be carried out according to any emergency which might arise. As the moon was bright, I had no fear of an attack from wild beasts.

We had gone on for about three hours, when Mango stopped. "See, massa, see!" he exclaimed. I looked ahead, and observed a ruddy glow in the sky. The ox at the same time poked out his head, as if he also saw something that interested him. Presently the light increased, and I could distinctly make out fires burning in the distance. "If those are camp-fires, they must have been lighted by a somewhat large party,' I observed. The further we advanced, the more distinct did the fires become. We proceeded eagerly. At length, to my surprise, the ox seemed unwilling to move on. In spite of

Mango's coaxing voice, it proceeded more and more slowly. At length I could distinguish not only the fires, but objects moving about; a waggon and numerous oxen tethered near, and horses and men, gradually came in sight. Then the barking of dogs reached our ears. This made me still more surprised at the unwillingness of the ox to proceed. Then I distinguished some water, on which the light of the fire was reflected. Between us and it, however, several dark objects appeared. In vain Mango now tried to urge on the ox. He stopped altogether. "Ah, massa, look dere!" he exclaimed in a terrified tone. He had cause for alarm. The fires just then blazing up more completely, exhibited the dark outlines of several lions and other creatures, which I took to be hyenas, standing on our side of the stream, watching the camp, while the dogs we had heard ran backwards and forwards, barking at them from the opposite side. My fear now was that the savage brutes might turn and attack us. Even if they did not do so, it might take us some time to find a ford and get round to the camp, unless we could make the travellers hear us and come to our assistance. Mango and I shouted again and again with all our might. Though our friends might not have heard our voices, the wild beasts did, for suddenly turning round, the whole pack, with angry roars, came bounding towards us.

CHAPTER XXIII.

DOINGS AT THE CAMP.

T was a nervous thing to stand in front of a dozen or more lions and hyenas bounding over the plain. I thought the ox would have bolted, in spite of Mango's efforts to hold him. To fly would have been more dangerous than standing still, so we remained firm, and shouted our utmost. The moon, which had before been behind a cloud, came out brightly, when the savage creatures, awed, if not terrified, by our cries, separated as they approached us, and bounded off on either hand into the wilds. The ox, recovering from his alarm, no longer refused to move on.

Reaching the banks of the stream, we again cried out, hoping to attract the attention of the travellers.

"Who are you? What is it you want?" shouted a voice from the other side.

"Andrew Crawford and Leonard Hyslop with the black Mango. We want to cross the river and join you," I shouted in return.

"Welcome! welcome! Move to the right! There is an easy passage. We will go that way and show you. Captain Hyslop and several of his party are here."

The last words which reached my ears were the first certain

intimation I had that my cousin Stanley was in the camp near us. I earnestly hoped that his sisters and David were there also. As we rode along we heard a number of voices, and saw men with torches moving rapidly along the side of the stream. Presently we came to a somewhat wider part, where the banks were very low, and where I should have expected to find a ford. At the same time several people were seen with torches crossing it. We went on to meet them, Mango leading the ox, which advanced without hesitation. We were already in the water when I heard Stanley's voice.

"Andrew, my dear fellow, is it you? and have you really brought poor Leo?" he exclaimed. "We had given you all up for lost!"

"I have brought him," I said; "but where is David?"

"He is in the camp; but having turned in, I suppose was not dressed in time to join us," he replied.

We had not time to exchange many words while crossing the stream; but as soon as we had got safe on dry ground I gave him a brief account of our adventures, and expressed my anxiety to have Leo placed under David's care without delay.

"And Kate and Bella!" I asked. "Are they with you, and well?"

"Yes, I am thankful to say so," he answered, "though they have had to go through much hardship, no little danger, and great fatigue; indeed, I do not know what would have occurred had not our friend Silva, and a party he had collected, arrived sooner than we expected. He had fallen in with a trader making an exploring expedition further north than any of his calling have hitherto reached, and, offering him a handsome remuneration, induced him to come on with his waggon and several good horses, in the hope of meeting us. . The trader—Donald Fraser by name, a Scotchman—having got into this unknown region, would not consent to proceed further, and was on the point of turning south again, when Silva induced him to remain another week, while Chickango went on to try and get tidings of us. We had, meantime, started south, and happily fell in with him, when reduced to extremities, about two days' journey from the camp. I am not surprised at our friend Donald's unwillingness to proceed, for he had fallen in with some rough customers, who were more likely to rob him of his goods than pay for them. However, by the exertion of the diplomatic talents of our friend Silva, they got free, and now, I am thankful to say, we are all well, and ready to march southward. Kate and Bella have been dreadfully cut up about Leo's loss, and yours, too, Andrew. But what has become of Natty? I hope the poor boy is not dead?"

I satisfied Stanley on that point.

"We must go back, then, for him at once," he remarked. "Though the Kabomba people may treat him well, we must not desert the poor lad."

By this time we had reached the camp. Although the rest of the party had been asleep, they had been aroused, and now appeared out of their respective huts to receive us. Kate and Bella greeted me kindly, but were too much occupied with poor Leo to exchange more than a few words. He was at once carried into their hut, where David went to attend to him. Senhor Silva, Jack, Timbo, and the other blacks, greeted me warmly.

"So glad, Massa Andrew, you come back; so glad," exclaimed Timbo. "Me pray always for you. Neber t'ought you lost. Knew you come back some day, dough me not den know de way."

Though I felt somewhat fatigued, my friends insisted on getting a substantial supper ready; and the relief I felt from the idea that my cares had now come to an end, contributed to give me a good appetite. I was introduced to Mr. Donald Fraser, a tall, gaunt, red-haired Scotchman.

"I am very glad to welcome you, Mr. Andrew Crawford," he said, putting out his horny-palmed hand. "You come from the North, I know, by your name, and you are none the less welcome so far from the old country, out in these southern and heathenish lands. Your stout arm and rifle will be a pleasant addition, too, to our party; for they are rough fellows we are travelling amongst, and I shouldn't be surprised if we had to fight our way out from their midst."

"My father came from Scotland, and though I have never been in it, I love the country for his sake," I answered. "Though I hope we shall have no fighting, I am ready to take my part if we have to defend ourselves."

"No doubt you would, Mr. Crawford," he said. "We are men of peace, and should never wish to fight, unless in cases of urgent necessity. I hope, now you are come, we shall begin our journey southward forthwith."

"I am afraid not, Mr. Fraser," said Stanley. "My brother, who has just arrived, will scarcely yet be able to move, and we have a young friend, I find, lying ill at a village some days' journey to the north of us; and until we get him we cannot leave this spot."

This information did not seem very palatable to our friend Donald; but after taking a glass of real Glenlivet, a flask of which stood in our midst, his countenance relaxed.

"Ay, to be sure. I had once a young brother of my own, a delicate boy. I had few else to love in the world. He is gone; but I know how you feel about this little fellow; we must not risk his life. And the other lad, the son of poor Captain Page—I knew him—made a voyage aboard his ship —and should like to do the boy a good turn for his sake. I don't greatly esteem the gratitude of this world, and yet it's pleasant to have the opportunity of repaying a debt for kindness received."

I was glad to hear these remarks, and trusting that Natty would find a friend in Mr. Fraser, I lay down to enjoy a sounder rest than I had for very long obtained.

Leo was much better in the morning, and David told me that though he was seriously ill, yet he trusted that he would shortly regain sufficient strength to travel. I begged of Stanley that he would allow me to accompany him to convey Natty to the camp. To this he willingly agreed, and it was arranged that Timbo was to take a third horse and act as interpreter, and that we were to travel during the bright moonlight hours of night.

I was anxious to set off immediately; but the horses were so tired with their hunting expedition of the previous day, that Stanley considered it was necessary to give them a couple of days' rest before they would be fit to start.

"When did you ride last, Mr. Crawford?" inquired Donald Fraser of me the following morning. "Because it strikes me that, unless you are a good horseman, you'll be little fit to take the journey the captain proposes, at the rate he goes over the ground."

I confessed that some years had passed since I had mounted a horse, though in my father's prosperous days I had owned one, and was then a fair rider.

"Well, then, we'll just take a canter across the plain this afternoon. It will not tire the horses, and it will help to get your muscles into play for the exertion you'll have to make by-and-by," he said.

I was very glad to accept his offer. After dinner, with our rifles at our backs—to be ready for any lion, panther, elephant, or rhinoceros which might cross our path—we set out for an honr's ride towards the south, Stanley cautioning us not to go far and fatigue the horses.

"Never fear, captain," answered Mr. Fraser. "We'll just go far enough to stretch our steeds' legs, and see how our young friend here sticks to his saddle."

As we rode along my companion gave me many valuable hints with regard to the journey I was about to undertake.

"Keep your horse well in hand," he observed, "your eyes about you, and your ears open; never press him unnecessarily; and then, should you meet a lion or be attacked by savages, you will be ready for action, and do what in my opinion is the wisest thing under such circumstances—get out of their way."

We had not gone far when an exclamation of pleasure burst from Donald, and I saw to the southward a vast herd of spring-boks crossing from east to west. Numerous as were the wild animals we had met with, I had never seen so many of one species together. They formed an immense herd extending for a full mile across our path, and, as far as we could judge, of the same width. On they went, bounding and leaping. "On! on!" cried my companion, forgetting all about our tired steeds; and putting spurs to the flanks of his, away he galloped, calling on me to keep up with him. The wary animals saw us coming, and, apprehending danger, immediately began to scour over the plain, turning, however, to the south-west. This placed us directly behind them. They would lead us a long chase, of that there was no doubt; but Donald was too eager to think of letting them escape. Mile after mile was passed over. We were approaching the herd. They now, however, began to scatter to the right and left, though still keeping in considerable bodies. We followed the centre one. At length we found ourselves in a rocky country, which com-pelled us to turn aside. Twice Donald fired, and each time brought down an animal. I also killed one; but could with difficulty rein in my horse while I reloaded my rifle. Away the springboks went, leaping over the rocks with wonderful agility. We had been gradually ascending, when Donald disappeared among the rocks and trees to the right, and shortly afterwards I found myself going down the somewhat steep side of a hill, with a number of springboks directly ahead of me. I again fired, but missed, when I stopped to reload; and just then looking up, I saw a high precipice, towards which several of the springboks were making. Rushing on, regardless of the height of the cliff, they leaped over it. I thought they must have broken their legs; but they alighted unhurt. Just then I saw Donald coming on at full speed, directly after another herd. They, too, made for the precipice. I shouted out to him, fearing that he might not see it, and that he and his horse would fall over and be killed. I shouted and waved again and again. Just before he reached the edge he saw me, and though he could not have heard what I said, he guessed there was danger, and reined in his steed; not, however, till they were both on the point of rushing over. Scrambling up the hill, I rejoined him. He had killed four antelopes—a welcome supply for our camp. We might have slaughtered many more, but those we had got we could not carry home. Cutting up four animals, we loaded our horses with the meat, and then drew the remaining two into a hollow of a rock, and filled up the entrance with stones and sand, hoping to send for them in the evening.

The springboks are so called from their wonderful agility. They are found in all parts of Southern Africa, and are more numerous than any other variety of the antelope. In form they are very graceful—not unlike the lovely gazelle of the north of Africa.

We had a somewhat fatiguing trudge towards the camp, though we had less to complain of than our steeds. The sup-

ply of venison was very welcome, though I was afraid, in consequence of our long chase, the intended journey might be
delayed another day.   Donald complimented me on my
horsemanship; indeed, I had not been five minutes in the
saddle before I found myself perfectly at home.   I was somewhat stiff, I must confess; but the horses were not much the
worse for their unexpected gallop.   We therefore prepared to
set off the following afternoon.

No time was lost in sending for the rest of the venison,
which the hyenas would soon have found out had it been
allowed to remain during the night.   Late in the evening
Chickango and one of the Hottentots, who had been sent to
bring it in, returned.   As they were approaching the camp,
one of the oxen, which had been allowed to feed for a moment,
was seen suddenly to stop, and begin to roar with pain, its
countenance exhibiting the utmost helplessness.   I, with
others, ran forward to see what was the matter, supposing that
it must have been bitten by some venomous insect or snake.
Donald soon followed, when, telling the men to hold the poor
ox's mouth, he took out of it a curious woody-looking substance, covered with sharp thorns.

" The poor creature has got this seed-vessel of the grapple
plant into his mouth," he said, exhibiting it.   " I suspect
that any of you who had taken the same between your jaws
would have roared too, if not so loudly."

He told us that if an animal lies down upon these seed-
vessels, they stick to his skin, so that he cannot possibly get
rid of them.   David, who examined it, said it came from the
plant *Uncaria procumbens*, or grapple plant.

I had gone out the next morning soon after sunrise to look
round the camp, when I saw several birds of a grayish colour,
about the size of a common thrush.   Their notes, too, reminded
me, as they sang their morning song, of the mistletoe thrush,

Presently they flew off together, some way up the stream. Turning round, I saw Chickango, Igubo, and several of Mr. Fraser's blacks following, with guns in their hands, accompanied by a pack of dogs. I pointed out the birds to them. "'Noceros not far off," observed Chickango. Presently we saw the birds pitch behind a neighbouring bush, and getting on one side of it, what was my surprise to find that they were standing on the back of a huge rhinoceros, sticking their bills into his head, and even into his ears, and uttering a loud harsh grating cry. The rhinoceros, we could see even at that distance, was a huge white monster, with a couple of horns, a short one placed on the head behind the front, and pointed—a formidable looking weapon. The object, probably, of these rhinoceros-birds, as they may be called, in thus pitching on his body, was to feed upon the ticks, and other parasitic insects, which swarm upon those animals. They also attend upon the hippopotamus, and, whether intentionally or not I cannot say, often thus give him warning of danger. Presently up rose the rhinoceros and looked about him. I, unfortunately, not intending to go far from the camp, had left my rifle behind. The dogs at that instant started off, rushing with loud barks towards the monster. They had better have kept at a distance, for, lowering his head, he caught the first which leaped towards him on his horn, and threw him back dead among the reeds. Then turning round, he charged directly towards us. The unarmed blacks immediately took to the water. Unable to escape by flight, I thought that my last moments had come; but, providentially, the dogs attracting his attention, diverted it from me. Chickango, rifle in hand, boldly ran up to face the monster, who at that instant seemed to catch sight of the waggon and cattle in the distance. He probably thought it an enemy worthy of his courage, for, to my great horror and dismay, in spite of our shouts and the barking of the dogs, he rushed off towards it. I could only hope that our friends saw him coming, though when I left the camp they were still asleep. I thought he would have struck Chickango, who was directly in his course; but the active black sprang out of his way, and then turning round, fired at his head. Though I was sure the bullet had struck, yet it did not stop his course. On he dashed towards the waggon. I shouted and shouted to Stanley, hoping that he might possibly hear my voice. In vain. The brute went on, and seemed to be almost in the midst of the camp. Aiming directly at the waggon, he struck it, and, heavy as it was, so great was the impetus of his huge body that he sent it on several feet. Fortunately he came against it in the rear, otherwise it must inevitably have been upset. Just then another shot was fired, and, greatly to my relief, over rolled the huge creature. Never have I heard such shouting, barking, and yelping of dogs, as immediately arose.

When I got to the camp I found our friends, as may be supposed, in a state of no small alarm; but that quickly subsided, and the blacks especially gave way to their delight at the prospect of so bountiful a supply of meat as the creature's carcass would afford them. We calculated that it was fully equal to three good-sized oxen. It was an enormous creature. David likened it to an immense gray hog shorn of its bristles. With the exception of a tuft at the extremity of the ears and tail, it had no hair on its body. Its eyes were absurdly small; indeed, at a little distance one could scarcely see them. We agreed that, what with its giant body, misshapen head, ungainly legs and feet, and absurdly small eyes, it was, according to our notions, the very image of ugliness. Next to the elephant, the white rhinoceros is the largest animal in existence, and scarcely inferior to it in strength, as this one had proved by the way in which it pushed on the huge waggon. Notwithstanding its ungainly appearance, it had shown us how active it could be, by the way it had turned about when assailed by the dogs, and the rapid charge it made towards the camp; indeed, I believe even a fast horse, with a rider on his back, could only keep pace with it. Senhor Silva told us it cannot go long without water, and it is, therefore, always found in the neighbourhood of somo pond or fountain, which it seeks at least once during the day, both to quench its thirst and to wallow in the mud, in which amusement it delights. Probably it is thus able to get rid of the insects which cling to its hide. We measured the animal, and found that it was nearly sixteen feet in length, from the snout to the end of the tail, and twelve feet in circumference. It is said to attain the age of one hundred years; indeed, judging from its horns, the old fellow we killed must have been nearly as old. The body was long and thick; the belly hanging nearly to the ground, and of great size. Its legs were short, round, and very strong; and its hoofs were divided into three parts, each pointing forward. The head was especially large, the ears long and erect, and its small eyes deeply sunk. The horns of the rhinoceros are composed of a mass of fine longitudinal threads, forming a hard solid substance, not secured to the skull, but merely attached to the skin. They rest, however, on a bony protuberance near the nostrils. The white rhinoceros, of which I have been speaking, has an extraordinary prolongation of the head, which we found to be nearly one-third of the length of the whole body. Its nose was square, and the after horn of considerable length. The horn of the black rhinoceros is much shorter, and the animal itself is smaller than the white species. There are, however, four species of rhinoceros—two black, or of a dark colour; and two of a whitish hue. The black is supposed to be of a wilder and more morose disposition than the white. It has a peculiar upper lip, which is capable of extension, and is extremely pliable, so that it can move it from side to side, and twist it round a stick. It in this way collects its food, and carries it to its mouth, making use of it somewhat as an elephant does his trunk. The black species are very fierce, and probably, next to the buffalo, are the most dangerous beasts in Southern Africa to encounter; for the lion gives notice of his approach by his roar, and can easily be driven off, while even the elephant is less pertinacious in assailing an enemy.

Senhor Silva said he had heard of rhinoceroses with three horns, but he had never seen them, and rather doubted their existence. One species known as the cobaba has a front horn frequently upwards of four feet in length, pointing slightly forward from the snout, at an angle of 45°. It can easily be conceived how fearful is a charge from an animal with such a weapon, active and determined as it is. Although the rhino-ceros sees but badly, it has a peculiarly acute sense of hearing and smell. It winds an enemy at a great distance; but the hunter may approach to leeward of it within a few paces, if he walks with care, without being discovered, though at the same time any noise will instantly arouse it. Ugly as the rhino-ceros is, the female is a very affectionate mother, and guards her young with the tenderest care. The calf also clings to its dam; and Senhor Silva told us that he had seen a calf watching by the side of the carcass two days after the mother had been killed. Until aroused, the rhinoceros looks the most stupid and inoffensive of animals; but woe betide the unwary traveller who offends him! If on horseback, he will have to scamper for his life; if on foot, his only chance of safety is to climb a tree, or hide on the opposite side of the thick trunk of one. A lion will never attack a rhinoceros, and slinks out of his way if he meets one. Even the elephant avoids an encounter, if he can, with so formidable an opponent, who, careless of the blows of his trunk or the thrusts of his tusks, will charge him with his sharp horn, and pierce him to the heart.

Senhor Silva told us that he once saw a battle between a large male elephant and a rhinoceros, when, after an encounter of some minutes, the elephant, who had at first shown great courage and activity, turned tail and fled, the blood flowing from the wounds he had received. He once also saw a battle between four enormous rhinoceroses. Again and again they charged each other, uttering the most horrible grunts, and digging their horns into each other's sides. So fiercely engaged were the monsters, that they did not observe the approach of his hunters, who succeeded in killing two of them, while the others escaped. Those killed were utterly unfit for food, their flesh being quite rotten from the wounds they had received on previous occasions.

The black rhinoceros feeds on a species of thorn known in Cape Colony as wait-a-bit, which gives it a somewhat acrid and bitter flavour. The white species, however, feeds chiefly on grass. The flesh has in consequence a pleasant taste, and is usually very fat. A high polish can be given to the horns of the rhinoceros, and they are valuable articles of commerce. They fetch, indeed, half as much as common elephant ivory. They are formed into drinking-cups, handles for swords, ramrods for rifles, and are used for many other purposes.

" When you speak of drinking-cups," said David to our Portuguese friend, who had given us this account, " I have heard that they are believed to possess the virtue of detecting poison. It is said that if wine is poured into them it forthwith rises and bubbles up as if it were boiling; and if poison is mixed with it, immediately the cup splits. It is said, also, that if poison by itself is poured into one of these cups, that the cup

will instantly fly to pieces. I confess, however, that I am inclined to doubt that such is the case."

"I also have no belief in the account," remarked Senhor
Silva.

The ordinary way of killing the rhinoceros is to stalk him
either when feeding or asleep. By approaching to leeward,
a good shot will kill him before he moves. Some hunters
prefer hiding themselves in huts or pits, as he comes to
drink in the stream at the morning or evening. Sometimes,
however, the animals are taken in pitfalls, such as are used to
capture elephants or other large game. Englishmen (for I
have not heard of any one else who has done so) occasionally
hunt the rhinoceros on horseback. Though their horses have
been able to keep up with the chase, the infuriated beasts
have been known to charge the hunter. In two instances I
heard of, the horses were completely run through by the creature's horns; and in two others, the unfortunate huntsmen
themselves were killed, being fearfully gored by the savage
brutes.

I was very anxious to set off to bring back Natty; and in the
afternoon Stanley pronounced the horses fit to proceed. Mr.
Donald Fraser proposed accompanying us; but when Stanley
promised to try and induce some of the blacks to come south and
trade with him, he abandoned his intentions, hoping to do a
stroke of business in the meantime with any natives who might
come to the camp. Timbo therefore took the third horse, and
I mounted the one he would have ridden. They were all
three fine strong animals, fleet and active; and we hoped on
their backs to bid defiance to any human beings or wild beasts
we might encounter. Stanley did not fail to urge on those
who remained behind the importance of keeping bright fires
burning round the camp at night, and being ever on the watch,
lest the wild beasts we had encountered might be tempted

to swim across the stream and attack either them or the oxen.

"Do you, my dear brother, be careful of yourself," said Kate, as she wished us good-bye. "You seem to forget that though you have attacked so many of them successfully, some day they may turn round and treat you in the same way."

CHAPTER XXIV.

HOUGH, after the wild life I had been so long living, I would gladly have remained behind in the society of my young cousins, I was so anxious to learn how Natty was going on that I felt very glad when I found myself in the saddle, with saddle-bags well stored with rhinoceros' meat and other eatables, and my rifle by my side. We had tethers for our horses, hooks for cutting grass for them, and axes for supplying ourselves with firewood to keep up blazing fires at night.

As we rode along, Stanley gave me fuller details of the attack which had been made on our village, and which had resulted in the party being compelled to quit it and seek safety at Kabomba. Soon after we had left our home on our unfortunate expedition, Timbo had set off to Kabomba, in the hope, as he said, of telling the natives about the Bible, showing them how much superior is the white man's religion to their foolish idolatry. They had listened more readily than he had expected; and his great wish now was to return there at some future day with missionaries, who might teach them to read about the matter themselves. He had just got back, when one morning Jack Handspike, who was on guard, observed a body of blacks approaching. At first he thought that they were the villagers for whose benefit Stanley had killed the man-eating lions. They, however, very soon exhibited their hostile intentions, by letting fly a shower of arrows into the enclosure. Happily no one was hit. Jack instantly roused the inmates, and fired his rifle at their assailants, while Stanley and the rest seized their arms and rushed out to defend the fortress. Their assailants were, however, too well acquainted with its construction, and were now seen rushing on, each man with a torch in his hand. These they threw among the prickly-pear hedge, which, dried by the hot sun, was as combustible as tinder. In an instant the whole was in a blaze. Stanley had collected his party, each one being loaded with as much property as could be carried. Then, sallying forth, they fired a volley, which drove the blacks to a distance. They were thus able to secure several of their animals, and to save a few more of their effects. They now retreated to some rising ground, where they witnessed the utter destruction of our habitation. The blacks had probably not expected so brave a defence. They once more came on ; but a volley killed three of their number, and the rest, disappointed of their expected plunder, took to flight. Timbo on this urged Stanley to set out without delay for Kabomba. They were happily able to reach it, though my young cousins had undergone great fatigue on the journey. After a stay of a week at Kabomba, they had received information that a party of white travellers had appeared at some distance to the south. Scarcely expecting that Senhor Silva could have returned so soon, they set off in the hope of falling in with the strangers, accompanied by an escort of the Kabomba people, who were anxious to show their gratitude by guarding them on their way. They had fallen in, as I have mentioned, with Chickango, and arrived safely at Donald Fraser's camp. Timbo supposed that the attack had been made by a tribe from the border of the lake, who had heard of the wealth possessed by the white men. It occurred to me that they had possibly come from the very village which our friends had advised us to avoid; and such I found was the case. Had we fallen into their hands, our fate would have been sealed.

Soon after leaving the camp, we saw before us a grove of tall palm-trees. At first they appeared to form a part of an extensive wood. As we drew nearer, we discovered that the trees grew at considerable distances from each other. They were tall and extremely graceful, each branch having the appearance of a beautiful fan; and as the wind waved them to and fro, the effect was peculiarly pleasing. They are known as " fan-palms "—the most beautiful, perhaps, of their tribe. We found fruit growing on them about the size of an apple, of a deep brown colour. Timbo begged us to stop, and he would try and get some. He accordingly climbed up one of the trees, helping himself with a band round his waist, and soon came down with a number of the fruit. They contained kernels as hard as a stone, which put us in mind of vegetable ivory. We found the fruit very palatable and refreshing. Most of the trees, however, were so tall, that it was evident the fruit could not be obtained without difficulty. I should have said we took a couple of dogs with us which had attached themselves to Stanley. They might prove useful at night in giving us warning of the approach of any wild beast; and we were therefore glad of their company. The country was tolerably open, but in some parts we had to pass through dense forests. In most of these, however, we could generally find an elephant path from one side to the other, always broad enough to allow two horsemen to ride abreast. Frequently Stanley rode ahead; while I rode alongside Timbo, who was more communicative than my cousin. He, I have already said, was a man rather of action than words; and would, for an hour together, ride without speaking, unless something attracted his attention. He had gone some way ahead, with the two dogs at his side; we following at a little distance, though, of course, always keeping him in sight. Timbo was recounting, with considerable animation, some of the adventures of his youth, when suddenly his narrative was interrupted by a loud trumpeting sound, and we saw Stanley wheel round and gallop towards us. At the same moment, a huge elephant, the largest monster I ever saw, with trunk projected, vast ears spread out, and tail erect, burst from the thicket, and in hot haste pursued my cousin.

"Fly! fly!" shouted Stanley; "gallop off for your lives!" We required no second order to obey him. Stanley was looking round at the monster; but, situated as he was in a pathway between thick trees, among which he could not force a passage, he was unable to fire. Flight was our only resource. We were already deep in the forest, and I had remarked no other way except the one by which we had come. Had we stopped and attempted to fire, we might too likely have shot Stanley, who was directly between us and the elephant. Had we missed, Stanley would certainly have been trampled upon; and so probably should we, as by the delay we should have impeded his progress, and prevented him from escaping. Very unwillingly, therefore, we turned our horses' heads and galloped on, hoping to keep ahead of him. His horse was, fortunately, the fleetest and strongest animal of the three. It seemed also to know its danger, and flew along over the ground at a rapid rate; but still the cumbrous monster came as fast, trumpeting and shrieking with rage. His huge feet almost touched the horse's hinder hoofs, so it seemed; while his trunk, in the glance I had got of him, appeared to be about to descend upon Stanley's head. So dangerous was the position in which he was placed, that I scarcely dared hope he would escape. "On! on!" he shouted. "On! on!" we shrieked in return, trying to urge forward our steeds at a little faster rate. The dogs, aware of their danger, scampered off, with tongues hanging out, watching for an opening in the thicket through which they might bolt. We had passed over several fallen trees and other impediments in the path; and I dreaded lest, coming against such, our horses might stumble. Now a trunk appeared before us. Our horses leaped boldly over it. I hoped that Stanley's would follow, and that it might offer some impediment to the elephant. Glancing for a moment anxiously round over my shoulder, I saw that the monster had also got over it without stopping. Could we once gain the open country, I knew that we should have a better prospect of escaping; because by separating the elephant would hesitate which to pursue, and while he followed one of us, the others would be able to fire at him. Still we had a considerable distance to go, for I calculated that we had penetrated a mile or more into the forest. It was indeed a gallop for life, and the elephant seemed determined to wreak his fury on us. What had offended him so much it was difficult to say—perhaps the sight of a horse, strange probably to him.

I think I have mentioned that when a troop of elephants are passing leisurely onwards, feeding as they go, their footfall is unheard; but when angry, the case is very different. The monster seemed to make the very ground quake beneath his feet, as he came trumpeting on behind us, adding not a little, I suspect, to the terror of our horses, which, with manes and tails streaming out, like some demon-pursued steeds of German legend, dashed through the wood. There was no need of whip or spur to urge them on. How thankful I felt when at length, under the tall arched trees, I caught sight of the open plain! Still our steeds dashed on. I turned my head to learn how it fared with Stanley. He was sitting his horse as composedly as ever, though the elephant was close behind him. "Andrew, turn to the right!—Timbo, keep ahead!" he shouted. We obeyed, and the elephant dashed out of the cover. The huge animal was coming on at even greater speed than at first, in no way out of breath with its long and tremendous charge. Stanley wheeled his horse to the left, while the elephant dashed forward, and seeing Timbo, pursued him. This was exactly what Stanley wanted. Again wheeling his horse, he followed, keeping on the quarter of the animal. I saw he was getting his rifle ready to fire. I imitated his example. The dogs, too, breaking from the cover, came in pursuit, and assisted us. With difficulty could Stanley curb in his horse. The elephant, hearing noises behind him, stopped. The instant he did so, Stanley's rifle was at his shoulder. There was a report, and the animal, a moment before so terror-inspiring by its bulk and powers of destruction, sank upon the grass. Its trunk fell, its mighty limbs stretched out, and before one of the yelping dogs could reach it, life was extinct.

Our escape had indeed been providential. It was some minutes before Timbo could rein in his horse, and we had to shout and shout to him to return. At length, however, he arrived, and was as delighted as we were to see our enemy overcome.

Timbo proposed that we should return to the camp and get our friends to come and carry off the tusks and flesh; but as I was anxious to get assistance for Natty as soon as possible, I begged Stanley to proceed, hoping that we might find the tusks on our way back.

"Dat bery unlikely," said Timbo; "but we cut dem out and hide dem, and den if black fellows come to take de meat, dey no find de tusks."

We accordingly set to work to cut out the tusks, which

Timbo then hid in the wood and covered them up with branches. I asked Stanley whether we should proceed by the pathway, or take the route outside the forest.

"There is but little fear of our encountering another fellow like the one we have killed," he answered. "He was evidently a solitary beast, by his savage disposition; and the chances are we shall get through without further interruption. If not, we can but have another gallop for it, Andrew. I rather enjoyed mine; though, to be sure, it was a neck or nothing affair."

This was the chief difficulty we met on our journey. We formed our camp at night, as we had proposed. With the aid of the dogs and the watch-fires, we were uninterrupted, although the roars of lions were heard in the distance, and we had visits from jackals and hyena-dogs, who came prowling round, attracted by the scent of our roasting meat; Stanley's unerring rifle supplying us amply with game. We had a pleasant addition one day in a large bustard which he shot. Though very abundant, the bird is shy, so that a good sportsman alone can hope to kill it. It weighed about fifteen pounds. The flesh was very tender and palatable, and we agreed that it was the best flavoured of the game birds we had met with. After each day's journey, Timbo generally went in search of small game or birds' eggs, of which he brought us a plentiful supply; so that we lived in abundance.

At length we recognized the reed-covered habitations of our Kabomba friends, the whole population apparently turning out to welcome us. The chief men, and those who had accompanied Stanley to the camp, hurried forward to grasp his hands, while the rest stood at a distance, gazing at the strange animals which our horses appeared to them ; indeed, those only who had been to the camp had ever seen a horse before. Our first inquiries were, of course, for Natty.

"Chief say better, but not like walk much," answered Timbo.

"Beg them to let me see him at once," I said, riding on.

It was difficult, however, to get through the dense mass who came to shake our hands and embrace Timbo—a ceremony to which they knew we objected. At length we reached the chief's house, at the entrance of which Natty was standing. Poor fellow! he still looked very pale and thin, and I was afraid from his appearance that his days were numbered.

"I shall get better now you have come for me," he said, looking up in my face. "I have been so longing for your return, and began to dread that some accident had happened. Do not be anxious about me, Andrew. I know—I am sure I shall get better."

I trusted so. "The food on which he has been living probably has not suited him," I thought; "and when he is placed under David's care, he may begin to improve." This hope prevented my spirits sinking, as they would otherwise have done. We told the Kabomba people that we were anxious to return immediately to our friends; and as I saw that it would be dangerous for Natty to ride behind one of us, as we had proposed, I begged the chief to allow some of his young men to carry him. To this he agreed; and forthwith I set to work, aided by Timbo, to form a litter. There were plenty of bamboos in the neighbourhood, and with these we constructed a light and very convenient conveyance, with a roof, back, and sides. The greater part was formed of bamboo, and matting served as a cover to keep off the sun's rays in the day-time, and the damp at night. We then had to train some bearers; for the people were unaccustomed to bear loads in the way a litter must be carried. Timbo employed his time, when not assisting me, in addressing his countrymen. When I asked him if he had succeeded in impressing on their minds any gospel truths—"Yes," he said; "I sow leetle seed, but it grow up and bear fruit some of dese days. No fear; dat seed I sow nebber rot."

Among the inhabitants of the village I recognized my three faithless attendants. The chief expressed himself very much ashamed at their having deserted me. They excused themselves by saying that they thought I had been made prisoner, and that they had run away to avoid sharing my fate. I replied that I was very glad they had got home safely, and that I harboured no ill-will towards them.

"I tell dem dat Christians ought to do good to deir enemies, so dey understand why you no beg de chief to kill dem," observed Timbo.

At break of day we commenced our return journey. Our style of travelling was very different from what it had been during my former adventures. We had bearers for Natty, and also a party of armed men with shields and spears as a body-guard, and others carrying provisions, while we ourselves were mounted on strong steeds. For most of the time I rode near Natty, anxious to keep up his spirits. Now and then Timbo took my place. Stanley generally rode ahead; as, however, we had to proceed slowly, he frequently started off with the dogs to get some sport. He was, as usual, successful, and kept our pots well supplied. I told him he must look out, and not be caught by another rogue elephant.

"No fear of that," he answered. "I keep my eye about me; and, in truth, I should rather enjoy being again chased. It is but fair, considering how fond I am of hunting animals, that I should occasionally be hunted in return."

We had accomplished four days of our journey, when, early in the morning, Stanley was riding some distance ahead, and Timbo and I were keeping at the side of Natty's litter. Natty was, I hoped, decidedly better. He was able to walk about every evening in the cool, and would sit at the camp-fire and join in conversation as well as any of us. We were passing along the edge of a wood, of which there were several scattered about in sight, though the country was generally open. A shorter way might have been found, perhaps, through the wood ; but our black friends declined entering it, declaring that many lions lurked there, and urging us to be on the watch for them.

" I only wish some of them would come out," observed Stanley. " I should like to carry home the hide of one, for I have lost all those I have killed."

Stanley, as I have said, was a little in advance, keeping close to the wood, looking apparently into it in search of game, for he was as good a shot on horseback as on foot. Presently I saw his horse swerve on one side. With whip and spur he brought the animal again up to the wood. Just then there was a fearful roar. The horse again started on one side, the suddenness of his movement almost unseating his rider, whose cap was knocked off. The next moment a huge lion, breaking cover, sprang out of the wood with a tremendous bound, and alighted on the back of the horse, grasping Stanley with one of his tremendous claws. Stanley, leaning over his horse's neck to avoid him, in vain attempted with his rifle to beat off the savage brute. To gallop to his rescue was the impulse of the moment. In another instant my cousin might be killed ; for had he once been dragged from his horse, nothing could have prevented the lion seizing him between his powerful jaws, wide open at that moment to grasp him. The risk Stanley had run in the adventure with the elephant seemed as nothing compared to the awful danger in which he was now placed. Our horses, though not unaccustomed to carry their riders in chase of lions, trembled in every limb. The frightened blacks were about to fly, leaving Natty on the ground. I shouted to them to come back, when Timbo and I spurred on our horses towards my cousin. He caught sight of us coming.

"Fire! fire!" he shouted. "Kill the brute! Never mind though you hit me!"

I sprang from my horse, and just as I got my rifle to my shoulder, Stanley, with the lion still clinging to him, dashed by. It was not a moment to hesitate. If I failed to hit the lion, my cousin must be killed. I fired, and he and the lion fell from the back of the horse. My heart felt sick, for I thought he had been killed. The horse, freed from the grasp of the mighty brute, galloped off across the plain. My cousin lay on the ground, and I saw that the lion's paw was still on him. I instantly began to reload. Timbo in the meantime had come up. What was my horror to see the lion, though wounded, working his way on towards Stanley's body. I was afraid if I now fired of hitting him. Without a moment's delay Timbo bravely rushed forward, shouting loudly, when the lion, raising himself on his fore-feet, and crouching down, prepared to make his deadly spring. Timbo stood firm as a rock. I fired. For an instant I saw the lion in the air; but the next he rolled over, not two feet from the brave black. I rushed up to Stanley. As I approached, he lifted himself on his arm, greatly to my relief.

"He nearly did for me; but I believe I am less hurt than I supposed!" he exclaimed.

However, even as he spoke, he sank back again. I knelt down by his side. The lion's claws had inflicted a fearful wound on his shoulder, and his hip also appeared to be greatly torn. Timbo, having ascertained that the lion was dead, now came up to assist me in supporting his master. Fortunately we had brought some spirits. I shouted to the blacks to come on with Natty and our goods, and as soon as possible poured a good portion of spirits and water down Stanley's throat.

Natty had got out of his palanquin and came towards us. Some of the blacks had, in the meantime, gone off to catch the horses. Poor Natty's concern was very great at seeing what had occurred.

"O Captain Hyslop, you must be put into my litter!" he said; "I am sure I shall be able to ride, for I feel quite strong now."

This indeed seemed the only way of conveying Stanley.

"But suppose I go on, and bring up Massa David," said Timbo. "Dat is de best t'ing."

I agreed with him. Having washed Stanley's wounds, and bound them up as well as I was able, with Timbo's assistance, we placed him in the litter; while Natty mounted my horse, I agreeing to walk by his side. The blacks having caught the horses, Timbo set off, leading Stanley's steed, in order that David might ride back on it to his brother's assistance. We then proceeded at a somewhat slower pace than before, the bearers finding a great difference between my strongly-built cousin and poor young Natty. As may be supposed, we kept a very strict watch at night, lest we might be visited by another lion. Stanley did his utmost to keep up his spirits; but from the fearful laceration he had suffered, his nervous system was greatly shaken, and he often relapsed into a state almost of unconsciousness. Natty, however, with the air and exercise, recovered his strength, and every day looked better. I was very thankful when, towards the end of the next day, I caught sight of two objects moving over the plain towards us. Gradually, as they approached, I made out two horsemen, and in a little time David and Timbo galloped up to our camp. Timbo's anxiety about his master had probably made him describe the wound as worse than it was, and David was in a state of great agitation when he arrived. He, however, after examining his brother's hurts, expressed a hope that they

would soon get well, and complimented me greatly on the way
I had treated him. Still, I was very glad that David had
arrived; for, in consequence of the constant state of stupor
into which Stanley had fallen, I began to feel very anxious
about him.

We continued to travel on at a very slow pace, as Stanley
could not bear any shaking. Three days more therefore
passed away before we came in sight of the camp.

I had never before seen my cousin Kate so much out of
spirits, and it was not till two or three days after our arrival,
when Stanley was found to be progressing favourably, that
she was at all herself again. To me, however, she was always
kind and gentle.

CHAPTER XXV.

N spite of Mr. Donald Fraser's expostulations, Senhor Silva would not consent to break up the camp till Stanley was in a fit state to travel. The honest trader, however, had no cause to complain, for he was driving a brisk trade, not only with our friends from Kabomba, but with the people of a number of neighbouring villages. Some he visited in a light cart, which accompanied the waggons, and a considerable number came to us. We had not forgotten the elephant's tusks, which Timbo had hid, and as soon as he believed his master was out of danger, he set off with one of the horses to bring them into camp. The elephant itself had long since disappeared, its skeleton alone whitened the prairie; but the tusks were safe, and were safely bestowed in the waggon, in part payment of our debt.

One morning our oxen, which were feeding near, suddenly started off in every direction, leaping, twisting, and turning about, and cutting the most ridiculous capers. They looked as if they had been seized with Saint-Vitus's-dance. On running towards them I discovered that a large flock of birds were clinging to their backs—three or four on each animal. Having my gun in my hand, I shot one, when I found that it

was the same bird which I had seen on the back of the rhinoceros. Senhor Silva, who arrived laughing heartily at the
commotion among our animals, told me that the bird is called
the *Buphaga Africana*. Its object in thus taking possession
of the backs of the cattle is for the purpose of feeding on the
ticks with which they are covered. They have particularly
long claws and elastic tails, which enable them thus to cling to
the hide, and to search every part of the beast, in spite of its
efforts to get rid of them. When animals are accustomed to
these birds they appear rather grateful for the visitation; but
Mr. Fraser's oxen had apparently never experienced a similar
visitation, and were therefore considerably astonished at being
thus unceremoniously assailed. By degrees, however, when
they found that they could not throw them off, and that
nothing very terrible happened, the oxen remained quiet, and
were probably much more comfortable from being delivered
from their parasitic pests.

The necessity of supplying our camp with meat compelled
us frequently to go out shooting. We greatly missed, on
these occasions, Stanley's unerring rifle. Our party generally
consisted of Senhor Silva, Timbo, and myself; but sometimes
Mr. Fraser took Senhor Silva's place; and he was, I must say,
the best shot of the party. We had been unsuccessful, however, on several occasions, and though there was no famine in
the camp, we had very little meat fit to eat; while our black
attendants were beginning to grumble greatly at being placed
on short commons. This made us more than ever anxious to
get some game. We had scoured the country towards the
south for some distance, and falling in with no animals, we
were induced to proceed further off than usual. The country
over which we were passing was a fine undulating plain. Now
and then there were dips of sufficient depth to conceal us from
each other, for we rode apart in order to cover a wider extent

of ground. My companions were not in sight. I had reached a slight elevation, when I saw in the distance a herd of large animals. At first I took them for buffaloes; but their movements soon convinced me that they were of the antelope species. The wind fortunately came from them, and I determined, without waiting for my friends, to endeavour to bring one of them down. I galloped on, till, to my delight, I saw before me an immense herd of the large eland, as they are called, or, more properly speaking, "cana." In stature they are equal to a good-sized horse. Their horns are long and spiral. The form of the creatures before me was massive, their tails terminating in tufts. I had never possessed much of the spirit of a hunter; but the necessity of obtaining food made me as eager as any professional hunter could be to bring down one of the fine animals. I put spurs to my horse, and galloped on, getting my rifle ready in the meantime to fire immediately I could obtain a fair shot. The creatures for some time did not see me, and not till I was close upon them did they take the alarm. Near me was a fine large buck. I had seldom fired from horseback; but my animal was steady, and I determined to make the attempt. I took aim, and, greatly to my satisfaction, struck the creature near the shoulder, and over he went. Seeing that he was utterly disabled, I dismounted from my horse, and gave him a merciful thrust, which deprived him of life. Immediately reloading, I again leaped on my horse's back, and made chase after the herd, which had now got to some distance. However, I found that I was coming up fast with them, and in a short time another fat animal lay rolling on the turf.

Wishing for the assistance of my companions in cutting up my prize, I rode to the nearest height in the hope of seeing them. I cast my eyes round in every direction. They were nowhere visible. I began to fear that they had gone in a

different direction. I shouted with all my might, thinking
that one or the other of them might be concealed in some
hollow, and that my voice might reach them. I could only
carry a part of the elands. After waiting awhile, I rode back
to where I had killed the last. Already several birds of prey
were hovering about. I scared them off, however, by my
shouts; and then passing the bridle of my horse round my
arm, I began in a very unscientific way to dismember the noble
beast I had killed. I did not like the employment; at the
same time, it was necessary to secure the meat. I had been
for some time thus employed, when I heard the sound of wings
close above me, and looking up, saw, with a feeling of no small
alarm, a flight of kites hovering near my head. My horse, too,
not liking their appearance, started back; and not without
reason, for they might quickly have torn out his eyes with
their powerful beaks and claws. I shouted, and waved and
clapped my hands. They retired to a short distance, but only
to come on again with renewed fierceness, seizing pieces of the
meat and flying off with them. I determined, however, not
to be defeated; and standing by the body of the eland, struck
out right and left with my knife. Some literally fell back on
the ground, spreading out their wings and talons and opening
their beaks to defend themselves. My determined onslaught
on them, however, compelled the first batch to beat a retreat;
but another immediately took their place, pouncing down as
the others had done on the carcass. I knocked over two or
three, and the second party retreated, a third, strange to say,
immediately afterwards coming on to the attack; but they
had become so wary that I was unable to reach them. Still,
as they kept about me, I expected every moment that they
would assail my head, and I could not help feeling how fearful
would be my position if they did so. At last I determined to
try the effect of my rifle, which I had not loaded after my last

shot — a neglect which might have proved extremely disastrous had any savage beast appeared. I loaded with shot. In consequence of my shouts and cries, and repeated blows made at the birds, they retired once more to a short distance. The next time they approached I fired into their midst, and a couple fell to the ground, and others were wounded. Still the army kept their ground. Seeing the effects of the first shot, I loaded again, and as they came hovering close to me, I fired once more, with the same success. Greatly to my satisfaction, on discovering that they could not obtain their feast without greater loss than it was worth, the whole army flew off, not appearing to stop while they remained in sight.

Thus being rid of my unwelcome visitors, I returned to my occupation; remembering, however, first to reload my rifle with ball, lest a hyena, panther, or lion might scent the dead eland and come to banquet off it. I had some leathern straps with me for the purpose of securing any animal I might kill, and with these I fastened to my saddle as much meat as my horse could carry. I was sorry to leave any part behind, knowing how much it was wanted in the camp. I now turned the horse's head towards the camp, intending to pass by the eland I had shot. As I approached, I saw some objects moving over the ground towards it. At that distance I could not tell what they were. They might be lions or panthers. If lions, I might probably have to do battle for my prize. I could not help thinking, too, of the way Stanley had been handled. It was not impossible that they would attack me, and get me and my horse, and the meat into the bargain. Knowing that on such occasions boldness is always the best policy, I rode forward, and in a short time distinguished three spotted hyenas stealing up towards the body of the eland. I determined to prevent them having their feast, or spoil it if I could not. So eager were they to seize their prize that they did not notice me. As they drew nearer they hastened their pace, and then made a dash at the carcass. At that moment, putting spurs to my horse, I dashed on towards them shouting and shrieking. They received me with loud snarls, appearing in no way disposed to take their departure. Not till I got close up to them did they retreat, snarling and grinning horribly at me. The scent of the meat had undoubtedly sharpened their appetites, and they certainly looked capable of making a spring and trying to carry away some of the joints. On this I charged them, and they retreated still further off.

I saw that unless I could find my friends, we should have no prospect of saving any of the meat. I therefore looked round again, and thinking that the sound of my rifle might attract them, I fired it at the nearest brute. Over the animal fell, and his companions scampered off to avoid a similar fate. As there was no object in delaying longer, I once more directed my horse's head towards the camp. Not till I had got to some distance did I catch sight of Donald Fraser and Timbo. They instantly galloped back, in the hope of being in time to secure the venison, and I proceeded at a slow pace, which the heavy weight my horse carried made necessary. I was still at some little distance when they overtook me, saying they had been too late. A number of birds and beasts of prey had set on the carcass, and devoured the greater portion. However, the supply I brought was doubly welcome. As it would only afford enough food for a day's consumption, we agreed to set out again immediately, in the hope of falling in with another herd of elands. The importance of obtaining food was very great. Mr. Fraser's attendants were already grumbling at their short allowance, and he was afraid that they would desert him, and leave us to make our way alone. He also was glad of an excuse for moving southward. We had been out a considerable part of the day without being able to get up to any

herd, though we saw one or two in the distance. I was talking to my companions, when, looking up, I saw before us
what seemed like a dark cloud moving through the air at no
great distance above the earth.

"What can that be?" I exclaimed, pointing it out to Mr.
Fraser.

"I will tell you presently," he said. "I fear it bodes us no
good!"

The cloud, as I may call it, now seemed to rise higher
in the air, in the same compact body as it at first appeared.
Then it suddenly sank and dispersed into smaller portions.
Now again it united, again to spread and to rise, very much
with the appearance of huge columns of sand whirled up by
the wind. On it came towards us.

"I will tell you what it is now," said Mr. Fraser. "That
is a flight of locusts. Woe betide the spot they select as a
resting-place!"

As he spoke it appeared as if a heavy snow-storm had begun,
for the locusts, as they alighted on the ground, looked exactly
like huge snow-flakes. Several thousands might have fallen
round us; still the whole mass seemed in no way diminished,
and on they flew, the noise of their wings sounding like that
produced by a gale of wind whistling through the rigging of
a ship at anchor. On and on we rode, but still the mighty
mass of winged insects advanced. Far as the eye could reach,
they appeared hovering in the air. We pushed on for some
miles, hoping to get beyond them; but the same dark cloud
appeared before us. Not an animal was to be seen. We
turned to the left and galloped on, but still could not get clear
of the mighty column.

"There will be small chance of our meeting with any game
to-day, I suspect," observed Donald, pulling up and looking
round him. "It will fare hard, too, with our poor cattle, I am

thinking, for these hungry creatures will make sad havoc in the camp if they pitch on it, and the surrounding country too."

Still, I urged that by pushing on we might fall in with a herd of deer, one or more of which would pay us for our long ride, and supply our larder with the much-needed flesh. We rode forward, hoping yet to fall in with some game. Still we were as unsuccessful as at first. We had gone some distance, when we came upon masses of creatures of a reddish colour with dark markings. In some places they covered the ground in layers two or three inches thick. Mr. Fraser told me they were the larvæ of the locusts, which the Dutch people at the Cape call *voet-gangers*—literally, foot-goers Some were seen hopping among the grass, devouring it with extraordinary rapidity.

"What do you think, Mr. Crawford, of the fruit on those bushes?" said Donald to me, pointing to some shrubs, from which hung what I took to be clusters of magnificent fruit.

Riding forward, I plucked some. My astonishment was great to find that they were merely the larvæ of the locusts hanging to the boughs. So thickly did they cover the branches, that they literally bowed them down to the ground. He told me that these creatures are especially dreaded by the colonists, as it is impossible to stop their progress, and they eat up every green thing in their way. They cross rivers or pools; for though the leaders are drowned, the others pass over the bridge thus formed by their bodies. Even fires, which are sometimes lighted in the hope of stopping their progress, are put out by the countless masses which crawl over them.

It was dark before we reached the camp. We rode together, keeping a sharp look-out, in case we might have been followed by any prowling inhabitant of the wilds. As we drew near the camp, a bright blaze appeared from one side to the other, and I could not help being alarmed at the thought that the waggon and tents, surrounded by dry grass, might have caught fire. Mr. Fraser, however, quickly calmed my fears.

" Our people, I suspect, are having a feast," he observed. " It is an ill wind that blows no one good; and if the locusts have eaten up the cattle's fodder, our people are engaged in eating up the locusts."

On entering the camp, we found all the blacks busily employed round the large fires which they had lighted. They were scraping together vast quantities of locusts which, passing through the flames, had scorched their wings, and fallen helplessly to the ground. Even Senhor Silva, and David, and the boys, were engaged in the work. In a short time they had collected enormous heaps of the insects, when the blacks began to roast and eat them as if they were the most delicate morsels. If they would support existence, there was no fear of our starving. I was tempted to taste some, but cannot say that I found them very palatable. As soon as the fires had burned low, the blacks set to work to dry the locusts in the hot ashes. Then obtaining some large flat stones from the river, they ground a number between them into powder, which they stowed away in all the receptacles capable of holding it.

Next morning, by daybreak, the neighbourhood of the camp was beset by numberless birds, among which I distinguished storks and kites, and soon perceived that they were employed in gobbling up the locusts. The most numerous, however, were small birds about the size of swallows. Mr. Fraser told me this was a species of thrush which constantly follows the locusts, and is said even to build its nest and rear its young in their midst. He called it by its Dutch name—*Springhaan oogel*. David said he knew it as the locust bird. We shot a number; and though we were not tempted to feed on locusts, we had no objection to breakfast off the result of our sport. The fires had partly saved the camp itself; fortunately so, for the locusts will not only eat grass, but every animal produce which comes in their way. Leathern thongs, even boots, shoes, and bags, have been destroyed by them. I saw a number also feeding on their own companions, which we had trampled on as we passed through them. No sooner had the warm rays of the sun been cast across the plain, than, with a loud whirring sound, the locusts rose in the air. Notwithstanding the number which had been destroyed, the mass appeared in no way diminished in size. Onwards they flew to their unknown destination; but what a scene of desolation met our eyes across the country on which they had rested! For miles, as Donald had feared, not a blade of grass was to be seen. As far as the eye could reach, where the evening before the ground had been green and smiling, it now looked brown and parched up, as if a fierce fire had passed over it.

"If we cannot move from here, we must just make up our minds to remain for ever," observed Donald, "for without food—and not a particle of grass do I see in any direction—the poor cattle will soon be starved to death. We must see Senhor Silva and your cousin David, and hear what they say."

Soon afterwards we were joined by David, and in reply to Mr. Fraser's remarks, he said he hoped that Stanley was sufficiently recovered to bear the fatigue of travelling in the waggon. I undertook to arrange a bed slung from the roof, by which any jolting might be avoided. Calling Jack Handspike to my assistance, we soon contrived a comfortable cot for my cousin.

Meantime every arrangement was made for starting. The oxen were harnessed to the huge machine; Kate and Bella took their seats by the side of their brother. The word was given to move on. The Hottentot drivers smacked their huge whips, and the lumbering waggon was put in motion. Natty and Leo had greatly recovered; but that they might not be fatigued, room was also made for them inside. Donald and

Senhor Silva mounted their horses, and David and I agreed to take one between us. When he rode, I was to walk, or mount an ox. The rest of the party proceeded on foot, or on the oxen. Far as the eye could reach, nothing but brown earth and leafless shrubs and trees could be seen; the ravaging hordes of locusts had cleared off every particle of verdure from the ground. North and south, east and west, the country had become a barren wilderness. The prospect for our poor oxen was a melancholy one. We could only push on therefore as fast as they could travel, in the hope of speedily getting beyond the ravaged country. Our friend Donald looked very grave. All day long we travelled on. The cattle began to show signs of thirst. We ourselves were also suffering from want of water, as we were afraid of exhausting the small supply we had brought in our bottles. At length some rocks appeared ahead, near which Donald told us was a pool. The cattle seemed to be aware of it, and eagerly moved on; but as we got near, no bright gleam, such as gladdens the sight of the thirsty traveller, played on the spot. On arriving at it, a mass of locusts and their larvæ were alone visible, completely filling the space where the water should have been.

Our men immediately set to work, and literally dug them out and threw them in masses on the shore. We ourselves could not have drunk a drop. The pool seemed filled with a thick, muddy, and putrid liquid. The cattle, however, when let loose, rushed down to it and drank eagerly, though I was afraid it would produce disease among them. Poor creatures! if there was nourishment in it, it was the only food they got; for that night we had to camp without water for ourselves or fodder for them.

CHAPTER XXVI.

ANY days had passed by, during which the usual incidents of African travel had occurred; but I need not stop to describe them, except to say that Mr. Fraser had been successful in killing several elephants, which he did for the sake of their tusks, and also in purchasing a large quantity of ivory from the natives who visited our camp to trade, or inhabited the villages near which we passed. Thus he had no reason to complain of the long journey he had made to rescue us, although we were not the less inclined to be grateful to him.

The country ravaged by the locusts had been passed at last, but not till our cattle were almost starved, and we and they had suffered greatly from want of water. The dried and pounded locusts had assisted to support our people, but we were now greatly in want of provisions. Stanley had borne the journey remarkably well, and was rapidly recovering from the hurts inflicted upon him by the lion; while Leo and Natty were completely themselves again. Stanley was very anxious once more to mount his horse and to assist in hunting, in order to supply the camp with food; but of this David would not hear, and declared that it would be equivalent to fratricide if he allowed it. Donald, Timbo, and I, and sometimes Senhor Silva, therefore scoured the country in every direction in search of game. Donald and I were riding on ahead one day, when he observed on a bush a fly somewhat smaller than the common blue-bottle fly—so annoying to the butcher—but with rather longer wings. Begging me to hold his horse, he jumped off and caught it. Instantly leaping into his saddle, he told me to turn and ride for my life, with an expression of consternation in his countenance which made me fancy that he had suddenly gone out of his mind. However, as we rode on he explained that the fly which he held in his fingers was the tsetse fly (David called it the *Glossina morsitans*), and that it was more dangerous to cattle and horses than all the lions and snakes in the country. Fortunately our horses had not been bitten by it. He told me that had such been the case their death would have been certain. It attacks, however, only domesticated animals, for wild beasts range over the country infested by it with impunity; while human beings are scarcely more annoyed by it than they are by flea-bites. It is confined to certain localities, and is never known to shift its haunts. He told me that it was found generally in the bush or among reeds. Though the insect is small, yet the poison it contains is of so virulent a nature that its bite is as deadly to horses and oxen as that of the most venomous serpent. Donald said he had ample reason to be afraid of it, for on one occasion, not believing in its power to injure him, he had attempted to pass through a district infested by it, when he lost all his oxen and horses, and very nearly his own life and that of his companions. They were in a wild and uninhabited district, and were barely able to secure provisions and water sufficient to support themselves, till they could obtain assistance. He said that four or five flies were sufficient to kill a full-grown ox. The animal, however, does not die so rapidly as when bitten by a snake. Sometimes, indeed, it exists for some weeks or months afterwards, gradually

losing its strength, and perishing ultimately of exhaustion.
Frequently, however, oxen die, especially should rain fall, soon
after they are bitten.   In the case of one of his horses which
had been bitten, the head and body swelled, its eyes became
so swollen that it could not see, and it was painful to hear it
neighing for its companions, who stood close to it while feeding.
A remarkable feature with regard to the poison of the tsetse
is that calves, and other young sucking animals, are safe so long
as they suck; but it has been remarked that dogs though reared
on milk die if bitten, while a dog which was reared on the
meat of game accompanied his master when hunting in the
districts infested by the fly without suffering.

We had now entered a far more desolate-looking country
than any we had yet passed through.   Vast sandy plains extended round us, broken here and there by clumps of low
bushes or coarse long grass, with occasional patches of more
nutritious verdure, from which our oxen plucked their scanty
meals.   Still, occasionally, herds of deer passed us in the distance, but they were so wary that we could not approach them.
The open nature of the country made stalking in the ordinary
way impossible.   Every night, however, Donald, accompanied
by Timbo, spent two or three hours, and sometimes longer, in
lying in ambush, hoping to get a shot at a passing animal, but
their success had hitherto not been sufficient to supply us with
as much meat as we required.   Water too was very scarce.
We had been travelling slowly all day, when our cattle began
to move on with greater rapidity than before, evidently believing that water was near.   Donald was riding ahead, looking
about him, when suddenly he and his horse disappeared.   I
was at no great distance behind him, and before I could pull
up I was very nearly following.   I found that he had slid
down into a large sand-well, some twenty feet in diameter at
the top, and upwards of twelve feet in depth.   As soon as he

was extricated from the pit, the men were called with spades to clear it out. However, after digging some time, no water flowed into it, though the bottom became thoroughly moist. We fortunately had some long reeds with us. These were sunk into the sand, and immediately water began to rise. We quickly got enough to quench the thirst of the people, but we had to wait some time before a supply could be procured for the cattle. As soon as water had been given to the horses, Timbo set out in search of game. We were as unfortunate as we had usually been of late. Perhaps this might have arisen from our want of skill. Donald was an inferior hunter to Stanley, and had he been well, we should have met with more success. Timbo was riding near me, when I found him eagerly examining the sand on one side. Without saying a word he jumped from his horse, and began scraping away. Presently he produced a huge egg, then another, and another.

"Dere!" he exclaimed; "we no want food now. See, here are anoder dozen! Dey eggs of ostrich!"

I looked into the nest, and saw that the eggs were arranged with their ends uppermost, to occupy, I concluded, as small a space as possible.

"But, Timbo," I said, "do you think they are fresh, for otherwise I fear they would be of little use?"

"Oh yes," he said; "de hen ostrich only just laid dem. See! see! dere she is, too, watching us!"

At that moment a loud roar saluted my ears. Instantly unslinging my rifle, I prepared to fire, believing that a lion was about to attack us, so similar was the voice to that of the king of beasts in a rage; but on looking round I could see no lion, but instead I caught sight, in the distance, of a huge long-necked bird, which I knew must be an ostrich, evidently observing with anxiety the visit we were paying to her nest. She had gone away, Timbo said, to feed, or otherwise we should probably have found her sitting, as the flamingoes do, with her legs astraddle above it. The poor bird did not attempt to fly, and accordingly gave us time to secure the eggs in a way which we hoped would prevent their being broken. Donald had by this time come up, and telling Timbo to take charge of the eggs, started with me in chase of the ostrich. As we approached the bird, under the idea perhaps of leading us from her eggs, or alarmed, more probably, she set off at full speed. It seemed hopeless, however, to me that we should ever catch her. Away she flew, at first with small strides, increasing every instant, and extending her wings like sails. Her feet scarcely seemed to touch the ground, and yet we could see huge stones thrown up behind her, flying into the air. " On, lad ! on !" shouted Donald. "We will weary her out this hot day. She will slacken her pace soon, and we may turn her maybe towards Timbo, if we do not run her down." Instead of pursuing directly in the wake of the bird, he turned on one side and I on the other; and at length she began, as he had expected, to slacken her tremendous speed. We were now moving up on parallel lines at some distance from her. At length we got ahead, when the bird, wheeling round, started back towards her nest. " Hurrah !" shouted Donald, " she is ours now !" Again we followed the mighty bird, never for a moment allowing her to stop. It seemed a question whether she or our horses would have to give in first. At length a patch of the candelabra-shaped tree euphorbia appeared in sight, and the hard-pressed ostrich darted towards it, endeavouring, it seemed, to force her way through. Pressing on, we were soon close to her, when Donald, raising his rifle, fired, and the bird fell over. I was galloping up, when he called to me. " Stand back ! You might as well get near a dying lion ! A kick from one of her feet would break your horse's leg, and kill you, if you got within her reach." In a few minutes the bird ceased to move, and jumping from our horses we approached. The ostrich must have been nearly eight feet in height, the feathers being of an ashy brown colour, slightly fringed with white. And now, for the first time, I saw those magnificent large plumes of beautiful white feathers which form the wings and tail of the bird. These the trader immediately began to pluck out with the greatest care, and having done so, secured them to our backs, where they were likely to be free from injury. He called me to assist him in hoisting the body up on his horse. It must have weighed upwards of two hundred pounds, no slight addition to the burden his tired steed had to bear.

On reaching Timbo we found that he had discovered another nest of eggs. With these I loaded myself, and well satisfied with our prizes we returned to the camp. "No starve now, Massa Andrew!" said Timbo, as he gave an affectionate glance at the huge eggs. As we rode up David and the two boys saluted us with shouts of laughter, at the extraordinary appearance we cut with the ostrich feathers sticking above our shoulders. Donald, I found, claimed them as his own property, and I did not wish to dispute the point, though I should have liked to have presented one to Kate and Bella. I could only hope to capture another bird without assistance. As soon as we had deposited our burdens, Timbo set to work to prepare the eggs. His process was a simple one. First, having made a hole at the end of the egg, he introduced into it salt, pepper, flour, and one or two other ingredients. He then shook the egg thoroughly, so as to mix what he had put in, as well as the white and yolk. He then placed the eggs he had thus prepared in the hot ashes, where they were soon perfectly baked. Meantime the other blacks, having skinned the bird, had cut it up, and began to roast it. We all quickly assembled round our usual supper-table—a cloth spread under an awning which projected a short distance from the waggon. The ostrich egg-omelets were pronounced excellent. Although it is said that the ostrich egg, prepared in the way I have described, is equal to that of two dozen common fowl eggs, Mr. Donald Fraser managed to eat a couple ; while I found no difficulty in swallowing the greater part of one of them. David, Kate, and Bella, however, expressed themselves perfectly satisfied with a single one divided among them.

As we were seated at our supper, various anecdotes were told of the ostrich. Donald said he had seen the Bushmen stalk them much in the same way that we had seen the blacks further north stalk the buffalo. The Bushman stuffs the head and neck of the ostrich, into which he introduces a stick, forming a sort of mantle for his shoulders with the feathers, so as greatly to resemble the bird. As his legs are black and the ostrich's white, he paints his legs with white, and taking his bow and arrow in his hand, sets off for the chase. It is extraordinary how admirably he mimics the ostrich—now stops, as if to feed, then turns his head as if keeping a look-out for enemies. Now he walks along slowly, then trots, just as the ostrich does, till he gets within bow-shot. With seldom erring aim he then pulls his bow. Instead of following the bird he has struck, however, when the others run away, he runs also. Should any wary old bird suspect that all is not right, and come towards him, he endeavours to escape; but if the bird approaches him, to avoid a stroke of its claws, or a blow from its wing, he sometimes throws off his disguise, which he leaves on the ground, and runs away to a distance to be prepared to pull his bow. He generally uses poisoned arrows, dipped in the milky juice of the tree euphorbia. A slight wound from his weapon quickly brings the ostrich to the ground. Formerly, he told us, it was supposed that the ostrich left its eggs to be hatched in the sand by the heat of the sun, as cold-blooded reptiles are known to do, but this is not the case. The hen-ostrich sits upon her eggs with great care, and as soon as the young are hatched, provides them with nourishment; and as broken eggs are generally found outside the nest, it is supposed that she keeps a certain number unhatched, that she may feed the young birds on them. She generally hatches about a dozen eggs; but the Hottentots play her a trick to induce her to lay a larger number. As soon as they find out a nest, they watch till the bird has left it to go in search of food, and then scrape out with a long stick two or three at a time. On returning and finding the number she expected deficient, she lays enough to supply their place, and thus goes on from day to day, till she has laid upwards of forty in the season. Timbo asserted that not only does man wage war against the ostrich, but that a white vulture is particularly fond of her eggs. As his beak is not sufficiently strong to break the shell, he seizes a large stone between his talons, and soaring with it high into the air, gets over the nest; he then lets it drop upon the eggs, seldom failing to break a sufficient number to afford himself a repast. The young ostriches, when they emerge from the nest, are about the size of pullets. They are quickly able to follow the mother, who supplies them for a considerable time with food. Their colour is a kind of pepper and salt, resembling the gravel and sand of the plain over which they roam; so that it is with the greatest difficulty they can be seen by the hunter, even when close to them. They are clothed with a kind of prickly stubble, which is neither down nor feathers, and which probably defends them from the coarse vegetation and gravel which covers the region where they exist. The Romans called the ostrich the *Struthio camelus*, in consequence of its resemblance in many respects to the camel of the desert. The ostrich, like the camel, is able, from the formation of its interior, to exist for a long time without water, feeding on the stunted and dried herbage of the desert. Its foot is formed curiously, like that of the camel; and it has also excrescences on its breast, on which it leans whilst sleeping. To complete the likeness, it has the same muscular neck, which rises high above the plain, and enables it to perceive the approach of an enemy, while its body is out of sight.

We had already witnessed the care which a hen-ostrich takes of her nest, and Donald told us that one day he was riding along, when he came near a bird evidently sitting. She remained quiet till he advanced, when instantly she sprang up and rushed towards him, hissing violently. When he turned round, she retreated a dozen paces or so; but directly he rode on she again rushed after him, endeavouring by her hisses to frighten him off.

"Did you kill her, Mr. Fraser, after her exhibition of maternal affection?"

"I did," was the answer; "and got her feathers and her eggs, and I and my people ate her up afterwards. Necessity has no law, I know; and if a trader in these regions were to give way to sentiment, he might have to go back with an empty waggon."

The ostrich has, properly speaking, only the rudiments of wings, which are utterly unable to lift it off the ground. It is, however, those magnificent white plumes in the tail and wings which assist it to run at the rapid rate I have described. Both male and female possess these white plumes. The body of the male differs from that of the female. It is of a deep glossy black, among which a few whitish feathers are mingled, but only visible when the plumage is ruffled.

While we were still talking about the ostrich, Leo started up, exclaiming, "See! see! there is one just outside the camp. Run for your gun, Andrew. You may get a shot at it."

There, sure enough, was an object moving slowly towards us, apparently utterly fearless of the fire. Now it began to run exactly as the ostrich does. Now it stopped and bent its head as if to feed. Presently it stretched out its neck, and a loud roar, which sounded very like that of a lion, burst from its throat.

"Do not fire, Mr. Crawford," exclaimed Donald; "for if you do, you will be apt to hit a friend;" and he and Stanley burst into a loud laugh, echoed by Timbo and some of the black boys near us, and directly afterwards the seeming ostrich came trotting merrily into the camp. Some of Donald's servants had been amusing themselves in forming such a disguise as I have already described, with the hope of catching a bird or more by means of it on the following day.

While the waggon proceeded onwards the next morning, our friend Donald again set out, accompanied this time by Chickango, to assist him in carrying home any game he might procure. They were to proceed on a line parallel with the caravan, while we ranged at a further distance. We went some little way together. We were about to separate, when, standing up, I caught sight of what I took to be the head of an ostrich in the distance, and we rode towards it. We had not got far when Donald exclaimed, "There is another! I hope there may be a family of them!" Directly afterwards we saw the female bird scampering away, and the male following at some little distance.

"I see no young birds," I observed. "I think you must have been mistaken."

"They are there, though, notwithstanding," observed my companion. "I know it by the way they run. Depend upon it, they would be going twice as fast as that if they were alone."

Putting our horses to their utmost speed, we at length nearly overtook the ostriches; and then I saw a number of little brown duckling-looking birds following at the heels of the female ostrich. Greatly to my surprise, the male ostrich at this moment stopped short, and then wheeling round, darted off on one side. As we were anxious to obtain the young as well as the mother, we continued our pursuit of her. On this he once more put on his utmost speed; but instead of going in a straight line, kept wheeling round and round us, using every effort to attract our attention. Instead of increasing, he decreased his circles, till he got within twenty yards of me, when, to my surprise, over he fell on the ground, and began to struggle desperately, and I thought he would easily be our prize. I therefore dashed forward; but quick as lightning he was on his legs again, running off in an opposite direction to that which the hen had taken. "You follow him, and I will go after the other," exclaimed Donald, perhaps thinking, from some remarks that I had made, that I should not have the heart to knock over the mother and her young brood. I had ridden some way in chase of the male ostrich, when a bird appeared in the distance, towards which he immediately directed his course, fancying, perhaps, that it was his own hen and her young ones. He was a long way ahead of me, and I had lost all hope of overtaking him, for my horse was already beginning to pant with exertion, when the report of a rifle came from the direction where I saw the other bird, and immediately my chase rolled over on the sand, the stranger rushing towards him, while three black heads appeared from some low rocks a little way beyond. Poor fellow! He deserved a better fate! The stranger bird turned out to be one of Donald's Hottentots, who, with his companions, had been fortunately in the right direction to intercept him. I insisted on appropriating the tail of the bird, asserting that the Hottentots would not have killed him had I not chased him up to them.

My horse being by this time well tired, we set off to overtake the waggon. Late in the day Donald arrived with the hen-ostrich over his saddle, his back and head ornamented with the feathers, and half a dozen young birds hanging from the crupper.

"Well, he does cut a curious figure!" exclaimed Leo, who saw Donald approaching. "If I had seen him for the first time, I should have taken him to be a fledged centaur — a mixture of man, quadruped, and bird."

Donald was inclined to claim the feathers I had appropriated; but Senhor Silva coming to my support, it was agreed that they were mine by right of conquest; and I had the satisfaction of presenting them to my fair cousins—the first trophies of the chase which I had deemed worthy of their acceptance.

We obtained, during the following days, a further supply of ostrich eggs, which, with the birds we had killed, gave us as much food as we required. We found it, when moving forward, very necessary to be careful not to deviate from the right course. Frequently over the country where there were no tracks, and often no landmarks, this was very difficult. Often it was a long day's journey from one fountain or pool to the next spot where water could be procured, and we knew well that without the necessary supply we and our cattle would suffer severely, even if we did not lose them altogether, in which case we should be involved in their destruction. Though I much enjoyed my gallops over the country, I was very thankful when Stanley was once more able to mount his horse; and I had, in consequence, generally to proceed on the back of one of the riding oxen, with Natty or Leo behind me. We were now able to carry far more water than usual. I should have said that the ostrich eggs were never broken. Their contents were extracted through a hole in one end, and were carefully surrounded by a basket-work of reeds, thus forming complete, and tolerably strong, bottles. At each fountain we came to they were re-filled.

"We have a long day's journey before us," observed Mr. Fraser one morning as we were inspanning, as the colonists call yoking, the oxen to the waggon; "and I wish I was sure that we should find water at the end of it. We have not enough left for the oxen, as we must keep all we have for the horses and ourselves."

He looked graver than usual, and not without reason. The heat was very great, and we had a wide extent of country before us, the soil consisting of light-coloured soft sand, which appeared incapable of producing any green thing for the support of animals. Pass it, however, we must, as it extended right across our path to the south, far away to the east, from the very coast of the Atlantic. Notwithstanding this, our party were in good spirits, from feeling that we were now making steady progress towards home.

"We have encountered so many dangers, and escaped them, that we should not mistrust the willingness of the kind hand of Providence to protect us to the end of our journey," observed Kate.

Her calm confidence gave us all courage, and we resolved not to allow any anticipation of evil to oppress us. Kate had never relaxed in her resolution to instruct Bella under all difficulties, and the greater part of the day they sat in the waggon with their books before them, or their work in their hands, labouring away as diligently as they would have done in their home in the colony. Leo and Natty were far more idle, though they occasionally took their seats near the young ladies, and either read to them, or listened to their reading. The Bible was their chief book, and happily its stores are inexhaustible. The other works they had read over and over again, till they declared that they could no longer look at them with patience. The heat was so great, that we were compelled to camp during the middle of the day, finding that we could make more progress by travelling early in the morning and again in the evening.

We had travelled on since daylight, when a group of trees, which are found here and there even on the desert, gladdened our eyes. We unyoked our weary oxen beneath them, and sought such shelter as their branches would afford; but not a drop or sign of water was to be seen round them. It seemed surprising how they could exist in that arid spot. Fires were lighted to cook the remnant of our provisions, though they also had fallen very short. We were seated at our meal, when Stanley started up, exclaiming, "We must have some of those fellows! Who will come with me?" He pointed eastward —the quarter whence the wind blew—and there I saw, moving slowly over the plain, and cropping the scanty herbage as they went, a large herd of antelopes.

"I will," I said, "if I can have a horse."

"You shall have mine," said Senhor Silva.

"I must go with you!" exclaimed Donald Fraser, gulping down the largest part of the contents of an ostrich egg.

Donald having giving directions for the caravan to move on, and appointed their halting-place, we mounted our horses, intending to meet it there at night, and galloped off towards the herd. I imitated my companions' attitude of leaning down, so as to conceal my head as much as possible, that we might get near without alarming the herd, keeping to leeward. Some time passed before they were aware of our approach.

"They are hartbeests," said Donald, "and will give us a good chase; but we may get within shot of them at last."

There was no shelter which would enable us to stalk them, and we therefore had to trust to their not taking alarm at the appearance of our horses. We rode on and on, and every instant I expected to see them start off, and scamper away fleet as the wind. They were noble-looking animals, with large horns rising on a line with their foreheads, and then bending curiously backwards. We rode on till we got within a hundred yards of them, when a wary old buck caught sight of us, and, suspicious of evil, gave the alarm to his companions.

"On, boys, on!" cried Donald, who had been watching for their expected start to put his horse at full speed. On we dashed, the hartbeests going away directly from the camp. They kept close together, somewhat impeding each other. They were now thoroughly alarmed, and away they went at a speed which it at first appeared would make it utterly impossible for us to come up with them. Not so, however, thought Stanley and Donald Fraser. Our horses seemed to enter into our wishes, and exerted themselves to the utmost. On kept the herd, throwing the dust up behind them, which rose in the air like clouds of smoke. After an hour's flight they began to slacken their speed, while our horses, urged on by our spurs, redoubled theirs. At last we got within a hundred yards of the hard-pressed herd. Stanley quickly threw himself from his horse, and firing, a fine buck flew up into the air; and the next moment, parting from the main body of the herd, darted off to the right; while Donald, aiming at another animal, brought it to the ground. I fired directly afterwards; but so excited had I become by the chase and ride, that I suspect my bullet flew over the heads of the animals.

"Mount and ride after that fellow, Andrew!" exclaimed Stanley, pointing to the hartbeest he had wounded.

I did as he directed me, while he and Donald Fraser, throwing themselves on their horses, again made chase after the herd. The wounded animal fled away by itself, and though evidently, by the flow of blood from its side, severely hurt, it yet continued springing forward at a rapid rate. Determined not to let it go, I urged on my horse in pursuit. At length, greatly to my satisfaction, for my horse was nearly done up, over the hartbeest rolled; and, springing from my saddle, 1 put an end to its sufferings. When I looked round, neither the herd nor my companions were to be seen. A long chase in that hot sun had made me very thirsty, and not a drop of water had I with me. I was hungry too, for I had only just begun my breakfast; though, if content to eat raw meat, I had the means of satisfying my appetite. The animal was so heavy that I could not lift it on my horse; and yet I did not like to leave it to be devoured by hyenas and jackals, or other beasts of prey, which it would, I knew, inevitably be very shortly, should I go away. I therefore waited and waited, hoping to see my companions return. I thought I remembered pretty accurately the direction I had come; but the clump of trees was but a small object to guide me over that extensive plain, on which, too, I knew that similar clumps existed. At length, not seeing my friends, I decided to load my horse with a portion of the antelope, and to try and find my way back to the camp. I had, as I mentioned, suffered greatly from thirst before, but it did not equal the pain I was now enduring. Not only did my mouth and throat feel dried up, but my whole stomach; and faint and hungry as I was, though I had an abundance of food with me, and might have collected grass and twigs enough to cook a portion, yet I could not swallow a particle. I felt growing weaker and weaker, and my head became so dizzy and my eyes so dim that I could not distinguish objects clearly before me. I began to fear that I had received a sunstroke, for the heat was greater than any I had yet experienced. I knew the fatal effects which might follow. Still, I managed to stick to my horse and ride on.

I had gone a considerable distance, and was trying to discover the wished-for clump of trees, when my eyes fell on a glittering pool of water, some way off to the left. I had not forgotten my experience when before wandering in search of water; but I was convinced that I could not be mistaken. By its side I saw several clumps of trees, and could even distinguish their reflection on the calm surface of the lake. The spectacle revived my spirits, and I urged on my horse, hoping soon to quench my thirst, and put an end to the suffering I was enduring. He too seemed equally eager to reach the lake. I was surprised that Donald had not known of it, as he certainly would have moved there instead of pushing on to the well, where he had doubts of finding water. I confess that had any one told me that what I saw before me was not water, I should have trusted my own senses rather than his assertion, and still gone on towards it. Bitter, therefore, was my disappointment, when in a short time I found myself standing on the margin of what I took to be a lake, but which was merely a dry basin incrusted with saline particles, which gave it, with the assistance of the existing mirage, thus exactly the appearance of water. I turned away, suffering even more than before from the fearful thirst which oppressed me. Still, I had been aroused, and I hoped to be able to return to the camp before being quite overcome. After going some distance, however, my spirits again sank, and I could scarcely sit my horse. In another moment I believe I should have fallen, when I saw a plant trailing along the ground, with large leaves, and among them a large melon-like fruit. Yes, there before me was a water-melon! I threw myself from my horse, and eagerly taking out my knife, cut a huge slice. Oh! how deliciously cool and refreshing it was! I let the juice trickle over my throat and down my mouth. I felt that I could never eat enough of it; it seemed to cool me even far more rapidly than water would have done. I did not forget my poor steed. He put down his head towards the fruit, part of which lay on the ground; and he seemed to relish it quite as much as I did. Having eaten my fill of the melon, I felt greatly relieved. My horse, too, had leisure to devour as much as he would. After riding on a little distance, I saw another fruit of the same appearance. I felt an inclination for a further supply; for when once the throat has become so dry as mine was, the sensation of thirst very quickly returns. I cut a slice, but the first mouthful I took made me throw it from me. It was perfectly bitter; so bitter, that even had I not tasted the previous one, I do not think I could have eaten it. My horse also, after licking it, refused to eat it. I tried another; that was equally bitter. I cut a third and a fourth; they had the same unpleasant taste. My horse also refused to touch them. I began to fancy that I had discovered the only sweet one. Still I persevered, and soon came upon another which was as delicious as the first. Three or four others were of the same character. My horse eagerly devoured them. Though loaded with meat, I could not refrain from adding several water-melons to my burden; and, thoroughly revived, set off in good spirits towards the camp.

CHAPTER XXVII.

RODE on and on, but still saw no signs of the camp Had it not been for the water-melons, I must inevitably have perished. Darkness came down over the dreary waste, making it appear still more desolate. I trusted that my steed would find his way to the camp, for I could no longer direct him with any degree of certainty. The stars shone brightly overhead; yet, as I did not know the exact bearings of the camp, they would only enable me to keep a direct line, and that might lead me far on one side or the other. Still, I should be prevented from going round in a circle, as travellers who have lost their way are apt to do, to find themselves after many hours at the spot from which they started. Every now and then I stood up in my stirrups, looking out eagerly for the camp-fire. Not a glimmer of light could I see. Dangers beset me also, I knew—old sand-wells and pitfalls might be in my path. Lions also, attracted by the smell of the meat I carried, might follow and seize me. I kept my rifle and hunting-knife ready for immediate use, while I cast an anxious look round me, every moment trying to pierce the gloom, lest some beast of prey might be stealthily approaching. It was a trying time. It would have been worse had I been suffering as before from thirst. At last I

began to fear that I must have passed the camp altogether. I
determined to halt, and was looking about for a bush or some
rock or slight elevation under which I might form my camp,
when I found my horse's fore-feet sinking into the ground. I
had great difficulty in keeping my seat; but immediately rear-
ing up, he sprang forward. The effort was vain, however, for
his feet alighted only on treacherous ground, and down he
sank into a large cavity. I made an attempt to spring off his
back, but the ground gave way, and I found myself sinking
down with him several feet below the surface. He kicked and
plunged, and very nearly struck me. I managed, however,
once more to get upon his back; and in a short time, finding
that his efforts to get out were hopeless, he remained quiet.
I had fallen into one of the holes I had dreaded. Even if I
could get out myself, there was no chance of extricating my
horse, and I should have to find my way to the camp on foot.
The loss, too, of the horse would be a serious matter. I was
as likely also to be attacked by a lion or hyena as I should
have been on the level ground; for though the wild beast,
if he got in, might not be able to get out again, I should
nevertheless become his prey. The poor horse could with
difficulty stand, and every now and then tried to change his
uncomfortable position. To relieve him, I got off and stood
as well as I was able, keeping my rifle ready for immediate
use.

Time went slowly by. Though tired and even drowsy, I
could not have gone to sleep, even had I wished it, in my uncomfortable position. I could see the stars overhead; but how
deep down I was I could not well judge. It was a depth, I
feared, from which I should have great difficulty, even in
daylight, in scrambling out. Now and then, indeed, the
dread came over me that I should be unable to do so; for the
sides were of such soft sand that when I attempted to climb

up it gave way below my feet, while there was sand alone at which I could grasp. I passed my time in devising plans for getting out; but I could not help acknowledging that the best of them were not likely to succeed. I might scrape the sand down, and thus, filling up the hollow, gradually rise to the surface; but there was danger of the mass above my head sliding down and overwhelming me.

These unpleasant reflections were presenting themselves to my mind, when I saw, against the sky, a huge head projecting over the edge. I could not be mistaken. It was that of a lion. In another instant he might spring down upon me. My only hope of escape was to fire immediately, and drive him off. Even then I dreaded that he might topple over when shot, and destroy me in his dying struggles. I raised my rifle and fired. A fearful roar was the answer to the sound of my piece. Scarcely daring to look up, I began loading again as rapidly as I could; for even then I feared that the monster would spring down. Roar succeeded roar. It seemed to me that it was echoed far and near by other lions. I waited for some time, but still no creature appeared. I began to hope that my shot had driven them away.

Once more I began to suffer from thirst; and cutting up a water-melon, I took part of it myself, and gave a portion to my poor steed, who showed his gratitude by licking my hand. I waited for some time, wishing for the return of day. Except when my horse made a movement, not a sound for many minutes together reached me. Then the distant roar of a lion might be heard, or the barks of hyenas or jackals. Suddenly I heard the sound of feet, as if a troop of antelopes was passing by. I hoped that they might not inadvertently tumble in on me. To scare them away I began to shout. I kept on, raising my voice to the utmost. To my surprise a shout came in return.

"Hillo! Where are you?"

I recognized Stanley's voice. I soon let him know. Presently I saw his head and Donald's projecting over the pit.

"A bad job for the horse, though we may soon get you out. But you must be almost dead of thirst, lad, as we pretty nearly are," observed Donald.

I told them of the water-melons I had found, and that I still had some remaining.

"Then hand them up, lad! hand them up!" cried Donald. "We shall have more strength to haul you up afterwards."

While he was speaking, he let down the tether-ropes. I fastened the water-melons to them.

"You will excuse us, Andrew," said Stanley, "if we satisfy our thirst before getting you up. You know from experience what it is."

Not waiting for my reply, their heads were in the melons, I suspected, before many seconds had passed. They did not keep me waiting long; but the next time the rope came down I fastened the meat to it. This was hauled up, Donald uttering exclamations of satisfaction at seeing it. By the aid of the rope I very quickly scrambled out; and as I did so I felt thankful that assistance had come, for from the depth of the hole and the nature of the sides I saw that I should not have got out without assistance. They had come upon the remains of the hartbeest, but had not discovered any water-melons, and their horses were, therefore, scarcely able to proceed. Even the small supply of the watery fruit we were able to give the poor animals greatly relieved them.

The next question was, how to get my horse up. I volunteered to descend again. With the aid of the tethers and all the straps we could muster, we managed to get a rope of sufficient length round his shoulders, so as to leave his limbs free, that he might help himself as much as possible. We then shuffled down the sand, making him leap up on it as it fell; and at length, by hard work, once more we got him on level ground.

My horse was heavily laden, but my friends remarked that could they have exchanged some of the meat for water-melons they would gladly have done so. We, however, could dis-cover none on the ground over which we passed. Fortunately they knew the bearings of the camp, and at length its fires appeared in sight. I was surprised to find in reality how short a time I had been in the pit; for I supposed I had passed the greater part of the night there.

We found our friends bitterly disappointed at having dis-covered no water, as they had expected, at their halting-place. Every one was complaining,—even Kate and Bella; for even the supply intended for the young ladies had been exhausted. My tidings of the water-melons was joyfully received; and it was arranged that a party should set out with oxen and baskets at daylight. I lay down, as did Stanley and Donald, to obtain a little sleep. I was to lead the party, as I fancied I knew the direction where I had found the juicy fruit. When Senhor Silva heard the account I gave, he expressed a hope that we should find not only an abundance of melons, but a root which he called *leroshúa*, which grows in the desert, and is of an excessively juicy nature.

While the waggons proceeded on southward, Senhor Silva and I scoured the plain in one direction, keeping sight of the oxen with the panniers, that we might summon them directly we discovered what we were in search of. Before going far, we saw the ground turned up as if some animal had been digging with its horns. Near it was a small plant, the stalk about the thickness of a crow's quill. It had apparently been broken off, and the root to which it had been attached had been consumed. Not far off, however, we saw several similar plants; and Igubo—who accompanied us with a spade—and the other blacks, who were not far off, were directed to dig. They had got down a little more than a foot, when a large tuber, twice the size of the ordinary turnip, was discovered ; and the rind being removed, we found it to consist of a mass of cellular tissue, filled with fluid like the root I have mentioned. We eagerly put it to our mouths, and found it deliciously cool. The poor oxen, as soon as it was given to them, ate it eagerly. We loaded one with the roots, and sent it on to overtake the caravan.

Senhor Silva said there was another root, of a similar nature, in other parts of the desert, called the *mokúri*. The tubers are far larger. It is a herbaceous creeper. The stem, rising out of the ground, sends out its branches horizontally to a distance of a yard or more on either side. They deposit underground a number of tubers, much larger than the first I have mentioned. The natives, when seeking them, strike the ground with a stone, and discover by the difference of sound when one is beneath the spot.

In half an hour, great was our delight to see the ground covered in all directions with the water-melons of which we were in search. Igubo and his sons, who had never before seen any, instantly set upon them. They spat out the first, with wry faces. They had seized upon a bitter one. The other blacks, more cautious, ran along, cutting a small piece off with their knives, tasting each in succession, leaving the bitter and only cutting the sweet.

As we had not more than a load for one buffalo, we pushed on further, hoping to find a larger supply. After going a few yards, I saw Donald, who was in front, standing up in his stirrups. On getting up with him, he pointed ahead, when we saw in the distance what looked like a number of black mounds.

"A troop of elephants!" he exclaimed. "But it will be no easy matter to get near enough for a shot in this open plain."

Riding on a little further, the elephants came more closely in sight; and near them were a number of rhinoceroses. It was soon evident that they were busily employed; and Senhor Silva said he had no doubt that they were eating the water-melons, a number of which probably grew there.

Eager as Donald was to obtain the tusks, we declined assisting him in so dangerous an undertaking. Turning southward to return to the caravan, we shortly afterwards caught sight of three lions, and a whole troop of hyenas and jackals, apparently quenching their thirst with the same juicy fruit. Scarcely had we lost sight of them, when several herds of antelopes appeared, scattered widely over the plain. They also were evidently feeding on the melons. We fortunately, however, fell in with a small patch of the fruit which had not yet been attacked, and were thus able to load our oxen. Continuing our course across the desert, we supported ourselves entirely by the watery fruit I have described.

We were pushing towards a village near a fountain, where Donald expected to find an ample supply of water. He and I were riding ahead. At length some circular, beehive-looking huts appeared in sight, with a few people moving about in front of them. The men were armed with spear and buckler, and wore the usual waist-cloth in front, and ornaments on their heads and arms. Several, when they saw us, came forward, and began to shake their spears and vociferate loudly. Before we could understand their meaning, they were joined by a tall oldish-looking man, who seemed to be their chief. After he had made a long harangue, Donald answered him; but I saw by his gestures and those of his followers that no satisfactory arrangement had been arrived at. Donald began to lose temper.

" The fellows guess the strait we are in, and refuse to give or sell us water," he exclaimed, " unless I deliver to them six of my best oxen, four muskets, and I do not know how many articles besides. I have told them I will do nothing of the sort ; and we shall soon see that they will come down in their demands. They know the country we have come through, and probably think we are harder up for water than is the case."

The chief waited to see if we would accede to his demands ; and Donald replied that as we could do very well without water they would get nothing, whereas we would have paid them liberally for what we took. Saying this, we turned round our horses and rode off. We had not got far when several arrows came whistling after us. Fortunately none struck us or our horses, for if they had, as they probably were poisoned, the result would have been serious. As we turned our heads for an instant, we saw a large number of people collecting from numerous huts scattered about in all directions.

We hastened back to the caravan to prepare for defence ; for the natives, it seemed, were too likely to attack us. Stanley at once proposed encamping and erecting a stockade, within which we might defend ourselves.

" Oh yes !" exclaimed Leo, " we could easily drive them off, as we should have done the natives of the north."

" But," observed Natty, " suppose they besiege us, what are we to do for water ?"

" You are right, Natty," said Stanley. " It would be better generalship to pass their village and try to gain another fountain further on."

This, indeed, was our only secure course ; for though our own blacks would certainly have fought well, Donald could not depend on his followers, who, he said, had shown the white feather on more than one occasion.

We therefore, instead of camping, as we had proposed, turned somewhat to the east, so as to leave the inhospitable village on our left hand, hoping to get a considerable distance to the south of it before daybreak. The country was tolerably level, and the moon was high enough to give us sufficient light to find our way. It was the first night we had attempted to travel without stopping, but it was absolutely necessary to do so to carry out our object. A battle with the natives was on every account to be avoided. Stanley and I rode as scouts on either hand, while Donald kept ahead to explore the way. We hoped thus to avoid being taken by surprise. We could see numerous animals moving around us. Once a vast herd of elephants hove in sight, another time one of buffaloes, while antelopes of various species bounded off as we came near. We could hear occasionally the muttering sound of lions and the cry of hyenas. Several, indeed, followed us, but as they did not approach, we refrained from firing at them, lest the sound of our rifles might betray our position. Timbo and Chickango brought up the rear on oxen, with directions only to fire in the case of any large body of natives being seen following, or should a wild beast threaten to attack them. Thus we travelled on hour after hour. We halted only once, to give the oxen some water-melons or leroshúa roots, and to take a little food and water-melon ourselves. I found that Kate and Bella had become very anxious, because one of the Hottentot boys, who spoke a little English, had been telling them all sorts of stories of the ferocity of the natives, and of the way they had attacked travellers and carried off their oxen. Donald, on hearing this, soundly rated the lad, assuring the ladies that, as he had never been in that part of the country before, he could know nothing of the matter.

After a short rest we again pushed on. The sun at length rose above the dry plain, shedding a brilliant glow of crimson over the whole eastern horizon, and lighting up the summits of the bare rocks, and clumps of trees here and there, with a red tinge. I could not help dreading, with the prospect of a burning day before us, that water might not be found. At length we arrived at one of the sand-wells I have before described. We eagerly rushed into it, and sank our reeds, in the hope of obtaining water. It came, but at a slow rate, which promised but a scanty supply for our thirsty cattle, even though we might obtain enough for our own wants. The blacks quenched their thirst by sucking narrow reeds, which they ran into the sand. Donald, after examining it, gave orders to them to dig, in the hope of obtaining a larger quantity. The result of the operation was satisfactory, and we accordingly resolved to encamp there for a day or two, till our cattle had obtained enough water to last them till we could get across the desert. There was an abundance of grass, growing in tufts, and a small group of trees near us, which would afford us shade and fire-wood. Stanley also hoped to kill some game. The poor cattle had to wait, though, till our horses had the water they required.

Leo and Natty had been amusing themselves outside the camp. "Here; see what we have got!" cried Leo, returning after they had wandered to a short distance. "Hillo!" he exclaimed, turning round as I went out to meet them. "Why, it was a long creature just now; and see, it has turned into a ball; and a big ball it is!"

The ball of which Leo spoke was covered with large black scales, somewhat the size and shape of the husk of the artichoke, which overlapped each other in a very curious and beautiful manner. David quickly solved the mystery of the scaly ball. Being allowed to remain quiet for a few minutes.

it unrolled itself, when it was seen to possess a head and a broad tail, likewise covered with scales. He pronounced it to be one of the manides or scaly ant-eaters—a rare animal, and seldom seen. It had a long extensile tongue, furnished with a glutinous mucous for securing its insect food. It was entirely destitute of teeth, so that it was evident it must suck in the creatures it caught, and swallow them whole.

David said that the manides are very inoffensive animals; that they live solely on ants and termites. They burrow to a great depth in the ground. For this, as also for extracting their food from ant-hills and decayed wood, we found that the creature's feet were armed with powerful claws, which it could double up when walking.

"We are getting into a thorny district," observed Donald, who had joined us, "very different to the thornless one we have passed through. What do you think of this?" he observed, stooping down and picking up a round disc with a sharp thorn in the centre. "Suppose this was to run into a poor animal's foot; it would take him months to get it out, even if he did not become lame for life."

Soon after we camped Donald started off on a hunt by himself. He had not been gone long when we saw him returning from the north, with a gemsbok, or oryx, as I have before called it, across his saddle. Considering the weight he carried, he came pressing on at a rapid rate. He was not a man much given to exhibiting his feelings, but I saw that something was the matter.

"Quick, lads!" he exclaimed. "We must get ready to defend our camp without loss of time. I thought as I rode along that I would just take a look at our inhospitable friends, and see what they were about. When I had got half-way between this and their village, I caught sight of a large number of them stealing along across the plain. I think they must have seen me, or perhaps they took me for a cameleopard or ostrich; for I only showed for a moment behind an ant-hill, then quickly again got under cover to reconnoitre them. There are some two or three hundred fellows at least, and by the way they were marching I am very certain that they intend to attack us. I had just shot this oryx, and I had no wish to leave it for them, or I might have been here sooner; but there is time to get ready, if we are sharp about it."

Stanley, on hearing this, was in his element. He immediately ordered out all hands to cut down the smaller trees from the group I spoke of, to form palisades. The waggons and carts were placed on one side, while palisades were fixed all round, and strong cross-beams secured to them. This done, we set to work to throw up an embankment, which, with the light sand, was easily accomplished, the upright posts keeping it in its place. We thus, in a wonderfully short time, had a little fortress which might have stood a siege against men armed with muskets. As we hoped our expected assailants had no fire-arms among them, we felt no apprehension as to the result. The chief danger was that they might try to starve us out, which there was a possibility of their doing should they persevere in surrounding us. We were working away till long after dark by the light of our fires. Scouts were sent out, but came back after some time stating that they could see nothing of the enemy. At length Stanley expressed his belief that Donald had been mistaken; at which our friend bristled up. No, he was certain he had seen an army of blacks; probably, however, when they caught sight of him, they might have thought better of the matter.

"But perhaps they were merely on a hunting expedition," said David, " or collecting water-melons."

" They were keeping too close together for hunting; and as they were following in our track, they would have found neither water-melons nor water-roots," answered Donald. " Do not be too sure that they will not come yet. These people, as I fancy you have experienced, like to take their enemies by surprise; and they will not come on in broad daylight with tom-toms and shouts, depend on that. It would be well that those who have the morning watch should keep a bright look-out, or we may be attacked when we least expect it."

Donald's advice was not thrown away on me. I had just relieved Stanley, who had taken what would at sea be called the middle watch, Jack and Timbo being my companions. The night was perfectly still. I could hear the low muttering of lions in the far distance, with an occasional roar as some other creature approached to dispute their prey with them. Now and then the trumpeting of elephants reached me, probably on their way to some distant fountain, or in search of roots or water-melons. I thought it was almost impossible that any enemies could approach without being discovered. Still, I had been too well accustomed to discipline at sea not to keep as bright a look-out as I should have done had I known they were near. I was standing with Timbo on the north side of the fort, when he asked me to let him go out to a little distance.

" If dey come, dey come soon; and we no see dem till dey close to de wall," he whispered.

Trusting to his judgment, I willingly let him do as he proposed. He accordingly slipped over the palisade on one side, and I could barely distinguish him as he crept along over the ground towards the north. He was soon lost to sight. Jack and I kept anxiously looking out for his return. I felt little alarm about the natives, but I was afraid that some prowling beast might attack him. I must have waited half an hour or more, when I distinguished a long object crawling along on the ground. In the gloom I could not make out whether it was Timbo, or a panther perhaps, or a huge snake, so noise-lessly and stealthily did it approach. It made, however, for the side of the fort, and in a short time Timbo came up to me, having been admitted by Jack through the sally-port in the rear.

"Dey come!" he whispered. "Dey no see me, dough. Dey t'ink dey find us all asleep. I go call de captain and de rest, and de black fellows; and we all get ready, and lie down and snore loud; and den, when de enemy come, we jump up wid loud shout, and dey run away."

Timbo's plan of action was simple, and I hoped might prove effective; so I begged him to carry it out. In a few seconds all our party, crawling out from their huts, or from beneath the waggon or lean-tos, assembled noiselessly, and took up their station at the palisades, kneeling down so as not to be seen by those approaching. Thus we all remained ready for the attack. Some time passed away, and no enemy appeared; and I could not help suspecting that Timbo might by some means have been mistaken. He, however, was positive that he had seen the enemy, and was rather indignant at my supposing that he could have been deceived. We kept watching on every side, not knowing on which the blacks, if they really were coming, might make their attack. At length I saw an object moving along the ground, exactly as Timbo had approached the fort; then another and another appeared. I found that Timbo had seen them too, and immediately he managed to give the information to our companions, when, somewhat to my amusement, a loud chorus of snores ascended from all parts of the camp. "Dat good," he whispered to me; "dey t'ink we all sleepy. Now, see!"

As he spoke, we could distinguish several black figures crawling on the ground close up to the fort. They stopped and listened, then rising to their feet, ran back to their companions, who yet, we supposed, remained concealed in the neighbouring bushes and long grass. Fearing, probably, that the snoring garrison might awake, the whole army of blacks now advanced, crouching down close to the ground, and had we not been watching for them, they might easily have got close up without being discovered. They advanced in a semicircle, closing gradually in on the fort. We lay still as death. The dogs, I should have said, had been muzzled, and stowed away under the waggon, where they remained quiet. Closer and closer the blacks advanced. "Dey t'ink dey climb over and we not know," whispered Timbo. "Now, see!"

We let the blacks get close to the palisades. They were touching them, expecting without difficulty to climb over, when at a word from Stanley up we all started, firing directly in their faces. The result was even more satisfactory than we could have anticipated, for in an instant the front ranks rushed away, knocking down those behind them in their terror, when the whole army instantly took to flight. The two boys gave vent to loud hurrahs, which were taken up by the rest of our party, when Kate and Bella, who had not been told of what was likely to take place, came rushing out of their tent to inquire what had occurred. We soon found, however, that we were not to gain so easy a victory as we had hoped. The blacks, recovering from their fright, and not acquainted with the effects our fire-arms were able to produce at a distance, once more assembled, and advanced bravely to the attack. We were consequently compelled to give them a volley, but except from the rifles of two or three of our best shots, very few of our bullets took effect. Seeing that we were not to be taken by surprise, the enemy again retired. We were in hopes that they had gone off altogether. To ascertain whether this was the case, Timbo volunteered once more to go out. He soon returned, saying that they had only retired under shelter, and from the sounds he had heard, he suspected that they proposed making another attack. We waited anxiously till daybreak. On looking out, we saw numerous blacks moving among the bushes. Then a large body appeared, apparently assembling to hold a consultation. After a time they separated, dividing into several small bodies. These marched forward, and posted themselves at equal distances round the camp. It was now clear that, having failed in their expectation of taking us by surprise, they had resolved on starving us out. Fortunately they could not interfere with our water, or they would have done so; indeed, they might possibly not have been aware of the supply we were gradually obtaining from the well.

The day passed away, but our pertinacious enemies made no signs of moving. Of course they kept us on the alert all night, not knowing at what moment they might again attack us. On the second day things began to look serious; for though we had water, provisions were growing scarce. Donald began to talk of cutting our way through the enemy; but as they could assail us at their pleasure as we marched along, this would have been a dangerous proceeding.

"It must be done," he said at last; "if we remain here another day we shall starve, and it is better to run the risk of fighting than to do that."

We had at length obtained a sufficient supply of water for the cattle, and had we been unmolested, we might now with confidence proceed on to the next fountain, after which Donald hoped to find each day a sufficient supply of water. Stanley however proposed, instead of risking an attack while moving on, to sally out with horse and foot and drive the enemy away. He, with Senhor Silva and Donald, were to form the cavalry, and I was to lead a party of infantry, consisting of Jack, Chickango, Igubo and his two sons, and four of Donald's Hottentots.

" We must go too, then ! " exclaimed Leo, when he heard the proposal.

" No," answered Stanley. " I have no doubt of your bravery, but you will show it better by remaining to assist David and Timbo in garrisoning the fort." After some hesitation Donald agreed to this plan.

At length, as evening drew on, the blacks appeared in greater numbers than before. Instead of allowing them to approach, however, we opened a warm fire upon them, when even at a considerable distance. This seemed to astonish them, as probably they were not aware that our bullets would reach so far, and once more they retreated under cover. Scarcely had they gone, when Donald gave us the unsatisfactory information that one meal alone remained for the party in the camp.

" Then, my friends," said Stanley, " let us lose no time in making our retreat. We may get to a long distance before the blacks discover that we have left the fort ; and if they follow us, we must turn round and drive them off."

The necessity of moving was so obvious, that no time was lost in preparing to start. The waggons were laden, the oxen yoked. The usual fires were lighted, to deceive the enemy. Then in perfect silence we quitted the camp, Stanley and I bringing up the rear, and Timbo and Jack and four other men, well-armed, on foot. We moved on slowly ; for neither we, our horses, nor cattle were capable of much exertion. Every now and then Stanley halted and faced round to ascertain whether we were pursued ; but some hours passed by, and we began to hope that the enemy had retreated before we commenced our march, or had not ventured to follow us. We knew well, however, that if the blacks did pursue us, they would come on stealthily, so that we should have but a short time to prepare for their reception. Leo and Natty were persuaded to remain in the waggon with their guns loaded, ready to do battle for Kate and Bella; while Donald had put arms into the hands of the most trustworthy of his men, who promised to fight bravely should we be attacked. However, he confessed that he had no confidence in their valour. After riding for some time at a distance from the waggon we once more joined them, hoping that we should be able to proceed without molestation.

I was very thankful when the sun rose; and though his beams were likely to be somewhat hot, they greatly cheered our spirits. I was on the point, at Stanley's request, of riding on to ask Donald Fraser when he proposed to camp, when, looking round, I saw away to the north, on the summit of an elevation we had passed over, a dark line moving towards us. I pointed it out to Stanley.

"It is the blacks! There can be no doubt about that," he answered. "We must be prepared for them. I did not suppose they would have ventured so far in pursuit."

"I say, Andrew, we must drive these fellows off, and have done with them," said Leo. "You will see how Natty and I will fight!"

I was sure from his determined look that he would be as good as his word, and that Natty would not be less courageous, though he made no remark. Stanley had given orders that not a shot was to be fired till he issued the word of command. We were standing in expectation of receiving it, when Timbo shouted out, "See! see! some horsemen come dis way!" We looked towards the west—the direction in which he pointed—where, under a cloud of dust, a herd of buffaloes were seen scampering across the plain, with several horsemen in close pursuit. On they came directly towards our black enemies, who did not perceive them till they were within a distance of four or five hundred yards. The herd of buffaloes dashed madly forward into the very midst of the blacks, whom they scattered in every direction. Numbers were knocked over. The rest, taken by surprise, attempted to escape by flight. Instantly Stanley threw himself upon his horse and galloped forward, shouting to the hunters. The buffaloes meantime continued their charge wherever they saw the negroes assembled, and in a few minutes swept half round the circle, raising the siege in the most effectual manner. Stanley's shouts soon attracted the attention of the hunters. A few words from him explained the state of affairs, and together they charged towards the remainder of the black army, who had hitherto stood their ground. The latter, without even stopping to draw their bows, took to flight towards the north, still followed, in the most extraordinary manner, by the buffaloes, who rushed in and out among them, urged on by the shouts and cries of the hunters in the rear. In a few minutes not a black was to be seen, except those who had been knocked over by the infuriated animals. All this time the only shots fired had been at the buffaloes, three of whom lay dead on the ground. At length the herd, after pursuing the blacks for a considerable distance, turned off to the east, leaving us possessors of the field.

As we were hurrying out to welcome the strangers, we saw Stanley warmly shaking hands with them, when what was my surprise as they rode up to recognize the Messrs. Rowley and Terence O'Brien!

"We will tell you all about it," said the latter, as we warmly shook hands. "But don't you know him?" and he pointed to the fourth horseman.

I could scarcely believe my eyes, when my friend, the worthy first mate of the *Osprey*—whom I thought had long been numbered with the dead—jumped off his steed and took me by the hand.

"I have not a very long yarn to spin," he said, "though it is a somewhat wonderful one. When I was washed off the deck, I found near me a topmast, which had probably been carried away and cut adrift from some craft ahead of us. I clung on to it, and was picked up a day or two afterwards by a vessel which had to touch at Walfish Bay on her way to the Cape. Finding a party settled there on a whaling speculation, I agreed to remain. However, after some time, as few whales were to be caught, I determined to go on to the Cape. Just as I was about to sail, I received an invitation from a gentleman—Mr. Ramsay—about to start into the interior on a hunting and trading expedition, to accompany him as an assistant. The life he proposed to lead was a new one to me; but I had had enough of salt water, and after a little consideration accepted it. Who should arrive directly afterwards but our friends here, who, after having been cast on shore and gone through all sorts of adventures as they travelled down the coast from the north, had at length reached Walfish Bay. But they will give you an account of themselves. Do not ask, though, about their poor sister," he whispered. "She is gone! Died soon after they landed; and that wretched fellow Kydd, he was washed off the raft in passing through the surf. These three young men alone remained, with scarce a rag on their backs, and not a sixpence in the world. They were therefore very glad to accept the offer made to them by my friend, to assist him in shooting elephants, and rhinoceroses, and other game. From what I have seen of them, they are better suited to that sort of work than the steady business of a colonist. We have now been out six months, and are on the point of turning westward; indeed, had the buffaloes not led us in this direction, we should not have come further to the east. The prospect of the desert is not over-inviting, and for my part, I have had enough of hunting. I have run a narrow chance of being killed a score of times by lions and elephants, not to speak of rhinoceroses and buffaloes, hyenas and snakes, and I do not know what other creatures. When my engagement is over, I have made up my mind not to accept another of the same sort, but to stick to the sea as long as I am fit for work, or till I can save enough to enable me to settle down in a snug cottage in old England."

After the hint I had received from my friend Gritton, I forebore to make too minute inquiries of the Rowleys as to their adventures. Terence O'Brien, however, gave me most of the particulars. They had undergone a fearful amount of suffering even before they were cast ashore, and a still greater amount afterwards. It is surprising, indeed, that poor Miss Rowley should have survived so long on the raft; and we all, indeed, had cause to be thankful that we had been preserved from similar sufferings.

As soon as part of the buffalo flesh had been divided among our half-famished party, and had been cooked and eaten, we inspanned and pushed on to join Mr. Ramsay's caravan. Though there was little chance of our being pursued by the hostile natives, I was very thankful when at length the fires of his camp appeared in sight. Terence O'Brien had galloped on to announce our coming, and he now came up with loud whoops and cries, followed by most of his party, from whom we received the warmest welcome. We had still, however, a long journey before us; but the road was known, the fountains were within moderate distances of each other, and the natives were friendly. Mr. Ramsay had been successful both in hunting and trading, and the large piles of huge elephant and hippopotamus tusks, lion and panther skins, and other articles, rather excited Donald Fraser's envy. "However," he observed to me, as he looked at his fellow-trader's well-filled waggons, " I have had the satisfaction of rescuing you and your friends, Mr. Crawford, out of as dangerous a position as travellers in Africa can well be placed in, and I have no reason to complain of the liberality of your generous friend, Senhor Silva."

We at length reached Walfish Bay, where we found a vessel, the *Flying Fish*, just on the point of sailing for the Cape. The Rowleys and Terence O'Brien were, however, so enamoured of their hunting life, that they determined to start off into the wilds again on their own account. Our kind, noble-minded, and generous friend, Senhor Silva, here bade us good-bye, intending to wait for a vessel which was expected to call in on her way to St. Paul de Loando. He shook my hand warmly.

"I am a widower, as you know, and I had a hope, I confess, which I must not speak of, for I see that it is vain," he said. "You will think of me, and so will your sweet cousin, I trust, sometimes; and I shall be truly glad to hear of your happiness."

We all embarked on board the *Flying Fish*, hoping at length, after all our adventures, to reach our destination in safety. I had made up my mind to settle on shore, and assist my cousins in cultivating their farm. Perhaps my cousin Kate had something to do with my resolution. At all events, when I proposed it she appeared very well pleased.

Leo, when he heard of it, exclaimed, "Oh, how delightful! because then, Andrew, you will not carry Natty away, as I was afraid you might have done; and he and I can manage to get on so capitally together. We have formed all sorts of plans already, and I only hope that you may marry Kate, and he, by-and-by, can marry Bella; and then we shall all be brothers, 'and live happily together to the end of our days,' as the story books say."

Though our voyage was a pleasant one, I was very thankful when at length the lofty height of Table Mountain appeared

ahead, covered with its table-cloth, and we dropped our anchor
off Cape Town. We had still a long journey before us; but
at length the anxiety which my uncle and aunt had been so
long suffering on account of the non-appearance of their children was relieved by our safe arrival at their farm.

After a few days' rest, we all set to work on the special
duties apportioned us. Kate did not neglect Bella's education, even though in the course of the following year she became the mistress of a house of her own, of which I was the
master. David settled down as the medical man of the district. Stanley, though he occasionally went out hunting, became a first-rate farmer, ably assisted by Timbo, Chickango,
and Igubo and his two sons, who expressed no desire to return
to their part of Africa. Jack Handspike accompanied Mr.
Gritton to sea, but lately came back again, saying that he had
had enough of it, and was determined henceforth to plough
the land instead of the ocean. I may say of myself and of all
my friends indeed, that "whatsoever our hands find to do,
we do it with all our might," humbly endeavouring to serve
God in our daily walk in life, and thereby enjoy that true
happiness which even in this world can be obtained by those
who know the right way to seek it.

THE END.

NELSON'S "ROYAL" LIBRARIES.

THE SHILLING SERIES.

ACADEMY BOYS IN CAMP.	*S. F. Spear.*
ALL'S WELL THAT ENDS WELL.	*Miss Gaye.*
ESTHER REID.	*Pansy.*
TIMOTHY TATTERS.	*J. M. Callwell.*
AMPTHILL TOWERS.	*A. J. Foster.*
IVY AND OAK.	
ARCHIE DIGBY.	*G. E. Wyatt.*
AS WE SWEEP THROUGH THE DEEP.	
	Gordon Stables, M.D.
AT THE BLACK ROCKS.	*Edward Rand.*
AUNT SALLY.	*Constance Milman.*
CYRIL'S PROMISE. A Temperance Tale.	*W. J. Lacey.*
GEORGIE MERTON.	*Florence Harrington.*
GREY HOUSE ON THE HILL.	*Hon. Mrs. Greene.*
HUDSON BAY.	*R. M. Ballantyne.*
JUBILEE HALL.	*Hon. Mrs. Greene.*
LOST SQUIRE OF INGLEWOOD.	*Dr. Jackson.*
MARK MARKSEN'S SECRET.	*Jessie Armstrong.*
MARTIN RATTLER.	*R. M. Ballantyne.*
RHODA'S REFORM.	*M. A. Paull.*
SHENAC. The Story of a Highland Family in Canada.	
SIR AYLMER'S HEIR.	*E. Everett-Green.*
SOLDIERS OF THE QUEEN.	*Harold Avery.*
THE CORAL ISLAND.	*R. M. Ballantyne.*
THE DOG CRUSOE.	*R. M. Ballantyne.*
THE GOLDEN HOUSE.	*Mrs. Woods Baker.*
THE GORILLA HUNTERS.	*R. M. Ballantyne.*
THE ROBBER BARON.	*A. J. Foster.*
THE WILLOUGHBY BOYS.	*Emily C. Hartley.*
UNGAVA.	*R. M. Ballantyne.*
WORLD OF ICE.	*R. M. Ballantyne.*
YOUNG FUR TRADERS.	*R. M. Ballantyne.*

T. NELSON AND SONS, London, Edinburgh, Dublin, and New York.

NELSON'S "ROYAL" LIBRARIES.

THE EIGHTEENPENCE SERIES.

T. NELSON AND SONS, London, Edinburgh, Dublin, and New York.

www.ingramcontent.com/pod-product-compliance
Lightning Source LLC
Chambersburg PA
CBHW020907210326

41598CB00018B/1797